Process Design Strategies for Biomass Conversion Systems

Process Design Strategies for Biomass Conversion Systems

Edited by

DENNY K. S. NG, RAYMOND R. TAN, DOMINIC C. Y. FOO, AND MAHMOUD M. EL-HALWAGI

WILEY

This edition first published 2016
© 2016 John Wiley & Sons, Ltd.

Registered Office
John Wiley & Sons, Ltd, The Atrium, Southern Gate, Chichester, West Sussex, PO19 8SQ, United Kingdom

For details of our global editorial offices, for customer services and for information about how to apply for permission to reuse the copyright material in this book please see our website at www.wiley.com.

The right of the author to be identified as the author of this work has been asserted in accordance with the Copyright, Designs and Patents Act 1988.

All rights reserved. No part of this publication may be reproduced, stored in a retrieval system, or transmitted, in any form or by any means, electronic, mechanical, photocopying, recording or otherwise, except as permitted by the UK Copyright, Designs and Patents Act 1988, without the prior permission of the publisher.

Wiley also publishes its books in a variety of electronic formats. Some content that appears in print may not be available in electronic books.

Designations used by companies to distinguish their products are often claimed as trademarks. All brand names and product names used in this book are trade names, service marks, trademarks or registered trademarks of their respective owners. The publisher is not associated with any product or vendor mentioned in this book.

Limit of Liability/Disclaimer of Warranty: While the publisher and author have used their best efforts in preparing this book, they make no representations or warranties with respect to the accuracy or completeness of the contents of this book and specifically disclaim any implied warranties of merchantability or fitness for a particular purpose. It is sold on the understanding that the publisher is not engaged in rendering professional services and neither the publisher nor the author shall be liable for damages arising herefrom. If professional advice or other expert assistance is required, the services of a competent professional should be sought.

The advice and strategies contained herein may not be suitable for every situation. In view of ongoing research, equipment modifications, changes in governmental regulations, and the constant flow of information relating to the use of experimental reagents, equipment, and devices, the reader is urged to review and evaluate the information provided in the package insert or instructions for each chemical, piece of equipment, reagent, or device for, among other things, any changes in the instructions or indication of usage and for added warnings and precautions. The fact that an organization or Website is referred to in this work as a citation and/or a potential source of further information does not mean that the author or the publisher endorses the information the organization or Website may provide or recommendations it may make. Further, readers should be aware that Internet Websites listed in this work may have changed or disappeared between when this work was written and when it is read. No warranty may be created or extended by any promotional statements for this work. Neither the publisher nor the author shall be liable for any damages arising herefrom.

Library of Congress Cataloging-in-Publication data applied for

ISBN: 9781118699157

A catalogue record for this book is available from the British Library.

Cover Image: Courtesy of the Author

Set in 10/12pt Times by SPi Global, Pondicherry, India
Printed and bound in Singapore by Markono Print Media Pte Ltd

1 2016

Contents

List of Contributors

Bawadi Abdullah, Department of Chemical Engineering, Biomass Processing Laboratory, Center of Biofuel and Biochemical, Green Technology (MOR), Universiti Teknologi PETRONAS, Malaysia

Viknesh Andiappan, Department of Chemical and Environmental Engineering/Centre of Sustainable Palm Oil Research (CESPOR), The University of Nottingham, Malaysia

Kathleen B. Aviso, Chemical Engineering Department, De La Salle University, Philippines

Mustafa Kamal Abdul Aziz, Department of Chemical and Environmental Engineering/ Centre of Excellence for Green Technologies, The University of Nottingham, Malaysia

Santanu Bandyopadhyay, Department of Energy Science and Engineering, Indian Institute of Technology Bombay, India

Paul Blowers, Department of Chemical and Environmental Engineering, The University of Arizona, Tucson, USA

Christina E. Canter, Department of Mechanical Engineering, University of Alberta, Canada

Peam Cheali, CAPEC-PROCESS Research Center, Department of Chemical and Biochemical Engineering, Technical University of Denmark (DTU), Denmark

Nishanth G. Chemmangattuvalappil, Department of Chemical and Environmental Engineering/Centre of Sustainable Palm Oil Research (CESPOR), The University of Nottingham, Malaysia

Carolina Conde-Mejía, Departamento de Ingeniería Química, Instituto Tecnológico de Celaya, Mexico

Rosa E. Del Río, Institute for Chemical and Biological Researches, Universidad Michoacana de San Nicolás de Hidalgo, Mexico

Nishith B. Desai, Department of Energy Science Engineering, Indian Institute of Technology Bombay, India

Mahmoud M. El-Halwagi, Chemical Engineering Department, Texas A&M University, USA

Rafiqul Gani, CAPEC-PROCESS Research Center, Department of Chemical and Biochemical Engineering, Technical University of Denmark (DTU), Denmark

Carina L. Gargalo, CAPEC-PROCESS Research Center, Department of Chemical and Biochemical Engineering, Technical University of Denmark (DTU), Denmark

Krist V. Gernaey, CAPEC-PROCESS Research Center, Department of Chemical and Biochemical Engineering, Technical University of Denmark (DTU), Denmark

J. Betzabe González-Campos, Institute for Chemical and Biological Researches, Universidad Michoacana de San Nicolás de Hidalgo, Mexico

Roman Hackl, Department of Energy and Environment, Chalmers University of Technology, Sweden

Simon Harvey, Department of Energy and Environment, Chalmers University of Technology, Sweden

Mimi H. Hassim, Department of Chemical Engineering, Universiti Teknologi Malaysia, Malaysia

Arturo Jiménez-Gutiérrez, Departamento de Ingeniería Química, Instituto Tecnológico de Celaya, Mexico

Antonis C. Kokossis, School of Chemical Engineering, National Technical University of Athens, Greece

Konstantinos R. Koutsospyros, School of Chemical Engineering, National Technical University of Athens, Greece

Weng Hui Liew, Department of Chemical Engineering, Universiti Teknologi Malaysia, Malaysia

Sergio I. Martínez-Guido, Chemical Engineering Department, Universidad Michoacana de San Nicolás de Hidalgo, Mexico

Elias Martinez-Hernandez, Department of Engineering Science, University of Oxford, UK

Noor Azian Morad, Malaysia-Japan International Institute of Technology, Universiti Teknologi Malaysia, Malaysia

Aikaterini D. Mountraki, School of Chemical Engineering, National Technical University of Athens, Greece

Fabricio Nápoles-Rivera, Chemical Engineering Department, Universidad Michoacana de San Nicolás de Hidalgo, Mexico

Denny K. S. Ng, Department of Chemical and Environmental Engineering/Centre of Sustainable Palm Oil Research (CESPOR), The University of Nottingham, Malaysia

Kok Siew Ng, Centre for Environmental Strategy, University of Surrey, UK

Rex T. L. Ng, Department of Chemical and Biological Engineering, University of Wisconsin–Madison, WI, USA

Lik Yin Ng, Department of Chemical and Environmental Engineering/Centre of Sustainable Palm Oil Research (CESPOR), The University of Nottingham, Malaysia

José M. Ponce-Ortega, Chemical Engineering Department, Universidad Michoacana de San Nicolás de Hidalgo, Mexico

Michael Angelo B. Promentilla, Chemical Engineering Department, De La Salle University, Philippines

Alberto Quaglia, CAPEC-PROCESS Research Center, Department of Chemical and Biochemical Engineering, Technical University of Denmark (DTU), Denmark

Luis F. Razon, Department of Chemical Engineering, De La Salle University, Philippines

Jhuma Sadhukhan, Centre for Environmental Strategy, University of Surrey, UK

Joost R. Santos, Engineering Management and Systems Engineering Department, The George Washington University, USA

Medardo Serna-González, Chemical Engineering Department, Universidad Michoacana de San Nicolás de Hidalgo, Mexico

Gürkan Sin, CAPEC-PROCESS Research Center, Department of Chemical and Biochemical Engineering, Technical University of Denmark (DTU), Denmark

Chiang Jinn Tan, Department of Chemical Engineering, Biomass Processing Laboratory, Center of Biofuel and Biochemical, Green Technology (MOR), Universiti Teknologi PETRONAS, Malaysia

Raymond R. Tan, Chemical Engineering Department, De La Salle University, Philippines

Chung Loong Yiin, Department of Chemical Engineering, Biomass Processing Laboratory, Center of Biofuel and Biochemical, Green Technology (MOR), Universiti Teknologi PETRONAS, Malaysia

Krista Danielle S. Yu, School of Economics, De La Salle University, Philippines

Suzana Yusup, Department of Chemical Engineering, Biomass Processing Laboratory, Center of Biofuel and Biochemical, Green Technology (MOR), Universiti Teknologi PETRONAS, Malaysia

Preface

Major environmental issues, particularly climate change, have stimulated research activities focusing on enhancing the sustainability of industrial processes. In particular, significant effort has been placed on developing viable alternatives to challenge the dominance of fossil fuels. Among the available technology options, biomass offers the possibility of a renewable supply of low-carbon feedstock for the production of clean energy, chemicals, and other products. Historically, interest in biomass as an industrial resource has peaked and waned in response to energy market trends and in fact has recently been dampened by the availability of low-cost fossil energy from nonconventional reserves; nevertheless, biomass is still widely regarded as an essential component toward the long-term development of low-carbon industries in the twenty-first century.

Research on biomass conversion is a primary requisite to the development of sustainable energy and chemical production systems. Such work is needed at different scales in order to provide the necessary scientific foundations for the improvement of existing processes, the innovation of new manufacturing routes, and the commercial deployment of new technologies. For instance, recent laboratory-scale experiments have yielded a multitude of reaction pathways for transforming a wide variety of biomass feedstocks into value-added products. The challenge of creating economically viable systems requires systematic process development approaches that lead to the synthesis and design of efficient biomass conversion facilities. Integration of the biomass conversion steps with the rest of the processing facility and utility systems offers opportunities for enhancing the efficiency and sustainability of the whole process. Furthermore, supply chain considerations should be a major component in the planning of large-scale biomass processing.

This book covers recent developments in process engineering and resource conservation for biomass conversion systems at scales ranging from the molecular level all the way to macrolevel supply chains. It provides an overview of process development in biomass conversion systems, with focus on biorefineries involving the production and coproduction of fuels, heating, cooling, and chemicals. Various techniques for enhancing the efficiency of natural resource utilization are also covered as an essential element of developing competitive biomass-based industries. Technical, economic, environmental, and social aspects of biorefineries are discussed and integrated.

The book features 14 chapters written by leading experts from around the world and presents an integrated set of contributions that are categorized into three major sections. The first part of the book deals with *Process Design Tools for Biomass Conversion Systems* and includes five chapters. Chapters 1–3, entitled "Early-Stage Design and Analysis of Biorefinery Networks" (*Peam Cheali*, *Alberto Quaglia*, *Carina L. Gargalo*, *Krist V. Gernaey*, *Gürkan Sin*, and *Rafiqul Gani*), "Application of a Hierarchical Approach for the Synthesis of Biorefineries" (*Carolina Conde-Mejía*, *Arturo Jiménez-Gutiérrez*, and *Mahmoud M. El-Halwagi*), and "A Systematic Approach for Synthesis of an Integrated Palm Oil-Based Biorefinery" (*Rex T. L. Ng* and *Denny K. S. Ng*), offer systematic approaches to the conceptual design, process synthesis, and screening of alternatives in the early process development stages. Chapter 4, entitled "Design Strategies for Integration of Biorefinery Concepts at Existing Industrial Process Sites: Case Study of a Biorefinery Producing Ethylene from Lignocellulosic Feedstock as an Intermediate Platform for a Chemical Cluster" (*Roman Hackl* and *Simon Harvey*), focuses on the coupling of emerging biorefineries with existing industrial infrastructures. Chapter 5, entitled "Synthesis of Biomass-Based Tri-generation Systems with Variations in Biomass Supply and Energy Demand" (*Viknesh Andiappan*, *Denny K. S. Ng*, and *Santanu Bandyopadhyay*), gives a synthesis approach to the energy and mass aspects of a bioconversion system in the context of energy and mass variability in the market.

The second part of the book features three chapters on *Regional Biomass Supply Chains and Risk Management*. Chapter 6, entitled "Large-Scale Cultivation of Microalgae for Fuel" (*Christina E. Canter*, *Luis F. Razon*, and *Paul Blowers*), surveys recent developments in the commercial-scale production of microalgal biomass, which is considered to be one of the most promising next-generation feedstocks due to its inherently high photosynthetic efficiency. In Chapter 7, entitled "Optimal Planning Sustainable Supply Chains for the Production of *Ambrox®* based on *Ageratina jocotepecana* in Mexico" (*Sergio I. Martínez-Guido*, *J. Betzabe González-Campos*, *Rosa E. Del Río*, *José M. Ponce-Ortega*, *Fabricio Nápoles-Rivera*, and *Medardo Serna-González*), a process systems engineering approach to the systematic design of a large-scale biomass supply chain is described. Then, systematic risk analysis focusing on ripple effects is discussed in Chapter 8, entitled "Inoperability Input–Output Modeling Approach to Risk Analysis in Biomass Supply Chains" (*Krista Danielle S. Yu*, *Kathleen B. Aviso*, *Mustafa Kamal Abdul Aziz*, *Noor Azian Morad*, *Michael Angelo B. Promentilla*, *Joost R. Santos*, and *Raymond R. Tan*).

The third part of the book covers *Other Applications of Biomass Conversion Systems*. Chapter 9, entitled "Process Systems Engineering Tools for Biomass Polygeneration Systems with Carbon Capture and Reuse" (*Jhuma Sadhukhan*, *Kok Siew Ng*, and *Elias Martinez-Hernandez*), presents the use of techno-economic analysis and carbon dioxide (CO_2) pinch analysis techniques to develop integration configurations for CO_2 utilization and exchange in a biorefining system. Another work on cogeneration system is found in Chapter 10, entitled "Biomass-Fueled Organic Rankine Cycle-Based Cogeneration System" (*Nishith B. Desai* and *Santanu Bandyopadhyay*). Chapter 11, entitled "Novel Methodologies for Optimal Product Design from Biomass" (*Lik Yin Ng*, *Nishanth G. Chemmangattuvalappil*, and *Denny K. S. Ng*), discussed the use of computer-aided molecular design technique for product design. Next, a comparative study using process integration technique between biotechnological and catalytic processes was reported in Chapter 12, entitled "The Role of Process Integration in Reviewing and Comparing Biorefinery Processing Routes: The Case of Xylitol" (*Aikaterini D. Mountraki*, *Konstantinos R. Koutsospyros*, and *Antonis C. Kokossis*). An experimental work was reported in Chapter 13, entitled "Determination of Optimum Condition for the Production of Rice Husk-Derived Bio-Oil by Slow Pyrolysis Process" (*Suzana Yusup*, *Chung Loong Yiin*, *Chiang Jinn Tan*, and *Bawadi Abdullah*). In the last chapter of the book, two important aspects of safety and health are reviewed in the work entitled "Overview of Safety and Health Assessment for Biofuel Production Technologies" (*Mimi H. Hassim*, *Weng Hui Liew*, and *Denny K. S. Ng*).

Together, these 14 chapters cover some of the most recent and important developments in biomass conversion systems research. We hope the book will serve as a useful guidebook for researchers and industrial practitioners working in biomass systems.

Acknowledgments

First, we thank all the chapter contributors who dedicated their time and effort in sharing the excellent work and state-of-the-art developments in biomass conversion system. We also acknowledge Viknesh Andiappan and Goh Wui Seng for their great help in formatting the chapters. Besides, we would also like to acknowledge Eureka Synergy Sdn Bhd and Havys Oil Mill Sdn Bhd for providing the photo for the book cover. We are also grateful to Sarah Keegan, Shiji Sreejish and the editorial team of John Wiley & Sons who provided invaluable assistance throughout the publication process. Finally, we thank our family members for their support throughout our professional careers. Denny Ng would like to thank his wife Yoe Pick Ling and his mother Wong Chook Chan for their continuous support and for taking good care of their children Wing Hsuan, Hon Seng, and Chi Hsuan. Raymond Tan would like to thank his family for their continued support of his scientific work and would like to especially acknowledge his late father, Ramon Tan, for encouraging intellectual pursuits. Dominic Foo would like to thank his wife Cecilia for tremendous support, especially in taking good care of their young daughters Irene and Jessica. Mahmoud El-Halwagi would like to acknowledge his parents, his wife Amal, and sons Omar and Ali for their constant support and unlimited love.

Denny K. S. Ng
Raymond R. Tan
Dominic C. Y. Foo
Mahmoud M. El-Halwagi

Part 1
Process Design Tools for Biomass Conversion Systems

1

Early-Stage Design and Analysis of Biorefinery Networks

Peam Cheali, Alberto Quaglia, Carina L. Gargalo,
Krist V. Gernaey, Gürkan Sin, and Rafiqul Gani
CAPEC-PROCESS Research Center, Department of Chemical and Biochemical Engineering,
Technical University of Denmark (DTU), Kongens Lyngby, Denmark

1.1 Introduction

The limited resources of fossil fuel as well as other important driving forces (e.g., environmental, social, and sustainability concerns) are expected to shape the future development of the chemical processing industries. These challenges motivate the development of new and sustainable technologies for the production of fuel, chemicals, and materials from renewable feedstock instead of fossil fuel. An emerging technology in response to these challenges is the biorefinery concept. The biorefinery is defined as the set of processes converting a bio-based feedstock into products such as fuels, chemicals, materials, and/or heat and power.

The design of a biorefinery process is a challenging task. First, several different types of biomass feedstock and many alternative conversion technologies can be selected to match a range of products, and therefore, a large number of potential processing paths are available for biorefinery development. Furthermore, being based on a natural feedstock, the economic and environmental viability of these processes is deeply dependent on local factors such as weather conditions, availability of raw materials, national or regional subsidies and regulations, etc. Therefore, the replication of a standard process configuration is often not convenient or impossible. Designing a biorefinery, therefore, requires screening among

Process Design Strategies for Biomass Conversion Systems, First Edition. Edited by
Denny K. S. Ng, Raymond R. Tan, Dominic C. Y. Foo, and Mahmoud M. El-Halwagi.
© 2016 John Wiley & Sons, Ltd. Published 2016 by John Wiley & Sons, Ltd.

a set of potential configurations in order to identify the most convenient option for the given set of conditions.

Detailed evaluation of each process alternative requires a substantial amount of information such as conversions and efficiencies for the different steps involved. Moreover, considerable time and resources are needed to execute the analysis, and it is therefore not practically possible to consider more than a handful of candidate processing paths. In order to partially overcome these drawbacks, a second level of decomposition is often employed based on the so-called *development funnel* approach (see Figure 1.1). The basic idea of the development funnel approach is to progressively reduce the number of candidate alternatives by employing simplified model and shortcut evaluation methods to identify nonconvenient or nonfeasible options and eliminate those from the set of candidate configurations.

One of the challenges associated with this development funnel approach lies in the ability of performing the early-stage screening in a project phase characterized by lack of detailed data. As a consequence, it is important to simplify and manage the complexity related to the vast amount of data that needs to be processed prior to identifying the optimal biorefinery processing path with respect to economics, consumption of resources, sustainability, and environmental impact.

In order to manage the complexity and perform synthesis and design of biorefineries, several publications have focused on simplification and different aspects of the problem: the study of Voll and Marquardt (2012) explored the use of reaction flux network analysis for synthesis and design of biorefinery processing paths, Pham and El-Halwagi (2012) proposed a systematic two-stage methodology to reduce the number of processing steps, Martin and Grossmann (2012) evaluated the heat integration on a biorefinery process flowsheet producing FT-diesel, Baliban et al. (2012) studied the heat and water integration and supply chain optimization of thermochemical conversion of biomass, Zondervan et al. (2011) studied the identification of the optimal processing paths of the biochemical platform, and finally, Cheali et al. (2014) presented a generic modeling framework to manage the complexity of the multidisciplinary data needed for superstructure-based optimization of biorefinery

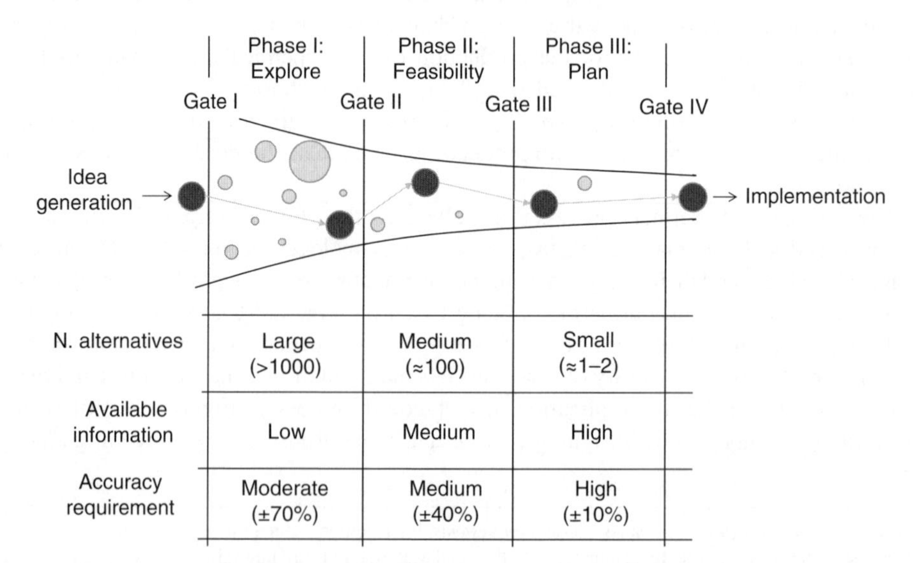

Figure 1.1 *A schematic representation of the development funnel for a project in the processing industries. Reproduced from Alberto Quaglia, Ph.D. thesis, with permission*

systems. A more detailed review of studies on the process synthesis of a biorefinery is given in Yuan et al. (2013).

While each of the abovementioned studies provided a valuable contribution, however, the scope of these studies was limited to one processing/conversion platform. Or, in other words, the studies focused either on biochemical, thermochemical, or biological platforms. In this contribution, as we focus on early-stage design and analysis of biorefinery systems, the scope of the biorefinery synthesis is broadened by considering a combination of thermochemical and biochemical platforms. In this way, the design space is extended significantly, meaning that more potential platforms and design alternatives can be compared resulting in a more robust and sustainable design solution. It is important to note that designing a biorefinery includes other challenges as well, such as the supply chain of the feedstock and land use, among others. These are beyond the scope of this study and will be considered in future work.

A methodology to generate and identify optimal biorefinery networks was developed earlier in our group (Zondervan et al., 2011; Quaglia et al., 2013). We present here the adaptation and extension of the methodology for the biorefinery problem. We expand the scope and the size of the biorefinery network problem by extending the database, the models, and the superstructure of the methodology with thermochemical biomass conversion routes. We then integrate the thermochemical superstructure with the superstructure of the biochemical conversion network. We then present a generic process modeling approach together with data collection and management for the multidisciplinary and multidimensional data related to different biorefinery processing steps. The optimal processing paths are then identified with respect to the given scenarios and specifications by formulating and solving an MILP/mixed-integer nonlinear programming problem (MINLP) problem using the GAMS optimization software. The resulting optimal biorefinery network is then further studied with respect to sustainability and environmental impact using two in-house software tools, SustainPro (Carvalho et al., 2013) and LCSoft (Piyarak, 2012), respectively.

1.2 Framework

This study uses the integrated business and engineering framework (Figure 1.2) which was successfully applied to synthesis and design of a wide range of different processes (Quaglia et al., 2013). The framework uses a superstructure optimization-based process synthesis combined with a generic modeling approach, thus allowing the possibility of generating a larger design space, of managing the data and model complexity, and of identifying the optimal processing path with respect to technical and economic feasibility. The framework is integrated with the analysis and evaluation of sustainability and environmental impact.

A schematic representation of the framework is reported in Figure 1.2. The description of the framework is presented step by step in this chapter:

Step 1: Problem definition

The first step includes the definition of the problem scope (i.e., design a biorefinery network, wastewater treatment plant network, a processing network for vegetable oil production), the selection of suitable objective functions (i.e., maximum profit of the biorefinery, minimum total annualized cost (TAC) of the wastewater treatment plant), and optimization scenarios with respect to either business strategy, engineering performance, sustainability, or a combination of such objectives.

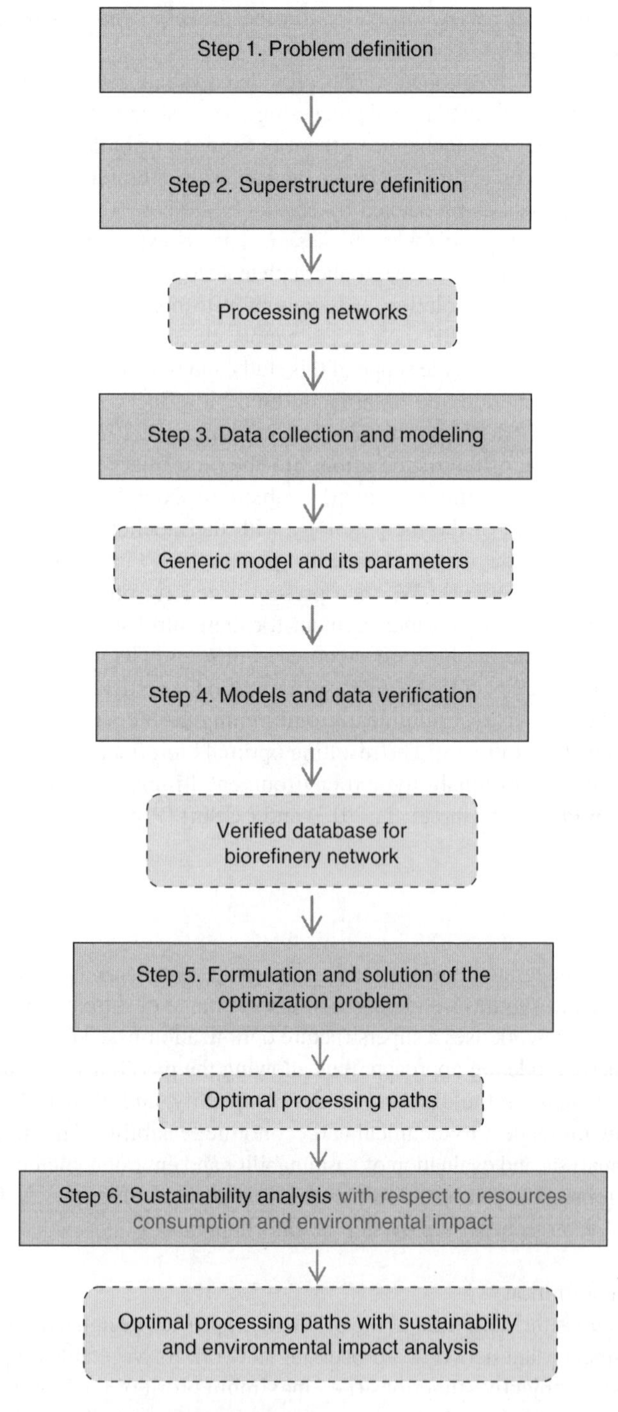

Figure 1.2 *The integrated business and engineering framework adapted: the dashed boxes indicate the outcome of each step of the workflow*

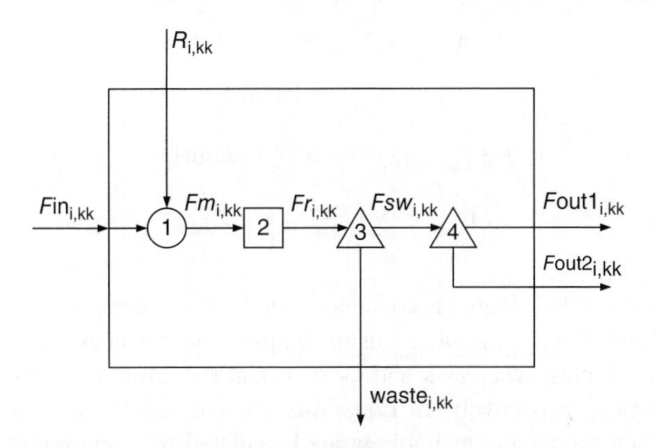

Figure 1.3 *The generic process model block. Reproduced from Cheali et al. (2014), © 2014, American Chemical Society*

Step 2: Superstructure definition
A superstructure representing different biorefinery concepts and networks is formulated by performing a literature review. A typical biorefinery network consists of a number of processing steps converting or connecting biomass feedstock to bioproducts such as pretreatment, primary conversion (gasification, pyrolysis), gas cleaning and conditioning, fuel synthesis, and product separation and purification. Each processing step is defined by one or several blocks depending on the number of unit operations considered in the step (several unit operations can be modeled using one process block). Each block incorporates the generic model to represent various tasks carried out in the block such as mixing, reaction, and separation (Figure 1.3).

Step 3: Data collection and modeling
Once the superstructure is defined, the data are collected and modeling is performed. Generally, the models for each processing technology are rigorous, nonlinear, and complex (e.g., kinetics, thermodynamics). However, in this step, a simple input–output-type generic model block is used, and this model is identified from the data generated from the aforementioned rigorous models. This generic model block thus consists of four parts of the typical simple mass balance equations: (i) mixing, (ii) reaction, (iii) waste separation, and (iv) product separation. The simple mass balance models representing each section of the generic model block are presented (Eqs. 1.1–1.7) and explained below:

$$Fm_{\mathrm{i,kk}} = Fin_{\mathrm{i,kk}} + R_{\mathrm{i,kk}} \tag{1.1}$$

$$R_{\mathrm{i,kk}} = \left(\mu_{\mathrm{i,j,kk}} \times Fin_{\mathrm{i,kk}} \times \alpha_{\mathrm{i,kk}}\right) \tag{1.2}$$

$$Fr_{\mathrm{i,kk}} = Fm_{\mathrm{i,kk}} + \mathrm{MW_i} \times \sum_{\mathrm{rr}} \left(\gamma_{\mathrm{i,rr}} \times \theta_{\mathrm{react,rr}} \times Fm_{\mathrm{i,kk}} / \mathrm{MW_{react}}\right) \tag{1.3}$$

$$F\mathrm{sw}_{\mathrm{i,kk}} = \left(1 - \mathrm{SW_{i,kk}}\right) \times Fr_{\mathrm{i,kk}} \tag{1.4}$$

$$\mathrm{waste_{i,kk}} = \mathrm{SW_{i,kk}} \times Fr_{\mathrm{i,kk}} \tag{1.5}$$

$$F\mathrm{out1}_{\mathrm{i,kk}} = \mathrm{split_{i,kk}} \times F\mathrm{sw}_{\mathrm{i,kk}} \tag{1.6}$$

$$Fout2_{i,kk} = (1 - split_{i,kk}) \times Fsw_{i,kk} \tag{1.7}$$

$$F1_{i,k,kk} \le S_p^{\ k,kk} \times Fout1_{i,kk} \tag{1.8}$$

$$F2_{i,k,kk} \le (S_p^{\ k,kk} - S^{k,kk}) \times Fout1_{i,kk} \tag{1.9}$$

$$Fin_{i,kk} = \sum_k \left(F1_{i,k,kk} + F2_{i,k,kk}\right) \tag{1.10}$$

The above equations (1.1–1.7) are the equations used for the generic model block to estimate the outlet mass flow ($Fout1_{i,kk}$, $Fout2_{i,kk}$) using simple mass balances. The subscripts i and j represent the components, whereas k and kk represent the upstream and downstream processing technologies, respectively. In Equations 1.1 and 1.2, the chemicals and utilities used ($R_{i,kk}$) for each processing technology are calculated by using the ratio ($\mu_{i,j,kk}$) to the inlet mass flow rate ($Fin_{i,kk}$). The parameter $\alpha_{i,kk}$ represents the amount of the utilities or chemicals carried to the outlets. In Equation 1.3, the reaction outlet mass stream ($Fr_{i,kk}$) is calculated based on stoichiometry, $\gamma_{i,rr}$, and conversion fraction, $\theta_{react,rr}$. In Equations 1.4 and 1.5, the waste stream ($waste_{i,kk}$) and the remaining stream ($Fsw_{i,kk}$) are calculated on the basis of the removal fraction, $SW_{i,kk}$. The product outlet streams are calculated in Equations 1.6 and 1.7 on the basis of a product separation fraction, $Split_{i,kk}$. Moreover, in order to connect each generic model block and thereby formulate the superstructure, Equations 1.8–1.10 are used. The mass outlet flows mentioned earlier ($Fout1_{i,kk}$, $Fout2_{i,kk}$) are called primary and secondary outlet flows, respectively. The primary and secondary outlet flows are connected to the next generic model blocks using binary variables (S_p, S), respectively. The outlet flows between the generic model blocks ($F1_{i,k,kk}$, $F2_{i,k,kk}$) of each stream (primary and secondary) are summed up as the input of the next generic model block. It is noted that recycle flows can be considered using Equations 1.8–1.10. There are two potential cases of recycle flows addressed: (i) recycle flows within the same processing step, that is, internal recirculation—the simulation of the recycle flows and their impact on process performance needs to be done prior to estimating the parameter values for the corresponding generic model block (e.g., processing step 2, 4)—and (ii) recycle between processing steps (e.g., processing step 4 to processing step 2 or 3), which is handled by using Equations 1.8 and 1.9.

The appropriate values for the aforementioned parameters can be collected in several ways including (i) literature sources or technical reports, (ii) experimental data, (iii) simulation results, or (iv) stream table or operating data of a designed flowsheet. The collected parameters are in the end organized in a multidimensional matrix form which represents the activities (chemicals/utilities used, reactions, separations, etc.) occurring in the processing alternatives.

Step 4: Models and data verification

After the superstructure is defined and the parameters are collected, a validation of the selected models and parameters needs to be performed for quality and consistency check. The verification can be performed in this step by fixing the decision variables in the MINLP problem formulation—that is, the vector y—and thereby to perform a simulation for each processing technology or path, followed by comparison of the simulation results against the available data. Such data can originate either from experiments or from the literature. All the necessary equations and constraints relevant to each processing technology are also formulated in this step, prior to being solved as MILP or MINLP problems in GAMS. The output of this step is a verified database representing the biorefinery superstructure formulated in step 2 and stored in an Excel worksheet.

Step 5: Formulation and solution of the optimization problem
In this step, the optimization problem is formulated as MILP or MINLP problem depending on the objective function definition and constraints using appropriate software, in this case GAMS. The output is the optimal biorefinery configuration. The generic models and structure of the optimization problem (MIP/MINLP) organized and used in this study are presented and explained in the following text.

The optimization formulation (presented in Eqs. 1.11–1.16) consists of the objective function (e.g., minimize TAC; Eq. 1.11) subjected to process constraints, the process models and constraints (Eqs. 1.1–1.10) of the generic model block mentioned earlier (*x* is a process variable, the mass flow rate), structural constraints (Eqs. 1.12 and 1.13) representing the superstructure which allows selection of only one process alternative in each step, and cost functions (Eqs. 1.14–1.16) to calculate the operating and capital costs using cost parameters ($P1_{\mathrm{i,kk}}^{\mathrm{waste}}$, waste treatment cost; $P2_{\mathrm{i,kk}}^{\mathrm{utilities/chemicals}}$, utility or chemicals cost; $P3_{\mathrm{a}}^{\mathrm{kk}}$, reactor investment cost; $P3_{\mathrm{b}}^{\mathrm{kk}}$, separation investment cost; $\mathrm{capex}_{\mathrm{kk}}$, capital expenditures). The details on a related optimization problem can be found in the previous studies (Zondervan et al., 2011; Quaglia et al., 2013).

As an example, the objective function is formulated such as to minimize the TAC:

$$\mathrm{OBJ} = \sum_{\mathrm{kk}} \mathrm{OPEX} - \left(\mathrm{CAPEX}_1 + \mathrm{CAPEX}_2\right)/t \tag{1.11}$$

Subject to the following constraints:

i. Process models of the generic model block $h\left(\mu_{\mathrm{i,j,kk}}, \alpha_{\mathrm{i,kk}}, \gamma_{\mathrm{i,rr}}, \theta_{\mathrm{react,rr}}, \mathrm{MW}_{\mathrm{i}}, \mathrm{SW}_{\mathrm{i,kk}}, \mathrm{split}_{\mathrm{i,kk}}\right) = 0$, as mentioned earlier, Equations 1.1–1.7 and 1.10
ii. Process constraints $g\left(S_{\mathrm{p}}^{\mathrm{k,kk}}, S^{\mathrm{k,kk}}\right) \leq 0$, as mentioned earlier, Equations 1.8 and 1.9
iii. Structural constraints:

$$\sum_{k} y_k \leq 1 \tag{1.12}$$

$$y \in \{0;1\}^n \tag{1.13}$$

iv. Cost constraints:

$$\mathrm{OPEX}_{\mathrm{kk}} = (P2_{\mathrm{i,kk}}^{\mathrm{utilities/chemicals}} \times R_{\mathrm{i,kk}}) + (P1_{\mathrm{i,kk}}^{\mathrm{waste}} \times F_{\mathrm{i,kk}}^{\mathrm{waste}}) \tag{1.14}$$

$$\mathrm{CAPEX}_{\mathrm{thermochem}} = \sum_{\mathrm{kk}} \mathrm{capex}_{\mathrm{kk}} \tag{1.15}$$

$$\mathrm{CAPEX}_{\mathrm{biochem}} = \sum_{\mathrm{kk}} \left[P3_a^{\mathrm{kk}} \times Fm_{\mathrm{i,kk}}^{\ \mathrm{n1}} + P3_b^{\mathrm{kk}} \times Fr_{\mathrm{i,kk}}^{\ \mathrm{n2}} \right] \tag{1.16}$$

v. Optimization constraints (big-*M* formulation):

$$\text{Process variables}\left(\text{e.g., } F_{\mathrm{i,kk}}\right) \leq M \times y_{\mathrm{kk}} \tag{1.17}$$

For the solution, GAMS retrieves the generic model parameters and other data appearing in the constraints (e.g., $\alpha_{i,kk}$, $\gamma_{i,rr}$, $\theta_{react,rr}$, $P1_{i,kk}^{waste}$, $P2_{i,kk}^{utilities/chemicals}$) from the database. In this way, the overall MINLP problem formulation is separated into two parts: (i) data handling and representation (as described in this contribution, with help of a generic process model and its parameters stored in a database) and (ii) solution and analysis of the problem. This separation of the problem in two parts helps with the management of the complexity of formulating an MINLP-based optimization problem for biorefinery networks.

Step 6: Sustainability and environmental impact analysis
The presented framework has been integrated with the sustainability and environmental impact assessment. Both sustainability and environmental impact analysis are performed using two in-house software tools, SustainPro (Carvalho et al., 2013) and LCSoft (Kalakul et al., 2014), respectively. As a prerequisite for this step, some extra data are needed such as rigorous mass and energy balance, connectivity among the unit operations within the flowsheet, duty, and reaction data which are explained in more detail in the following text.

1.2.1 Sustainability Analysis

A sustainable process is characterized by the use of renewable resources as raw materials and as energy sources and utilities to produce biodegradable products while minimizing the production of waste, the use of nonenvironmentally friendly chemicals and the external dependence on energy and water. In this study, a generic and systematic approach of sustainability analysis developed by Carvalho et al. (2013) has been integrated with the business and engineering framework already proposed (see Figure 1.2). This integrated tool enables the user to evaluate the process in terms of raw material, water, and energy usage through the calculation of sustainability metrics (Azapagic, 2002) and by a set of previously defined indicators (Uerdingen et al., 2003). The sustainability-related methodology was adapted into Excel VBA-based software, resulting in the so-called SustainPro software tool, which contains four main parts as follows (see Table 1.1): (i) flowsheet decomposition into open path (OP) and closed path (CP), (ii) path flow assessment in terms of mass and energy indicators analysis, (iii) sustainability metrics calculation, (iv) and generation and comparison of new retrofit alternatives—note that this last stage is beyond the scope of this study.

The models used to perform the indicators analysis in SustainPro are presented below (Eqs. 1.18–1.22) (Uerdingen et al., 2003):

Material value added (MVA): The indicator that reflects the value added between the entrance and exit of a given compound in a certain path, and therefore, this indicator is only estimated for OPs. A high negative value of MVA indicates that the specific compound in that particular path is losing value along the process. Ideally, MVA should be close to zero. As mentioned before, the software defines paths or routes for each and every compound in the system, classifying a compound as OP, meaning that the compound enters and leaves the system, or CP, meaning that the compound is being recycled:

$$\mathrm{MVA}_{o}^{i} = m_{o}^{i} \times \left[\mathrm{PP}_{o}^{i} - \mathrm{CA}_{0}^{i} \cdot \left(\sum_{rm=1}^{RM} \frac{\left| \upsilon_{0}^{(rm)} \right| \cdot M_{0}^{(rm)}}{\upsilon_{0}^{i} \cdot M_{0}^{i}} \times \mathrm{PR}_{0}^{(rm)} \right) \right] \tag{1.18}$$

Table 1.1 *The simplified explanation of each section in SustainPro*

Part	Name	Description
Prerequisite	Input	• Mass and energy balance (stream table) • Component properties • Reactions • Duty of unit operations
(i)	Flowsheet decomposition (path decomposition)	Process streams are decomposed into: • Open path (OP): refers to a certain compound that enters and leaves the system, therefore characterizing a specific path • Closed path (CP): refers to a certain compound that is being recycled within the system, therefore characterizing a specific path
(ii)	Path flow assessment	The performance of each path flow is assessed through a set of indicators (MVA, EWC, RQ, AF, TVA) which are explained in Eqs. 1.18–1.22. Rank the paths from top to bottom, where the top are the worst in terms of the process sustainability, therefore identifying the process critical points (bottlenecks) with respect to energy and mass (raw materials and water)
(iii)	Evaluation	Sustainability metrics calculation with respect to energy, water, and raw material usage
(iv)	Generation and comparison of new alternatives	• Generation of retrofit alternatives according to the critical points identified by the application of the general known retrofit rules

where $m_{\mathrm{o}}^{\mathrm{i}}$ is the flow rate of component i in the OP flow o, $\mathrm{PP}_{\mathrm{o}}^{\mathrm{i}}$ is the purchase price, $\mathrm{CA}_{0}^{\mathrm{i}}$ is the cost allocation factor calculated by dividing the purchase price of the component i by the total purchase price of product, $\upsilon_{0}^{\mathrm{i}}$ is the stoichiometric coefficient, M_{0}^{i} is the molecular weight, and $\mathrm{PR}_{0}^{(\mathrm{rm})}$ is the raw material price.

Energy and waste cost (EWC): The indicator of overall process costs related to utility consumption and waste treatment of a component path flow—for OP and CP. A high value of EWC means that the given path carries important units of energy that could be used for heat integration within the system but are being discarded instead:

$$\mathrm{EWC}_{\mathrm{k}}^{\mathrm{i}} = \sum_{\mathrm{u=1}}^{\mathrm{U}} \mathrm{PE}_{\mathrm{u}} \times Q_{\mathrm{u}} \times \frac{m_{\mathrm{k}}^{\mathrm{i}} \cdot A_{\mathrm{u,k}}^{\mathrm{i}}(T,p)}{\sum_{\mathrm{uk=1}}^{\mathrm{UK}} m_{\mathrm{uk}} \cdot A_{\mathrm{u,uk}}(T,p)} \tag{1.19}$$

where PE_{u} is the energy price of subunit operation u, Q_{u} is the subunit operation duty, $m_{\mathrm{k}}^{\mathrm{i}}$ is the mass flow rate along the path flow k of the component i, and $A_{\mathrm{u,k}}^{\mathrm{i}}$ is the specific heat capacity.

Reaction quality (RQ): The indicator of the process productivity. A positive value (0–1) shows that a certain compound in a given path has a positive contribution to the process productivity. On the other hand, a negative value means that the given path has a negative impact on the overall process productivity. This metric is also estimated for OP and CP:

$$\mathrm{RQ}_{\mathrm{k}}^{\mathrm{i}} = \sum_{\mathrm{r=1}}^{\mathrm{R}} \sum_{\mathrm{rk=1}}^{\mathrm{RK}} \frac{\xi_{\mathrm{r,rk,k}} \cdot E_{\mathrm{r,rk,k}}^{i}}{\sum_{\mathrm{fp=1}}^{\mathrm{FP}} n^{(\mathrm{fp})}} \tag{1.20}$$

where $\xi_{\mathrm{r,rk,k}}$ is the extent of reaction rk, $n^{(\mathrm{fp})}$ is the mole flow rate of a desired final product, and $E_{\mathrm{r,rk,k}}^{\mathrm{i}}$ represents the effect of the component i on reaction rk: +1 means component i is favored to the desired product, 0 means no effects, and −1 means that the compound *i* is inhibiting the formation of the desired product.

Accumulation factor (AF): The indicator that reflects the accumulation behavior of the compounds that are being recycled within the system, meaning that the recycle flow rate of a certain compound (a certain representative path in the system) is being evaluated regarding the amount of fresh compound that is being added to the system:

$$\mathrm{AF}_{\mathrm{z}}^{\mathrm{i}} = \frac{m_{\mathrm{z}}^{\mathrm{i}}}{\sum_{\mathrm{i}=l}^{\mathrm{I}} \cdot \left(\sum_{\mathrm{a=1}}^{\mathrm{EP}} f_{\mathrm{i,a}}^{\mathrm{i}} + \sum_{\mathrm{op=1}}^{\mathrm{OP}} d_{\mathrm{i,op}}^{\mathrm{i}} \right)} \tag{1.21}$$

where $m_{\mathrm{z}}^{\mathrm{i}}$ is the mass flow rate in cycle path flow z and $f_{\mathrm{i,a}}^{\mathrm{i}}$ and $d_{\mathrm{i,op}}^{\mathrm{i}}$ are the flow rates leaving the cycle path flow.

Total value added (TVA): The indicator that reflects the economic impact of a given path directly related to a certain compound in the system. Due to the aforementioned MVA and EWC constraints, a high negative value of TVA indicates that a given compound in a certain path is losing value along the process in terms of energy that is being wasted, raw material that is not being recycled, or even a valuable by-product that is being discarded:

$$\mathrm{TVA}_{\mathrm{k}}^{\mathrm{i}} = \mathrm{MVA}_{\mathrm{k}}^{\mathrm{i}} - \mathrm{EWC}_{\mathrm{k}}^{\mathrm{i}} \tag{1.22}$$

1.2.2 Environmental Impact Assessment

Life cycle assessment (LCA) is a method to quantify the environmental impact of the designed process throughout the product–process life cycle. LCA is defined by the ISO14040 and ISO14044 standards as the evaluation of inputs, outputs, and the system during the lifetime of the process. It consists of a systematic framework including the following steps: (i) goal and scope definition, (ii) inventory analysis or life cycle inventory (LCI), (iii) impact assessment or life cycle impact assessment (LCIA), and (iv) interpretation.

Earlier in our group, an Excel VBA-based software tool—LCSoft—was developed according to the standard for the environmental impact and carbon footprint analysis (Piyarak, 2012). The tool contains three main parts: (i) LCI knowledge management, (ii) calculation factor estimation, and (iii) LCA calculation. It is noted that US EPA and IPCC emission factors are used to calculate the environmental impact for a given process. Figure 1.4 and Table 1.2 present the framework and a short description of each step of LCSoft.

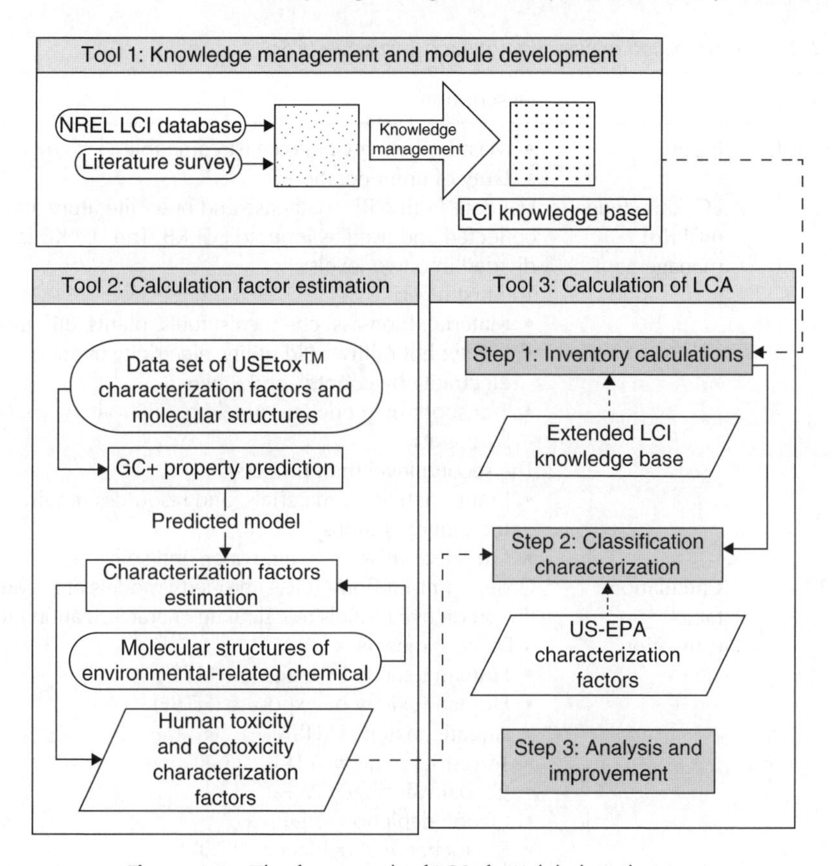

Figure 1.4 *The framework of LCSoft (Kalakul et al., 2014)*

In step 2 of Tool-3, the resources and energy consumption are calculated using the following equations (Eqs. 1.23–1.25):

$$R_{\text{total}} = R_{\text{renew}} + R_{\text{non-renew}} \tag{1.23}$$

$$R_{\text{renew}} = \frac{\left(\sum_{\text{r,i}} m_{\text{i}}^{\text{PI}} \times x_{\text{r,i}} \times \text{HV}_{\text{f}}\right)}{m_{\text{product}}} \tag{1.24}$$

$$R_{\text{non-renew}} = \frac{\left(\sum_{\text{nr,i}} m_{\text{i}}^{\text{PI}} \times x_{\text{nr,i}} \times \text{HV}_{\text{nr}}\right)}{m_{\text{product}}} \tag{1.25}$$

where *r* refers to the renewable resource used to produce input i, nr represents nonrenewable resource used to produce input i, $x_{\text{r,i}}$ is the mass of renewable resource *r* used to produce 1 kg of input i (kg), $x_{\text{nr,i}}$ is the mass of nonrenewable resource nr used to produce 1 kg of input i (kg), HV_{r} is the heating value of renewable resource r (MJ/kg_{r}), HV_{nr} is the heating value of nonrenewable resource nr (MJ/kg_{nr}), R_{total} is the total energy from resource consumption per

Table 1.2 *The simplified explanation of each section in LCSoft*

Part no.	Name	Description
Prerequisite	Input	• Mass and energy balance (stream table) • Duty of unit operations
Tool-1	LCI knowledge (LCI KB) management	LCI data from NREL database and open literature are collected and used as input to LCI KB. The LCI KB is divided into two levels: The first level: • Material: biomass, chemicals, fuels, plants, and others • Utility: hot utility, cold utility, electricity by fuel, electricity by country, and others • Transport: by mode (air, pipeline, rail, road, water) and by country The second level (unit process): • Inputs: activities, materials, and resources required in the unit operations • Outputs: emission to air, water, and soil
Tool-2	Calculation factor estimation	Group-contribution+ (GC+) property models are used as the predictive models to calculate characterization factors (CFs) of 8 impacts: • Human toxicity by ingestion (HTPI) • Human toxicity by exposure (HTPE) • Aquatic toxicity (ATP) • Terrestrial toxicity (TTP) • Global warming (GWP) • Ozone depletion (ODP) • Photochemical oxidation (PCOP) • Acidification (AP) The USEtox database is also used to develop the predictive models to calculate CFs for 3 other impacts: • Carcinogenic (HTC) • Noncarcinogenic (HTNC) • Freshwater ecotoxicity (ET)
Tool-3	Calculation of LCA	This tool contains 3 steps: • Step 1: to check the existence of LCI data and to retrieve LCI data from LCI KB • Step 2: to calculate resources and energy consumptions (Eqs. 1.22–1.24) and to calculate carbon footprint (Eq. 1.25) • Step 3: to assess and analyze the environmental impact (Eq. 1.27)

1 kg of product ($MJ/kg_{product}$), R_{renew} is the total energy from renewable resource consumption per 1 kg of product ($MJ/kg_{product}$), and $R_{non-renew}$ is the total energy from nonrenewable resource consumption per 1 kg of product ($MJ/kg_{product}$).

In this step, the carbon footprint is calculated by means of the following equation (Eq. 1.26):

$$CO_{2_{eq}} = \frac{\left(m_{GHG,air}^{PRO} \times CF_{GHG,air}^{GWP}\right)}{m_{product}} \tag{1.26}$$

$$\text{Carbon footprint} = \sum \text{CO}_{2\text{eq}} \tag{1.27}$$

where GHG represents the greenhouse gases emitted to air from the process, $m^{\text{PRO}}_{\text{GHG,air}}$ is the mass flow rate of GHG emitted to air from the process, $\text{CF}^{\text{GWP}}_{\text{GHG,air}}$ is the characterization factor (CF) for the global warming effect of the greenhouse gas (GHG), and $\text{CO}_{2\text{eq}}$ represents the carbon dioxide equivalent per 1 kg of product.

In step 3, the environmental impact is calculated by the following equation (Eq. 1.28):

$$I^k = \sum_{\text{i,c}} \text{EM}_{\text{i,c}} \times \text{CF}^{\text{k}}_{\text{i,c}} \tag{1.28}$$

where i refers to a chemical emitted to compartment c, k represents the impact category, $\text{CF}^{\text{k}}_{\text{i,c}}$ is the CF of chemical n emitted to compartment c for impact category k, $\text{EM}_{\text{i,c}}$ is the mass of chemical i emitted to compartment c per 1 kg of product, and I^k is the potential environmental impact (PEI) of chemical i for a specific impact category of concern k.

1.3 Application: Early-Stage Design and Analysis of a Lignocellulosic Biorefinery

In this section, the application of the framework is shown and discussed. The superstructure of the biochemical and thermochemical processing networks and the combined network is presented. The data collection and verification are briefly illustrated; a more detailed study on data collection was presented in the previous work (Cheali et al., 2014). The optimal solutions are then identified with different optimization scenarios under techno-economic constraints, hence reflecting the CAPEX and OPEX costs. After the optimal solutions are identified with respect to techno-economic criteria, the solutions are further analyzed with respect to sustainability and environmental impact.

1.3.1 Biorefinery Networks and Identification of the Optimal Processing Paths

Step 1: Problem definition
Four optimization scenarios were studied for technical and economic feasibility on the basis of the extended superstructure which has extended capabilities to compare more potential processing technologies and platforms. The scenarios selected were as follows: (i) to maximize the production of FT-gasoline/diesel; (ii) to maximize FT-gasoline/diesel sales, min. operating cost, and min. investment cost; (iii) to maximize production of ethanol; and (iv) to maximize ethanol sales, min. operating cost, and min. investment cost.

Step 2: Superstructure definition
The superstructure of the biochemical conversion platform of biomass (corn stover) using biomass pretreatment, hydrolysis, and fermentation technologies was developed earlier to produce ethanol, butanol, succinic acid, and acetone with and without gasoline blend as presented in Figure 1.5 (Zondervan et al., 2011). Note that the short description of the processing blocks can be found in Zondervan et al. (2011) as well.

To increase the design space and the potential number of scenarios, the superstructure was extended by combining the biochemical conversion platform with a thermochemical conversion platform (Cheali et al., 2014). A literature review was performed to formulate

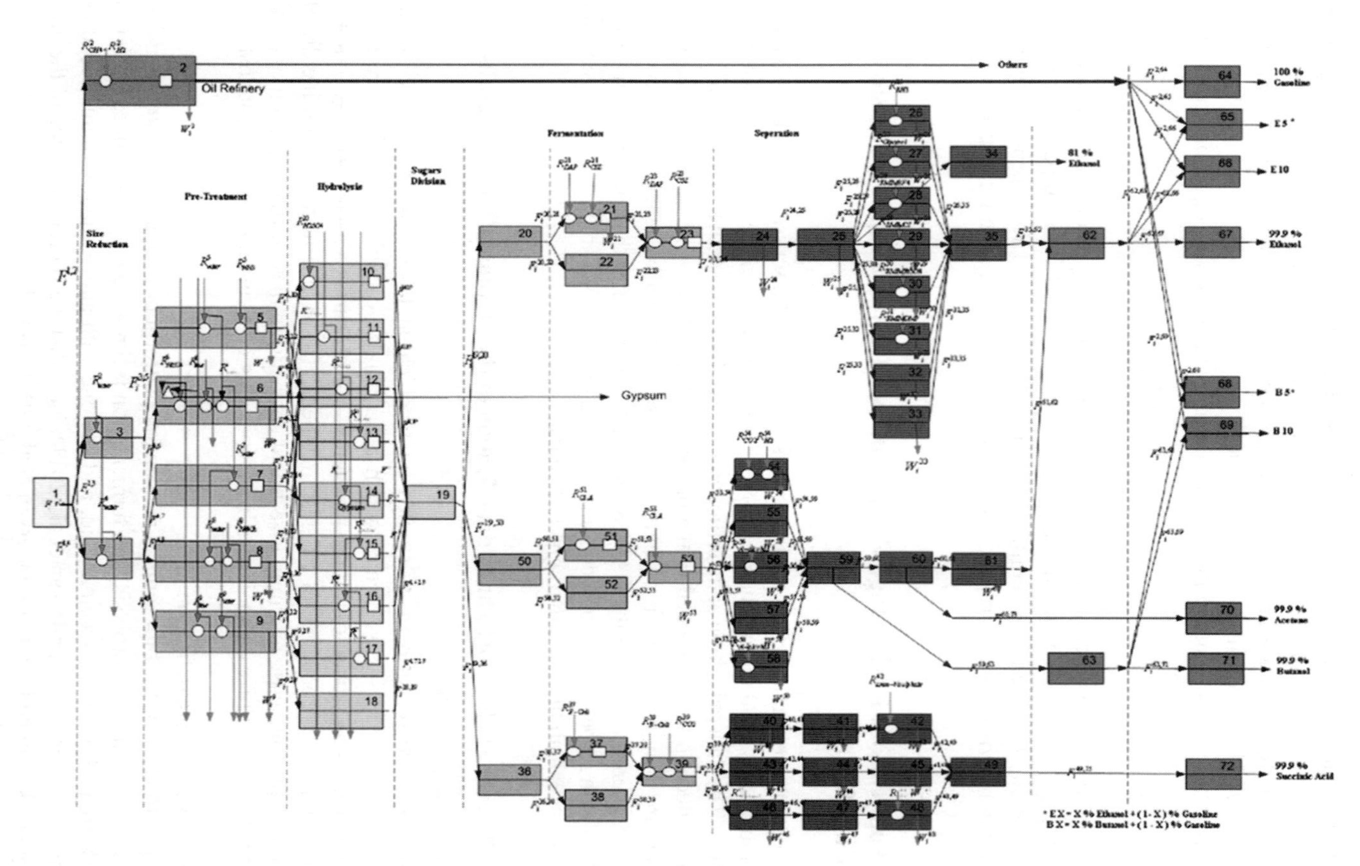

Figure 1.5 *The superstructure of the biochemical conversion platform of biomass*

the superstructure, and in particular, the simulation studies of the US National Renewable Energy Laboratory (NREL) (Phillips et al., 2007; Dutta and Philips, 2009; Dutta et al., 2011; Swanson et al., 2010; Wright et al., 2010) and the Pacific Northwest National Laboratory (PNNL) (Jones et al., 2009) were considered. Figure 1.6 illustrates the superstructure of the thermochemical conversion platform which is proposed based on NREL/PNNL studies. The superstructure consists of 7 sections: 1 feedstock section (source), 5 processing steps (processing tasks), and 1 product section (sink) resulting in a total of 27 process intervals.

As shown in Figure 1.6, corn stover (1) and woody biomass (2) were considered as alternative raw materials. For the thermochemical conversion platform, the processing techniques are generally divided into five processing tasks: (i) pretreatment (size reduction, dryer), (ii) primary conversion (gasification, pyrolysis), (iii) gas cleaning and conditioning (reformer, scrubber, acid gas removal, water–gas shift, PSA), (iv) product synthesis (Fischer–Tropsch (F-T), alcohol synthesis, hydroprocessing unit for pyrolysis oil), and (v) product separation and purification (hydroprocessing unit for F-T products, mol. sieve, and distillation). According to the considered raw materials and processing techniques, a number of products were considered: FT-gasoline, FT-diesel, ethanol, mixed alcohols, two waste heat streams from the gasification, and the reformer. Utilities and waste separation for each processing technology are presented in Figure 1.6. However, recycles which were also considered are not presented in the superstructure.

Then, the superstructure of thermochemical conversion was combined with the superstructure of biochemical conversion resulting in a superstructure with a total of 96 processing intervals, 3 raw materials, 79 processing technologies, and 14 products (main and by-products) as presented in Figure 1.7. The short description of each process interval of the combined superstructure can also be found in Table 1.3.

The extended biorefinery network (thermochemical and biochemical platforms) expands the design space significantly. This means that one can now screen among more potential processing paths and alternatives. The extended networks can also generate more scenarios and solutions and can serve more requirements and specifications of the end users (engineers,

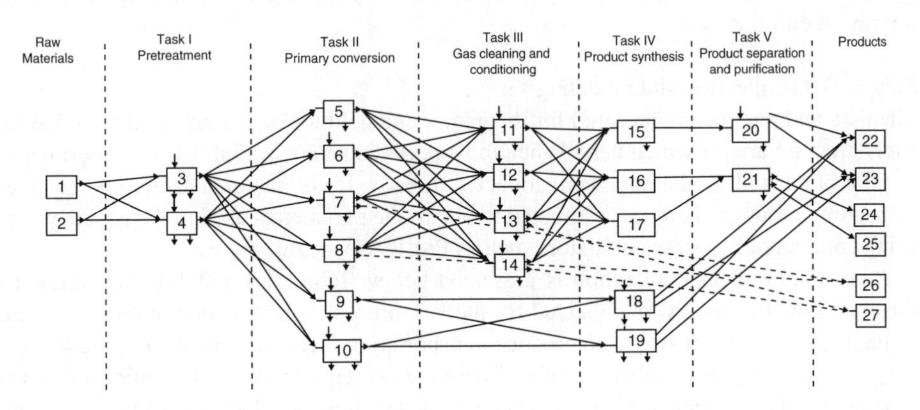

Figure 1.6 *Combined superstructure of two biorefinery conversion platforms: thermochemical (top) and biochemical platform (bottom). Reproduced from Cheali et al. (2014), © 2014, American Chemical Society*

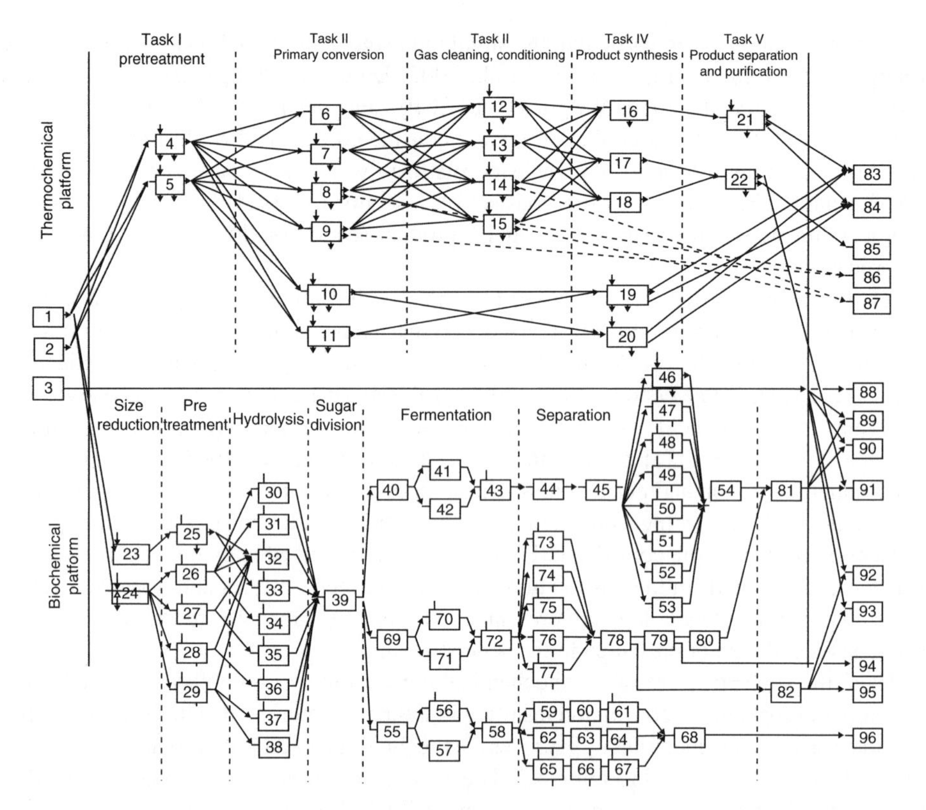

Figure 1.7 Combined superstructure of two biorefinery conversion platforms: thermochemical (top) and biochemical platform (bottom). Reproduced from Cheali et al. (2014), © 2014, American Chemical Society

researchers, managers, etc.). The expanded superstructure, alternatively, can also be used for the bottleneck studies in the existing processes, thereby helping end users (e.g., engineers) improve their processes.

Step 3: Data collection and modeling
The data and parameters required for the generic model blocks that are used to define the superstructure are presented here. When the reported data are available from experimental or pilot plant studies, the data were collected directly. If not, the data need to be obtained from simulations or should be estimated to obtain the parameters used in the general block using commercial process simulators such as ProII, Aspen, etc.

An example of data collection is presented below. Tables 1.4 and 1.5 and Figure 1.8 illustrate how the data were collected for the F-T process, which is one of the processing technologies to convert syngas to produce transportation fuels. The process requires a clean syngas to prolong the catalyst lifetime. There are no requirements for utilities and waste separation. The effluents are (i) unconverted gas which is recycled and (ii) liquid product which is discharged to a hydroprocessing unit. Generally, F-T catalytic synthesis is a well-known and commercial process producing a wide range of alkanes under a wide range of operating conditions and type of catalysts. Here, we have used the design data reported by

Table 1.3 *The description of the process intervals presented in Figure 1.7*

No.	Description	No.	Description
Raw materials			
1	Corn stover (33% moisture), 2000 tpd (dry)	3	Gasoline (for blending and comparing, 400 tpd)
2	Wood (35% moisture), 2000 tpd (dry)		
Thermochemical conversion platform (processing technologies)			
4	Size reduction, dryer (steam, indirect contact)	14	Steam reforming, WGS, acid gas removal—amine
5	Size reduction, dryer (flue gas, direct contact)	15	Steam reforming, WGS, acid gas removal—DEPG
6	Entrained-flow gasifier with size reduction	16	Fischer–Tropsch with special H_2S removal
7	Bubbling fluidized-bed gasifier	17	Alcohol synthesis (metal sulfide catalyst)
8	Indirectly heated with circulating gasifier	18	Alcohol synthesis (MoS_2 catalyst)
9	Directly heated with bubbling gasifier	19	Hydroprocessing (H_2 production)
10	Pyrolysis (bubbling fluidized bed)	20	Hydroprocessing (H_2 purchasing)
11	Fast pyrolysis (fluidized bed)	21	Decanter with hydroprocessing unit
12	SWGS, acid gas removal—amine, PSA-H_2	22	Molecular sieve, two distillation columns
13	Direct cooler, WGS, acid gas removal—amine		
Biochemical conversion platform (processing technologies)			
23	Size reduction by 60% water	53	Molecular sieve
24	Size reduction by 54% water	54	Anhydrous ethanol
25	Ammonia fiber explosion	55	Fermentation feed handling
26	Pretreatment dilute acid	56	Seed production
27	Controlled pH pretreatment	57	Seed production bypass
28	Aqueous ammonia recycle pretreatment	58	Succinic acid fermentation by *E. coli*
29	Lime pretreatment	59	Filtration
30	Dilute acid hydrolysis	60	Evaporation
31	Concentrated acid hydrolysis	61	Crystallization
32	NREL enzyme hydrolysis	62	Water splitting electrodialysis
33	Spyzyme hydrolysis from AFEX	63	Electrodialysis
34	Spyzyme hydrolysis from dilute acid	64	Crystallization
35	Spyzyme hydrolysis from controlled pH	65	Reactive distillation
36	Spyzyme hydrolysis from APR	66	Vacuum distillation
37	Spyzyme hydrolysis from lime	67	Crystallization
38	Hydrolysis bypass	68	Succinic acid storage
39	Sugar division	69	Fermentation feed handling
40	Fermentation feed handling	70	Seed production
41	Seed production	71	Seed production bypass
42	Seed production bypass	72	Butanol fermentation
43	Ethanol fermentation	73	Gas stripping
44	Flash	74	Adsorption

(*continued overleaf*)

Table 1.3 *(continued)*

45	Distillation column	75	Solvent extraction by oleyl alcohol
46	Solvent-based extraction by ethylene glycol	76	Pervaporation
47	Solvent-based extraction by ethylene glycerol	77	Membrane separation
48	Extraction with ionic liquid—EMIMBF4	78	Distillation for butanol
49	Extraction with ionic liquid—EMIMCl	79	Distillation for acetone
50	Extraction with ionic liquid—EMIM + $EtSO_4$	80	Distillation for ethanol
51	Extraction with ionic liquid—EMIM + DMP	81	Total ethanol production
52	Membrane separation	82	Butanol storage
Products and by-products			
83	FT-gasoline	90	E10 (ethanol–gasoline blend)
84	FT-diesel	91	Ethanol (100%)
85	Higher alcohols (C_3-ol, C_4-ol, C_5-ol)	92	B5 (butanol–gasoline blend)
86	Hot flue gas from gasifier combustor	93	B10 (butanol–gasoline blend)
87	Hot flue gas from tar reformer combustor	94	Acetone
88	Gasoline (100%)	95	Butanol (100%)
89	E5 (ethanol–gasoline blend)	96	Succinic acid

Reproduced from Cheali et al. (2014), © 2014, American Chemical Society.

Table 1.4 *The data collection example for Fischer–Tropsch reactor*

Descriptions	Raw data from NREL study		Generic model block parameters	
Utilities	—			
Reaction	Stoichiometry	N/A	$\gamma_{i,rr}$	(Eq. 1.29)
	Conversion fraction of CO (once through)	0.4	$\theta_{react,rr}$	1
Waste separation	—			
Product separation	Gas product	Recycled	$split_{i,kk}$	0
	Liquid product	Main product	$split_{i,kk}$	1

the NREL study of Swanson et al. (2010) which uses an Anderson–Schulz–Flory chain-growth probability model to describe the product distribution. Based on the model mentioned, the chain-growth value (α) was selected, resulting in the product distribution as a function of chain-growth value (α) for F-T reaction. Figure 1.8 and Table 1.5 show the stoichiometry values estimated from the model.

Chain Anderson–Schulz–Flory chain-growth probability model:

$$\gamma_{C_n} = \alpha^{n-1} * (1-\alpha) \tag{1.29}$$

The example mentioned in Table 1.4 shows how multidisciplinary data (simulation results, kinetics, separation efficiency, etc.) are converted into a generic form as a set of constant parameters. The collected data are then stored as a database in a multidimensional

Table 1.5 *Example of the stream table of the Fischer–Tropsch reactor*

Component	Chain-growth equation (Eq. 1.29)	$\gamma_{i,rr}$ (stoichiometric coefficient)	$\theta_{i,rr}$ (Conversion fraction)
CO	—	−1	1
H_2	—	−2.1	—
C_1	$\alpha^{1-1}\times(1-\alpha)$	0.010	—
C_2	$\alpha^{2-1}\times(1-\alpha)$	0.009	—
C_3	$\alpha^{3-1}\times(1-\alpha)$	0.0081	—
C_4	$\alpha^{4-1}\times(1-\alpha)$	0.00729	—
C_5	$\alpha^{5-1}\times(1-\alpha)$	0.00656	—
C_6	$\alpha^{6-1}\times(1-\alpha)$	0.005905	—
C_7	$\alpha^{7-1}\times(1-\alpha)$	0.005314	—
C_8	$\alpha^{8-1}\times(1-\alpha)$	0.004783	—
C_9	$\alpha^{9-1}\times(1-\alpha)$	0.004304	—
C_{10}	$\alpha^{10-1}\times(1-\alpha)$	0.003874	—
C_{11}	$\alpha^{11-1}\times(1-\alpha)$	0.003486	—
C_{12}	$\alpha^{12-1}\times(1-\alpha)$	0.003138	—
C_{13}	$\alpha^{13-1}\times(1-\alpha)$	0.002825	—
C_{14}	$\alpha^{14-1}\times(1-\alpha)$	0.002542	—
C_{15}	$\alpha^{15-1}\times(1-\alpha)$	0.002288	—
C_{16}	$\alpha^{16-1}\times(1-\alpha)$	0.002059	—
C_{17}	$\alpha^{17-1}\times(1-\alpha)$	0.001853	—
C_{18}	$\alpha^{18-1}\times(1-\alpha)$	0.001668	—
C_{19}	$\alpha^{19-1}\times(1-\alpha)$	0.001501	—
C_{20}	$\alpha^{20-1}\times(1-\alpha)$	0.001351	—
Wax	$1-\left(\sum_{N}^{20}\alpha^{N-1}\times(1-\alpha)\right)$	0.012158	—
H_2O	—	1	—

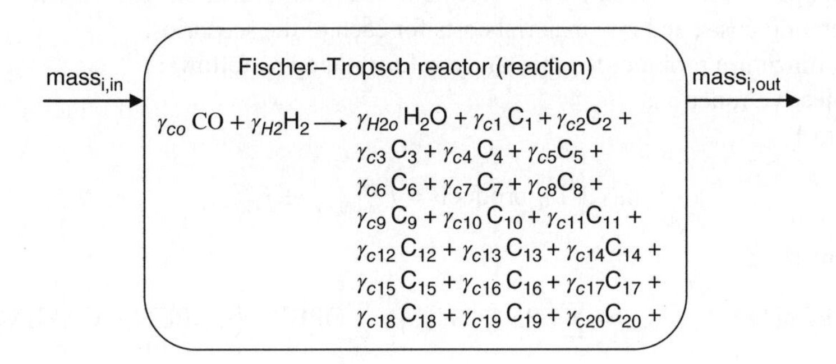

Figure 1.8 *Process diagram showing mass inlet/outlet, the reaction, and its stoichiometry for the Fischer–Tropsch reactor*

matrix (the database uses the Excel spreadsheet environment, but any other software environment would work, e.g., Matlab, MS Access, etc). In this way, storage of the data is flexible as it only requires simple column and row operations to add, modify, or update data in the database. At the same time, storing the data in matrix form provides a certain structure to organize the data and manage the complexity in a compact and efficient way.

The description and the data collection (plus parameter estimation where necessary) for the other process intervals included in the superstructure of the thermochemical platform (Figure 1.6) are summarized in Table 1.6 (Cheali et al., 2014). For each process interval, mixing parameters ($\mu_{i,j,kk}$, $\alpha_{i,kk}$), reaction parameters ($\gamma_{i,rr}$, $\theta_{react,rr}$), waste separation parameters ($SW_{i,kk}$), and a product separation parameter ($split_{i,kk}$) are provided.

Step 4: Models and data verification

After the data conversion (or estimation), the verification needs to be performed as a consistency check prior to performing the optimization. Seven processing paths based on five NREL reports and a PNNL report were used to verify the models and data used for each process interval and processing path. As explained earlier, the verification can be performed by fixing the processing path and comparing the simulation results with the NREL and PNNL studies. The full verification results were presented in the previous study (Cheali et al., 2014).

Here, an example of verification is presented. The data collected and modeled of the F-T reactor from the previous step were verified in this step. The simulation results of this study (implemented in GAMS) are necessary in order to verify the quality of the collected data and the models used in this study. In the previous section, the data collection was presented as examples for the entrained-flow gasifier. Here, the collected data for both examples are validated and presented in Table 1.7, respectively. The validation results confirm that the quality of the collected data is good and the data are consistent. The full simulation results (implemented in GAMS) can be found in the previous study (Cheali et al., 2014).

Step 5: Formulation and solution of the optimization problem

Four optimization scenarios selected in step 1 were formulated and solved in GAMS to identify the optimal biorefinery processing network. The optimization problem for scenario 2 resulted in 4,737,904 equations with 4,705,181 single variables (668 discrete variables). This problem was solved using the DICOPT solver using Windows 7 as operating system and an Intel® Core™ i7 CPU@ 3.4GHz, with 4GB RAM, resulting in an execution time of 12s. Table 1.8 presents the optimization results consisting of the processing paths; production rates; earnings before interest, taxes, depreciation, and amortization (EBITDA); investment costs; operating costs; and raw material costs for each of the scenarios.

The optimization problem for this study is formulated as follows:

The objective functions

Scenario 1:

$$\max.\mathrm{FT-products} = F^{out}_{gasoline,kk} + F^{out}_{diesel,kk} \tag{1.30}$$

Scenario 2:

$$\max.\mathrm{EBITDA}_{gasoline,\ diesel} = \sum_{i,kk}\left(P3_{i,kk} * F^{out}_{i,kk}\right) - \sum_{kk}\mathrm{OPEX} - \left(\mathrm{CAPEX}_1 + \mathrm{CAPEX}_2\right)/t \tag{1.31}$$

Table 1.6 *Summary table for the data collection (mixing, $\alpha_{i,kk}$, $\mu_{i,kk}$ reaction, $\gamma_{i,rr}$, $\theta_{react,rr}$, waste, $SW_{i,kk}$, and product, $split_{i,kk}$, separation) for thermochemical processing networks (Cheali et al., 2014)*

No.	Description	Mixing ($\alpha_{i,kk} = 1$)	$\mu_{i,kk}$	Reaction (stoichiometry, $\gamma_{i,rr}$)	$\theta_{react,rr}$	Waste separation	$SW_{i,kk}$	Product separation (in primary outlet)	$Split_{i,kk}$
4	Hammer mill, rotary dryer (indirectly contact with steam)	• Electricity to biomass ratio • Steam to %moist ratio	4 5.9	—	—	Moisture	0.6	Steam	0
5	Hammer mill and rotary dryer (directly contact with hot flue gas)	• Electricity to biomass ratio • Waste heat from process	4	—	—	Moisture	0.96	Waste heat	0
6	Entrained-flow (free-fall) gasifier	• O_2 to biomass ratio • Steam to %moist ratio	0.35 0.48	$C + 0.13H_2O + 0.6O_2 + 0.0017S$ $\rightarrow 0.13H_2 + 0.0007N_2 + 0.66CO +$ $0.34CO_2 + 0.002H_2S +$ $0.07SOOT + 1.3SLAG$	1	Ash, soot	0.99	One outlet stream	1
7	Bubbling fluidized-bed gasifier	• O_2 to biomass ratio • Steam to %moist ratio	0.26 0.17	$C + 0.11H_2O + 0.34H_2 + 0.5O_2 +$ $0.002S + 0.007N_2$ $\rightarrow 0.36CO + 0.41CO_2 + 0.002H_2S +$ $0.01NH_3 + 0.08CH_4 + 0.01C_2H_6 +$ $0.02C_2H_4 + 1.2CHAR$	1	Ash, char	0.99	One outlet stream	1
8	Indirectly heated with circulating FB gasifier	• Air to biomass ratio • Steam to %moist ratio • Fresh olivine to biomass ratio • MgO to ash ratio	2.2 0.38 0.003 0.004	$C + 0.48H_2 + 0.6O_2 +$ $0.0006S + 0.001N_2$ $\rightarrow 0.37CO + 0.39CO_2 +$ $0.12H_2O + 0.0006H_2S +$ $0.002NH_3 + 0.13CH_4 +$ $0.02C_2H_6 + 0.03C_2H_4 + 0.23TAR$	1	Ash	0.99	H_2O CO_2 O_2, N_2, Ar	0.75 0.25 0

(*continued overleaf*)

Table 1.6 (*continued*)

No.	Description	Mixing ($\alpha_{i,kk} = 1$)	$\mu_{i,kk}$	Reaction (stoichiometry, $\gamma_{i,rr}$)	$\theta_{react,rr}$	Waste separation	$SW_{i,kk}$	Product separation (in primary outlet)	$Split_{i,kk}$
9	Directly heated with bubbling gasifier	• O_2 to biomass ratio • Steam to %moist ratio • Fresh olivine to biomass ratio • MgO to ash ratio	0.22 0.2 0.003 0.004	$C + 0.47H_2 + 0.04H_2O + 0.4O_2 + 0.0006S + 0.001N_2 \rightarrow 0.19CO + 0.39CO_2 + 0.0006H_2S + 0.002NH_3 + 0.18CH_4 + 0.01C_2H_6 + 0.013C_6H_6 + 0.55TAR + CHAR$	1	Ash, char	0.99	Two product stream (same component)	0.6
10	Pyrolysis (bubbling fluidized bed)			$C + 0.3O_2 + 0.6H_2 + 0.007N_2 + 1.5ASH + -0.001S$ $\rightarrow 0.05H_2O + 0.05CO + 0.07CO_2 + 0.01CH_4 + 4.5CHAR + 15PYRO-OIL$	1	Ash, char	0.85	H_2, O_2, N_2, CO, CO_2, CH_4 H_2O	0 0.63
11	Fast pyrolysis (fluidized bed)			$C + 0.5O_2 + 0.7H_2 + 0.45ASH + -0.0001S \rightarrow 1.4H_2O + 0.002CO + 0.25CO_2 + 15PYRO-OIL$	1	Ash, char	1	H_2, O_2, N_2, CO_2 H_2O CO	0 0.15 0.01
12	SWGS, acid gas removal—amine, PSA-H_2	Steam to inlet flow	0.14	$CO + H_2O \rightarrow CO_2 + H_2$	0.3	H_2O, N_2, CO_2, NH_3, H_2S COS	1 0.7	One outlet stream	1
13	Direct cooler, SMR, WGS, acid gas removal—amine	Steam to inlet flow	0.36	• $CH_4 + 0.1C_2H_6 + 0.2C_2H_4 + 1.6H_2O \rightarrow 4.4H_2 + 1.6CO$ • $CO + H_2O \rightarrow CO_2 + H_2$	0.35 0.25	NH_3, TAR, Ash, Char, H_2O H_2S CO_2	0.99 0.95 0.9	One outlet stream	1

14	Tar reformer, scrubber, acid gas removal—amine	Air to inlet flow	1.2	$TAR + 0.7CH_4 + 0.015C_2H_6 + 0.15C_2H_4 + O_2 \rightarrow 1.5H_2 + CO + 0.3CO_2 + 0.4H_2O$	1	H_2O H_2S CO_2	0.98 0.8 0.6	Ar, N_2 H_2O, CO_2	0 0.5
15	Tar reforming, scrubber, acid gas removal—DEPG	Air to inlet flow	1	$TAR + 0.8CH_4 + 0.02C_2H_6 + 0.14C_2H_4 + 0.8O_2 + 0.2CH_4O \rightarrow 2.5H_2 + 1.3CO + 0.5CO_2 + 0.1H_2O$	1	NH_3, H_2O, TAR, Ash, Char H_2S CO_2	1 0.65 0.6	Ar, O_2, N_2 H_2O CO_2	0 0.6 0.4
16	Fischer–Tropsch	—	—	$CO + 2.1H_2 \rightarrow 10.8C_1 + 9.8C_2 + 8.8C_3 + 7.9C_4 + 7.1C_5 + 6.4C_6 + 5.7C_7 + 5.2C_8 + 4.6C_9 + 4.2C_{10} + 3.7C_{11} + 3.4C_{12} + 3C_{13} + 2.7C_{14} + 2.5C_{15} + 2.2C_{16} + 2C_{17} + 1.8C_{18} + 1.6C_{19} + 1.5C_{20} + 13Wax + H_2O$	0.4	—	—	One outlet stream	1
17	Alcohol synthesis (modified F-T catalyst, MoS2)	—	—	$CO + H_2 + 0.006H_2O \rightarrow 0.36CO_2 + 0.06CH_4 + 0.04CH_4O + 0.24C_2H_6O + 0.026C_3H_8O$	0.4	—	—	One outlet stream	1
18	Alcohol synthesis (metal sulfide synthesis catalyst)	—	—	$CO + 1.2H_2 \rightarrow 0.07H_2O + 0.07CO_2 + 0.1CH_4 + 0.1CH_4O + 0.2C_2H_6O + 0.017C_3H_8O$	0.26	—	—	H_2, N_2, CO, CH_4 H_2O, alcohols CO_2	0.01 1 0.08

(*continued overleaf*)

Table 1.6 *(continued)*

No.	Description	Mixing ($\alpha_{i,kk} = 1$)	$\mu_{i,kk}$	Reaction (stoichiometry, $\gamma_{i,rr}$)	$\theta_{react,rr}$	Waste separation	$SW_{i,kk}$	Product separation (in primary outlet)	$Split_{i,kk}$
19	Hydroprocessing (H2 production)	• O_2 to biomass ratio • Steam to %moist ratio	1.87 0.37	$Pyro-oil + 0.02H_2O + 0.01O_2 \rightarrow 0.02CO_2 + 0.0014gasoline + 0.0007diesel$	1	—	—	H_2O, O_2, CO_2	0
20	Hydroprocessing (H2 purchasing)	H_2 to inlet flow ratio	0.05	$Pyro-oil + 0.014H_2O + 0.027H_2 \rightarrow 0.003CO_2 + 0.003CH_4 + 0.002gasoline + 0.0012diesel$	1	—	—	H_2O, O_2, CO_2, CH_4	0
21	Hydroprocessing unit	H_2 to wax ratio	0.0257	$Wax + 5.4H_2 + 0.5C_5 + 0.48C_6 + 0.44C_7 + 0.35C_9 + 0.32C_{10} + 0.28C_{11} + 0.25C_{12} + 0.23C_{13} + 0.2C_{14} + 0.18C_{15} + 0.15C_{17} + 0.14C_{18} + 0.12C_{19} + 0.11C_{20} \rightarrow 2gasoline + 2.76diesel$	1	—	—	H_2O, H_2, CO, CO_2, light hydrocarbon	0
22	Mol.sieve and distillations	—	—	—	—	H_2, N_2, CO, CH_4 H_2O	 1 0.8	CH_4O C_2H_6O Higher alcohols	0.07 0.99 0.05

Reproduced from Cheali et al. (2014), © 2014, American Chemical Society.

Table 1.7 *Summary of the verification results for the Fischer–Tropsch reactor*

	Inlet flow	The reported results from NREL report (Swanson et al., 2010)					The simulation results of this study				
		Recycles	R(i)	Waste (i)	Fout1	Fout2	Recycles	R(i)	Waste (i)	Fout1	Fout2
Total (tpd)	3376	1292	0	0	427	4237	1292	0	0.03	427	4237
H_2O	45					642					642
H_2	288	81				225	81				225
O_2											
N_2											
S											
C											
ASH											
CO	1818	500				1391	500				1391
CO_2	190	106				296	106				296
H_2S	0.03								0.03		
NH_3											
COS	1.2	0,68				1,88	0.6				1.8
AR	544	305				850	305				850
CH_4	63.4	38.6				107	38.6				107
C_2H_6	106.9	65				180	65				180
C_2H_4		86					86				
C_6H_6		101.9					102				
C_3	141.56	1.5				239	1.5				239
C_4	167.55	0.7				283	0.7				283
C_5	2.29	0.72			15.49	3.87	0.7			15.4	3.8
C_6	2.46	0.72			16.65	4.16	0.7			16.6	4
C_7	1.15	0.72			17.54	1.95	0.7			17.5	1.9
C_8	1.18	0.71			17.99	2	0.7			18	2
C_9	1.11	0.33			18.1	2.01	0.3			18	2
C_{10}	1.16	0.32			18.12	2.01	0.3			18	2
C_{11}	0.83	0.31			17.62	1.96	0.3			17.6	1.9
C_{12}	0.13	0.29			17.25	0.91	0.25			17.2	0.9
C_{13}	0.02	0.28			16.7	0.88	0.25			16.5	0.8
C_{14}					16.16	0.85				16	0.8
C_{15}					15.57	0.82				15.5	0.8
C_{16}					14.93	0.79				15	0.8
C_{17}					14.75					14.7	
C_{18}					14.05					14	
C_{19}					13.35					13.3	
C_{20}					12.64					12.6	
Waxes					170					170	

Table 1.8 *The optimization results and comparison. Highlighted in bold are the processing networks for each task*

Scenario	Objective function	Processing path	EBITDA (MM$/a)	Fuel production (tpd)	Investment cost (MM$/a)	Operating cost (MM$/a)	Feedstock cost (MM$/a)
1	Max. FT-products	2 **4** 6 **15** 16 21 83 84	207	171 (gasoline), 403 (diesel)	11.5	18.6	60
2	Max. FT-product sales, min. investment, and operating cost	2 **5** 6 **14** 16 21 83 84	210	170 (gasoline), 400 (diesel)	9	17	60
3	Max. ethanol	2 **4** **6** **15** **18** 91	100	602	10.5	12	60
4	Max. ethanol sales, min. investment cost, and operating cost	2 **5** **9** **14** **17** 91	100.5	589	10.4	10	60

Note the number of process intervals refer to Figure 1.7.

Scenario 3:

$$\max.\text{Ethanol} = F^{\text{out}}_{\text{ethanol,kk}} \tag{1.32}$$

Scenario 4:

$$\max.\text{EBITDA}_{\text{ethanol}} = \sum_{\text{i,kk}}\left(P3_{\text{i,kk}} \times F^{\text{out}}_{\text{i,kk}}\right) - \sum_{\text{kk}}\text{OPEX} - \left(\text{CAPEX}_1 + \text{CAPEX}_2\right)/t \tag{1.33}$$

Subject to the following constraints are (i) process models, material balances of the generic model block (Eqs. 1.1–1.7); (ii) process constraints, rules defining the superstructure together with binary variables ($S_{\text{p}}^{\text{k,kk}}, S^{\text{k,kk}}$) and the flow constraints (Eqs. 1.8–1.10); (iii) structural constraints using $\sum_k y_k \le 1$ with binary variable ($y \in \{0;1\}^n$), to define the extended superstructure (Eqs. 1.12 and 1.13); (iv) cost constraints, to calculate capital and operating cost regarding the estimated mass flow rate (Eqs. 1.14–1.16); and (v) optimization constraints using big-M formulation, where M was set to 10,000 in this study which is about three times higher than the maximum flow rate (Eq. 1.17).

As can be seen, for the gasoline and diesel cases, a higher production, 171 tpd-gasoline and 403 tpd-diesel, was found (scenario 1) as well as the higher EBITDA in scenario 2. For the ethanol production cases, a higher ethanol production, 600 tpd, was also found

Table 1.9 *The top five ranked solutions identified in scenario 2: max. FT-product sales, min. operating cost, and investment cost. Highlighted in bold are the processing networks for each task*

Rank no.	Process interval selection	EBITDA (MM$/a)	Production (tpd)	EBITDA (MM $/year)	TAC (MM $/year)
1	2 **5 6 14** 16 21 83 84	210	170[a], 400[b]	170	86
2	2 **4 6 15** 16 21 83 84	206	171[a], 403[b]	158	90
3	2 **5 11 20** 83 84 (Jones et al., 2009)	148	245[a], 311[b]	148	133
4	2 **5 8 15** 16 21 83 84	93	141[a], 334[b]	93	108
5	2 **4 8 14** 16 21 83 84	93	138[a], 327[b]	93	101

[a] FT-gasoline.
[b] FT-diesel.

(scenario 3). The results show that the thermochemical conversion platform is able to produce a higher amount of ethanol compared to the biochemical platform under the considered design space of alternatives. Furthermore, the entrained-flow gasifier and reformer are found to be the best technical alternatives to produce higher FT-gasoline and FT-diesel. The results indicate a higher potential of the thermochemical platform due to a relatively lower investment cost (in agreement with Foust et al., 2009) and a higher ethanol production of the optimized thermochemical processing path.

We further analyzed the results obtained from scenario 2 with respect to the top five ranking solutions. The results are given in Table 1.9 which features the objective function value, production rates, EBITDA, and TAC for the top five solutions. Analyzing the solutions in Table 1.9 reveals the following: (i) wood and entrained-flow gasifier are the most favorable ones, and (ii) no processing paths of the biochemical platform were selected in this top five. The difference in the objective function value is reflected in the selection of particular processing technologies. For example, the differences between the first- and second-ranked solutions are the selection of a dryer using process waste heat (5) and a steam reformer with amine-based acid gas removal (14) instead of a dryer using steam (4) and a catalytic reformer with DEPG acid gas removal, respectively. The changes of these processing technologies cause the changes in operating and capital cost resulting in a better EBITDA. The simplified flow diagrams for the first and the second processing path are presented in Figures 1.9 and 1.10, respectively.

Moreover, the top two optimal processing paths were then selected to perform further analysis on sustainability and environmental impact in the next section in order to create sufficient data to perform multicriteria evaluation.

1.3.2 Sustainability Analysis with Respect to Resource Consumption and Environmental Impact

Step 6: Sustainability analysis with respect to resource consumption and environmental impact

In this section, the two highest ranking optimal processing paths identified with respect to the techno-economical metrics presented in Figures 1.9 and 1.10 were selected for further analysis with respect to the following sustainability metrics, resource consumption and environmental impact, using two software tools, SustainPro and LCSoft, respectively.

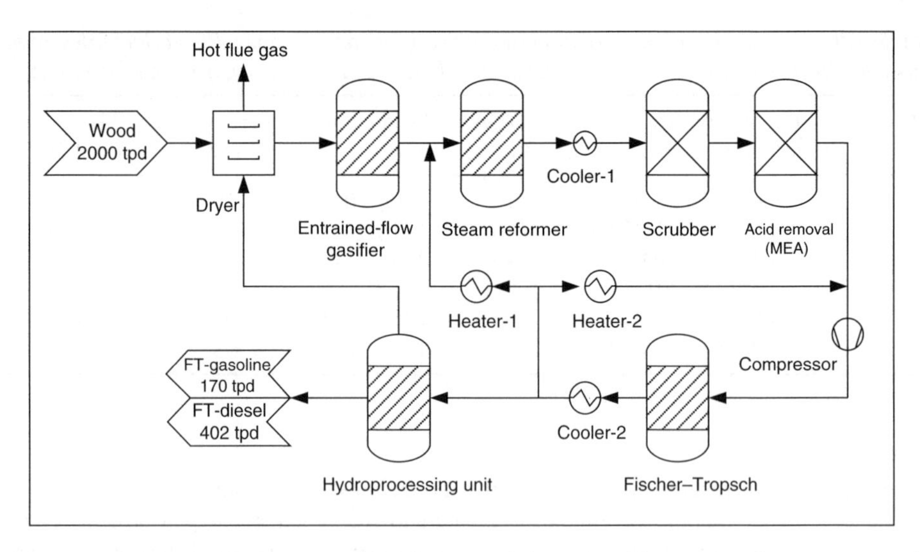

Figure 1.9 *The simplified process flow diagram of the second best optimal processing path of scenario 2 (as presented in Table 1.9)*

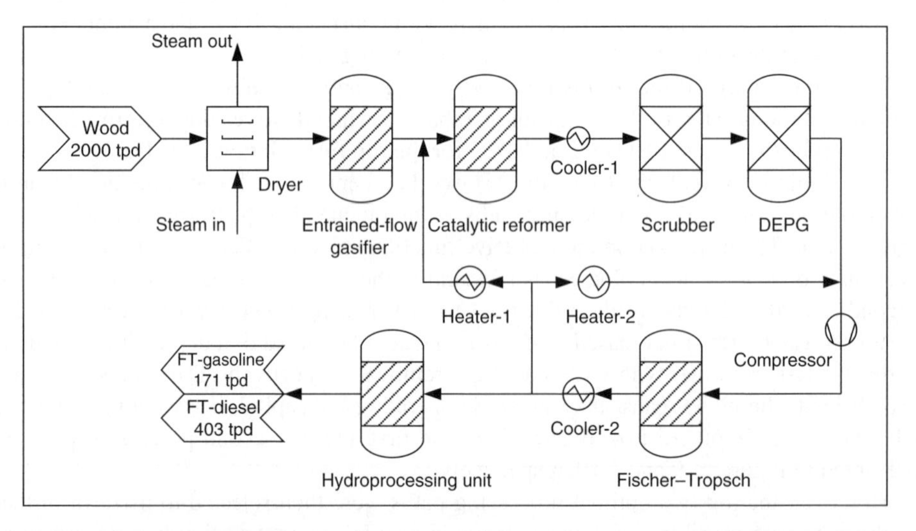

Figure 1.10 *The simplified process flow diagram of the best optimal processing path of scenario 2 (as presented in Table 1.9)*

1.3.2.1 Resource consumption analysis

The simulation results obtained from the sustainability analysis for two processing paths are presented in Tables 1.10, 1.11, 1.12, and 1.13. Tables 1.10 and 1.11 present the potential of improvement and the path flow details of the first optimal processing path (Figure 1.9), respectively. Tables 1.12 and 1.13 present the potential of improvement and the path flow details of the second optimal processing path (Figure 1.10). The potential of improvement is here related to the ability to change the process path flow with the aim of increasing the process sustainability.

Table 1.10 Identified process critical points for the first processing path, regarding the open and closed paths, respectively

Open path	MVA	EWC	TVA	Probability
OP 2	−40,951	3	−40,954	High
Closed path	EWC	Probability	AF	Probability
CP97	1125	Check AF	3,277	High
CP99	61,37	Check AF	0,6	High

Table 1.11 Path flow details on the critical points identified for the first processing path indicated in Table 1.10

Open path	Component	Path stream			Flow rate (kg/h)
OP 2	H_2O	FT—out	Hot flue gas—out	—	10,992
Closed path	Component	Path stream			Flow rate (kg/h)
CP97	H_2	Gasifier—out	Acid—out	FT-out	38,348
CP99	O_2	Gasifier—out	Acid—out	FT-out	30,737

Table 1.12 Identified process critical points for the second processing path, regarding the open and closed paths, respectively

Open path	MVA	EWC	TVA	Probability
OP 2	−1,952,512	2,956	−1,952,515	High
Closed path	EWC	Probability	AF	Probability
CP50	1,433	Check AF	3,892	High

Table 1.13 Path flow details on the critical points identified for the second processing path indicated in Table 1.12

Open path	Component	Path stream			Flow rate (kg/h)
OP 2	H_2O	Steam—in	Steam—out	—	636,038
Closed path	Component	Path stream			Flow rate (kg/h)
CP50	H_2	Gasifier—out	Acid—out	FT—out	45,539

As mentioned before (Table 1.1), SustainPro decomposes the flowsheet into OP and CP. An OP corresponds to the mass or energy entering and exiting the system boundaries, and a CP reflects the recycle routes for mass or energy. After establishing each path flow rate, the software estimates the mass and energy indicators that reflect a complete analysis of the process. The software tool then ranks the OP and CP, from top to bottom, according to the respective impact on the overall process sustainability. In other words, the top indicators stand for the paths (and respective compounds) that have the highest impact on the process with respect to the resource consumption and the respective economic impact.

Tables 1.10 and 1.11 show the top indicators obtained through SustainPro for the optimal processing path (read as biorefinery network) obtained from the previous steps. OP2 was identified as a possible bottleneck due to the high negative value of MVA. It reflects the water vapor flow rate that is discarded from the F-T reactor to the dryer, being discarded as hot flue gas (waste) instead of being used/recycled. A possible solution to this is to recycle this water flow, hence decreasing the net amount of freshwater that will be added to the system, and it needs to be analyzed in more detail. Moreover, CP97/CP99, which stand for the recycling of H_2 and O_2, respectively, are also identified as possible critical points, due to the fact that they carry high units of energy. This points toward the possibility of using these streams as potential heating source for heat integration. Therefore, by decreasing the external dependence on energy from fossil sources, the user achieves an improvement of the overall process sustainability and environmental impact.

For the second optimal processing path, there are two main paths (presented in Table 1.12 and 1.13, respectively) that were indicated as holding a high potential of improving the process sustainability if changes were performed in the mentioned paths. OP2 (steam used to dry biomass) presents a highly negative MVA. This is due to the fact that the process is discarding the steam after it is being used, and therefore, it is losing value along the way. In other words, this is a critical point in the process that clearly indicates that the process sustainability (in terms of natural resource usage) could be significantly improved if this steam was recycled, pressurized, and then reused resulting in an improvement on the process sustainability, decreasing the freshwater consumption and external dependence. Moreover, CP50, in which H_2 is being recycled from the F-T reactor to the gasifier, also indicates that this path carries more units of energy that are not being efficiently used. Therefore, there is a possibility of improving the process sustainability and of decreasing the external dependence by using this path as heating source through heat integration (HEN).

As mentioned previously, SustainPro also estimates the sustainability metrics. This is a valuable feature that allows the user to compare several processes in order to assess their degree of sustainability, presenting a complete evaluation with respect to water, energy, and raw material consumption. Table 1.14 presents both results from the first and the second processing path.

Sustainability metrics indicate that both the first and second processing paths have the same energy consumption per kg of product that is being produced. Nevertheless, the first process has a higher net primary energy usage per unit value added, which means that

Table 1.14 *Sustainability metrics for the first and second processing path*

	First path	Second path
Energy		
Total net primary energy usage rate (GJ/y)	414,301	414,301
% Total net primary energy sourced from renewables	0.9989	0.9989
Total net primary energy usage per Kg product (kJ/kg)	991.06	991.06
Total net primary energy usage per unit value added (kJ/$)	0.215	0.0738
Raw materials		
Total raw materials used per kg product (kg/kg)	3.256	16.85
Total raw materials used per unit value added (kg/$)	0.0007	0.00125
Fraction of raw materials recycled within company	0.9627	0.1936
Fraction of raw materials recycled from consumers	0	0
Hazardous raw material per kg product	0	0
Water		
Net water consumed per unit mass of product (kg/kg)	35.09	48.69
Net water consumed per unit value added (kg/$)	0.0076	0.0036

the total MVA (see Eq. 1.18) is lower than the one characterizing the second process. This means that the first process uses energy more efficiently.

With respect to water and raw material consumption, one can see from Table 1.14 that the first processing route shows a better performance, also including a higher percentage of raw materials that are being recycled. For this particular analysis, the optimal biorefinery concept identified with respect to techno-economic criteria was found to perform better with respect to utilization of resources too. In addition to the retrofitting options mentioned previously, further process improvements and integration scenarios can be generated using formal chemical engineering techniques such as pinch analysis. However, this is beyond the scope of this study.

1.3.2.2 Environmental impact assessment

For each optimal path previously selected, the output data from the different subprocesses in the network is combined and presented as total emissions of a compound. To perform that analysis, the impact assessment requires component-specific CFs, which are estimated based on a methodology implemented in LCSoft, in particular Tool-2 (Table 1.2) using group-contribution+ (GC+) property models together with the US EPA USEtox™ database.

The results of the environmental impact analysis for the two mentioned processing paths are presented in the following text. With respect to the total carbon footprint, one can see (Table 1.15) that the second optimal processing path emits approximately 7% more CO_2, expressed as kg of CO_2 eq. per kg of ethanol being produced, than the first optimal process. Table 1.16 presents the list of PEIs also estimated by LCSoft for each one of the options. Also for this analysis, the first path has also been reported as the best one, having the lower environmental impact.

After the environmental impact assessment, one can see (Tables 1.15 and 1.16) that the first processing path is a better solution with respect to environmental impact since it has a lower total carbon footprint and lower values for the PEI metrics. This is due to the fact that the first process uses a waste heat stream as internal utility instead of using an external utility. Moreover, even when including a high electricity consumer such as the compressor, the first processing path includes coproduction of electricity that fully satisfies the electricity demands.

Table 1.15 *Total carbon footprint for first and second processing path*

Unit operations/processes	Duty/work (GJ/h)	First path	%	Second path	%
Biomass dryer	0.069	1.62E−09	0	7.58E−04	0
Entrained-flow gasifier	0.291	6.83E−09	0	0.0032	0.1
Cooler-1 prior to acid gas removal	1.981	0.0086	0.2	0.0029	0.1
Compressor to F-T	15.03	1.742	44.3	1.738	41.1
Fischer–Tropsch	0.146	0.0006	0,0	0.0002	0
Cooler-2 after Fischer–Tropsch	484.5	2.108	53.6	2.199	52
Heater-1 for recycles	17	0.0739	1.9	0.0248	0.6
Heater-2 for recycles	20	4.69E−07	0	0.2196	5.2
Hydroprocessing unit	3.88	9.11E−08	0	0.0426	1
Total carbon footprint (kg CO_2 eq.)	—	3.93	100	4.23	100

Table 1.16 *Potential environmental impacts for first and second processing path*

Environmental impact	Unit	First path	Second path
Human toxicity by ingestion (HTPI)	1/Lethal dosage (LD_{50})	5.23E−07	4.75E−05
Human toxicity by exposure (HTPE)	1/Time-weighted average (TWA)	2.93E−04	3.65E−04
Global warming potential (GWP)	CO_2 eq.	3.93	4.23
Ozone depletion potential (ODP)	CFC-11 eq.	5.24E−12	1.8E−06
Photochemical oxidation (PCOP)	C_2H_2 eq.	5.86E−09	1.05E−05
Acidification (AP)	H^+ eq.	0.033	7.02E−03
Aquatic toxicity (ATP)	$1/LC_{50}$	1.89E−06	1.98E−05
Terrestrial toxicity (TTP)	$1/LD_{50}$	5.23E−07	4.75E−05
Carcinogenics (HTC)	kg benzene eq.	0.042	0.049
Noncarcinogenics (HTNC)	kg toluene eq.	5.37E−05	0.071
Freshwater ecotoxicity (ET)	kg 2,4-dichlorophenol eq.	3.35E−07	7.43E−06

In summary, the framework presented is a promising tool to represent the ever-increasing number of biorefinery alternatives with their competing technologies and routes and help evaluate them at their optimality for early-stage design and analysis purposes in terms of techno-economic analysis. Sustainability analysis and environmental impact assessment are included in the framework which enables a more detailed and comprehensive analysis of the biorefinery alternatives. The framework helps formulate a multicriteria evaluation (techno-economic, environmental impact, and sustainability analysis) of the biorefinery concept.

1.4 Conclusion

A systematic framework with a superstructure-based optimization approach was presented for the early-stage design and analysis of biorefinery alternatives. In the description of the framework, we have especially highlighted the use of the superstructure to represent many competing biorefinery alternatives (from thermochemical to biochemical), and we have used generic and simple models to describe the processing steps in the biorefinery. These generic models are coupled to a data structured to manage the multidisciplinary data needed to solve the problem, and we cast the resulting optimization problem as an MINLP and solve the problem to identify optimal processing paths for a given objective function definition. To further complement the analysis, we then analyze the results in terms of sustainability and environmental impact.

The integration of the sustainability and environmental impact analysis to the framework provides a more comprehensive decision making and multicriteria evaluation to enable the users to screen for economic but sustainable biorefinery concepts early on in the project development life cycle.

The framework was evaluated with a case study focusing on design of a lignocellulosic biorefinery. The results showed that the tool could find new optimal processing paths for different scenarios (objective functions) which provide a better, more efficient production process and less utility, waste, investment, and operating costs using the expanded biorefinery network. The optimal biorefinery pathways were further analyzed with respect to sustainability and environmental impact, which revealed further options for sustainability improvement in the optimal biorefinery pathway. The results presented in this chapter convincingly demonstrate the promising potential of the framework as a decision support tool for analyzing biorefinery concepts early on for technical, economical, as well as sustainability and environmental impact-related aspects.

Nomenclature

Indexes

i	Component
k	Process interval (origin)—optimization model
kk	Process interval (destination)
react	Key reactant
rr	Reaction

Sustainability analysis model

o	Open path flow—sustainability model
rm	Raw material—sustainability model
RM	Total number of raw materials—sustainability model
k	Component path flow—sustainability model
z	Component cycle path flow—sustainability model
u	Subunit operation—sustainability model
U	Total number of subunit operations—sustainability model
uk	All component path flows in a given subunit operation—sustainability model
m	Mean value—sustainability model
rk	Reaction—sustainability model
r	Reactive unit operations—sustainability model

LCA model

c	Compartment (air, water, and soil)

Parameters

MW_i	Molecular weight
$P1_{i,kk}$	Raw material prices
$P2_{i,kk}$	Utility prices
$P3_{i,kk}$	Product prices
$SW_{i,kk}$	Waste fraction
$S_{k,kk}$	Superstructure (binary)
$S^{p}_{k,kk}$	Superstructure (binary)
a_{kk}	Coefficient for capital cost estimation
n_{kk}	Coefficient for capital cost estimation
$\alpha_{i,kk}$	Specific utility consumption
$\gamma_{i,kk,rr}$	Reaction stoichiometry
$Split_{i,kk}$	Split factors

$\theta_{react,kk,rr}$	Conversion of key reactant
$\mu_{i,kk}$	Fraction of utility mixed with process stream
t	Time

Sustainability analysis model

PP_o^i	Purchase price of component i in the open path flow
$PR_0^{(rm)}$	Raw material price of component i in the open path flow
CA_0^i	Cost allocation factor of component i in the open path flow
υ_0^i	Stoichiometry of component i in the open path flow
M_0^i	Molecular weight of component i in the open path flow
PE_u	Energy price of subunit operation u
Q_u	The duty of subunit operations
$A_{u,k}^i$	Specific heat capacity of component i in the path flow k in a given subunit operation
$\xi_{r,\,rk,k}$	The extent of reaction rk
$n^{(fp)}$	Mole flow rate of a desired final product
$E_{r,rk,k}^i$	Effect of the component i to reaction rk
AF_z^i	Accumulation factor of component i in the cycle path flow z

LCA model

HV_r	Heating value of renewable resource r
HV_{nr}	Heating value of nonrenewable resource nr
$CF_{GHG,air}^{GWP}$	Characterization factor value regarding the global warming of greenhouse gas (GHG)
$EM_{i,c}$	Mass of chemical i emitted to compartment c per 1 kg of product
$CF_{i,c}^k$	Characterization factor of chemical n emitted to compartment c for impact category k

Variables

$F_{i,k,kk}$	Component i flow from process intervals k to process intervals kk
$F_{i,kk}^M$	Component flow after mixing
$R_{i,kk}$	Utility flow
$F_{i,kk}^R$	Component flow after reaction
$F_{i,kk}^{out}$	Component flow after waste separation
$waste_{i,kk}$	Component flow of waste stream after waste separation
$F_{i,kk}^{out1}$	Component flow leaving process intervals kk through primary outlet
$F_{i,kk}^{out2}$	Component flow leaving process intervals kk through secondary outlet
y_{kk}	Selection of process intervals (binary)
$w_{j,kk}$	Selection of a piece of the piecewise linearization (linear)

Sustainability analysis model

MVA_o^i	Mass Value Added (MVA) of component i in the open path flow o
m_o^i	Flow rate of component i in the open path flow o

EWC_k^i	Energy and waste cost of component i in the path flow k
m_k^i	Mass flow rate along the path flow k of the component i
RQ_k^i	Reaction quality of component i in the path flow k
$f_{i,a}^i$, $d_{i,op}^i$	The flow rate leaving the cycle path flow

LCA model

R_{renew}	Total energy from renewable resource consumption per 1 kg of product
$R_{non\text{-}renew}$	Total energy from nonrenewable resource consumption per 1 kg of product
R_{total}	Total energy from resource consumption per 1 kg of product
$x_{r,i}$	Mass of renewable resource r used to produce 1 kg of input *i*
$x_{nr,i}$	Mass of nonrenewable resource nr used to produce 1 kg of input *i*
CO_{2eq}	Carbon dioxide equivalent per 1 kg of product
$m_{GHG,air}^{PRO}$	Mass flow rate of greenhouse gas (GHG) emitted to air from the process
I^k	The potential environmental impact of chemical i for a specific impact category of concern k

Abbreviations

CAPEX	Capital investment
EBITDA	Earnings before interest, taxes, depreciation, and amortization
LB	Lower bound of the objective function
UB	Upper bound of the objective function

References

Azapagic, A. (2002). *Sustainable Development Progress Metric*. IChemE Sustainable Development Working Group, Rugby.

Baliban, R.C., Elia, J.A., Weekman, V., Floudas, C.A. (2012). Process synthesis of hybrid coal, biomass, and natural gas to liquids via Fischer–Tropsch synthesis, ZSM-5 catalytic conversion, methanol synthesis, methanol-to-gasoline, and methanol-to-olefins/distillate technologies, *Computers & Chemical Engineering*, **47**, 29–56.

Carvalho, A., Matos, H., Gani, R. (2013). SustainPro—A tool for systematic process analysis, generation and evaluation of sustainable design alternatives, *Computer & Chemical Engineering*, **50**, 8–27.

Cheali, P., Gernaey, K.V., Sin, G. (2014). Towards a computer-aided synthesis and design of biorefinery networks—Data collection and management using a generic modeling approach, *ACS Sustainable Chemistry & Engineering*, **2** (1), 19–29.

Dutta, A., Phillips, S. (2009). Thermochemical Ethanol via Direct Gasification and Mixed Alcohol Synthesis of Lignocellulosic Biomass, NREL Technical Report, No. TP-510-45913, National Renewable Energy Laboratory, Colorado, USA.

Dutta, A., Talmadge, M., Hensley, J., Worley, M., Dudgeon, D., Barton, D., Groenendijk, P., Ferrari, D., Stears, B., Searcy, E.M., Wright, C.T., Hess, J.R. (2011). Process Design and Economics for Conversion of Lignocellulosic Biomass to Ethanol, NREL Technical Report, No. NREL/TP-5100-51400, National Renewable Energy Laboratory, Colorado, USA.

Foust, T.D., Aden, A., Dutta, A., Phillip, S. (2009). An economic and environmental comparison of a biochemical and a thermochemical lignocellulosic ethanol conversion processes, *Cellulose*, **16**, 547–565.

Jones, S.B., Valkenburg, C., Walton, C.W. & Elliott, D.C. (2009). Production of Gasoline and Diesel from Biomass via fast Pyrolysis, Hydrotreating and Hydrocracking: A Design Case. PNNL-18284, U.S. Department of Energy, Pacific Northwest National Laboratory, Richland, WA.

Kalakul, S., Malakul, P., Siemanond, K., Gani, R. (2014). Integration of life cycle assessment software with tools for economic and sustainability analyses and process simulation for sustainable process design, *Journal of Cleaner Production*, **17**, 98–109.

Martin, M., Grossmann, I.E. (2012). Energy optimization of bioethanol production via hydrolysis of switchgrass, *AIChE Journal*, **58**, 1538–1549.

Pham, V., El-Halwagi, M. (2012). Process synthesis and optimization of biorefinery configurations, *AIChE Journal*, **58**, 1212–1221.

Phillips, S., Aden, A., Jechura, J., Dayton, D., Eggeman, T. (2007). Thermochemical Ethanol via Indirect Gasification and Mixed Alcohol Synthesis of Lignocellulosic Biomass, NREL Technical Report, No. NREL/TP-510-41168, National Renewable Energy Laboratory, Colorado, USA.

Piyarak, S. (2012). Development of Software for Life Cycle Assessment. Master thesis. The Petroleum and Petrochemical College, Chulalongkorn University, Bangkok.

Quaglia, Alberto et al. (2012). Integrated Business and Engineering Framework for Synthesis and Design of Enterprise-Wide Processing Networks. *Computers & Chemical Engineering*, **38**, 213–223. Available: 10.1016/j.compchemeng.2011.12.011.

Quaglia, A., Sarup, B., Sin, G., Gani, R. (2013). A systematic framework for enterprise-wide optimization: Synthesis and design of processing networks under uncertainty, *Computers & Chemical Engineering*, **59**, 47–62.

Swanson, R.M., Satrio, J.A., Brown, R.C., Platon, A., Hsu, D.D. (2010). Techno-Economic Analysis of Biofuels Production Based on Gasification, NREL Technical Report, No. NREL/TP-6A20-46587, National Renewable Energy Laboratory, Colorado, USA.

Uerdingen, E., Fischer, U., Hungerbuler, K., Gani, R. (2003). A new screening methodology for the identification of economically beneficial retrofit options in chemical process, *AIChE Journal*, **49**, 2400–2418.

Voll, A., Marquardt, W. (2012). Reaction network flux analysis: Optimization-based evaluation of reaction pathways for biorenewables processing, *AIChE Journal*, **58**, 1788–1801.

Wright, M.M., Satrio, J.A., Brown, R.C., Daugaad, D.E. & Hsu, D.D. (2010). Techno-Economic Analysis of Biomass Fast Pyrolysis to Transportation Fuels. NREL Technical Report, No. NREL/TP-6A20-46586, National Renewable Energy Laboratory, Colorado, USA.

Yuan, Z., Chen, B., Gani, R., (2013). Applications of process synthesis: Moving from conventional chemical processes towards biorefinery processes, *Computers & Chemical Engineering*, **49**, 217–229.

Zondervan, E., Nawaz, M., de Haan, A.B., Woodley, J., Gani, R. (2011). Optimal design of a multiproduct biorefinery system, *Computers & Chemical Engineering*, **35**, 1752–1766.

2

Application of a Hierarchical Approach for the Synthesis of Biorefineries

Carolina Conde-Mejía[1], Arturo Jiménez-Gutiérrez[1], and Mahmoud M. El-Halwagi[2]

[1] *Departamento de Ingeniería Química, Instituto Tecnológico de Celaya, Celaya, Mexico*
[2] *Chemical Engineering Department, Texas A&M University, College Station, TX, USA*

2.1 Introduction

A biorefinery has been defined as a facility which transforms biomass into fuels, chemicals, and energy (Naik et al., 2010). Biorefinery implementation became a reality with the so-called first-generation biorefineries (FGB), mainly for the production of bioethanol and biodiesel. Generally, FGB use corn, sugarcane, and soya bean as feedstocks, which is a major issue because of the competition with food sources. A second generation of biorefineries (SGB) has based its production on materials such as agricultural residuals and industrial residuals and grows without feed use; such types of materials are known as lignocellulosic materials (LCM). There are five principal platform conversions for a biorefinery: thermochemical, biogas, biochemical, bio-oil, and carboxylate (Dimian and Bildea 2008; Holtzapple and Granda 2009); in some cases, it is easy to select a type of platform, only depending on the characteristics of feedstock and products. For example, if biodiesel production is of interest, the proper platform will be bio-oil; however, there are cases such as bioethanol whose production can be based on thermochemical, biochemical, or carboxylate platforms (Piccolo and Bezzo, 2007, 2009; Granda et al., 2009; Holtzapple and Granda, 2009; Anex et al., 2010). The scope of this work is limited to the biochemical platform and is focused on bioethanol production.

Process Design Strategies for Biomass Conversion Systems, First Edition. Edited by Denny K. S. Ng, Raymond R. Tan, Dominic C. Y. Foo, and Mahmoud M. El-Halwagi.
© 2016 John Wiley & Sons, Ltd. Published 2016 by John Wiley & Sons, Ltd.

For bioproducts from LCM, the transformation process shows several technical barriers. For example, a type of feedstock is only available during one season of the year, the feedstock generally requires some type of pretreatment, the inhibition of the main product may occur because of by-product formation, and the energy consumption for the separation step is usually high. Several research works have focused on biorefinery technical problems. Bowling et al. (2011) developed a supply chain optimization problem for a biorefinery, involving plant location. Another approach for supply chain problem was presented by Stephen et al. (2010), who took the Peace River region of Alberta to locate potential facilities, assessing variations in availability of residual biomass from agricultural crops. Thorsell et al. (2004) analyzed the cost to harvest lignocellulosic feedstock for biorefineries. A pretreatment step is implemented before the conversion stage to improve the sugar yield, which represents an additional cost with respect to a conventional process. Some pretreatment methods are based on steam explosion (SE), liquid hot water (LHW), dilute acid (DA), organosolvent (OS), and irradiation. Several experimental works have been done to evaluate the pretreatment methods (Negro et al., 2003; Lloyd and Wyman, 2005; Mosier et al., 2005; Silverstein et al., 2007; Gupta et al., 2011; Binod et al., 2012); a valuable effort has been reported by Wyman et al. (2005a, 2005b), who compared several pretreatment methods based on sugar yields for LCM such as corn stover and poplar wood. One important aspect of LCM bioconversion is that a hydrolysis step is required in order to obtain sugars from cellulose; acid hydrolysis (AH) or enzymatic hydrolysis (EH) can be used. Cellulose AH can be implemented at high acid concentration and low temperature or low acid concentration and high temperature. Equipment corrosion and product degradation are the main disadvantages of high acid concentration, and high energy requirement is the main disadvantage of low acid concentration (Rinaldi and Schüth, 2009). The EH is characterized by the use of low temperatures and low energy requirements, but it is limited by the enzyme cost (Steele et al., 2005; Aden, 2008; Xu et al., 2009; Kazi et al., 2010). One alternative to make this stage more efficient is the recovery and recycle of the enzymes (Gregg and Saddler, 1996; Gregg et al., 1998; Lu et al., 2002); the correct implementation for enzyme recovery highly depends on the reactor design. Also, the correct microorganism and conversion configuration selection can contribute to achieve high ethanol yields (Wingren et al., 2003; Öhgren et al., 2007; Morales-Rodriguez et al., 2011a). Although alcoholic fermentation has been known for many years, there are several topics on bioconversions that need to be improved. In order to increase the ethanol yields, it is needed to modify the microorganism performance to allow higher tolerance to increases in ethanol concentration (Krishnan et al., 1999). Another aspect is the inhibition of the main product. Ethanol is typically inhibited by furfural and acetic acid. Furfural is a xylose degradation product that can be produced during pretreatment and hydrolysis steps, while production of acetic acid can occur in pretreatment, hydrolysis, and fermentation steps. Experimental reports have contributed to understand the conditions that minimize inhibition (Larsson et al., 1999; Palmqvist and Hahn-Hägerdal 2000a, 2000b; Klinke et al., 2004; Thomsen et al., 2009). Based on these data, it is possible to establish an operation range to improve ethanol production. For example, one can aim to control the dilution level of the fermentation mixture. For the bioethanol separation from a biochemical platform, the two main problems are the initial low composition (2–6 wt%) and the azeotrope mixture of ethanol and water. In industry, conventional distillation is probably the most widely used separation system for hydrous ethanol production; azeotropic distillation, extractive distillation, and adsorption onto molecular sieves are the more popular processes for anhydrous ethanol production (Dias et al., 2009b; Kumar et al., 2010). However, there are other alternatives that can preconcentrate or dehydrate ethanol, such as liquid–liquid extraction, pervaporation, or vapor permeation. Some analysis has been done to compare separation alternatives (Vane and Alvarez, 2008; Álvarez et al., 2009; Dias et al., 2009b; Junqueira et al., 2009; Huang et al., 2010; Avilés Martínez et al., 2011; Neves et al., 2011). Energy consumption is one of the main factors that have been considered in these studies. There are several investigation opportunities in order to achieve the implementation of these separation processes. For example, new solvents have to be considered for liquid–liquid extraction and extractive distillation processes, and new materials have to be identified for membrane separation. It is important to highlight that integration technics can substantially improve the efficient use of basic resources for a biorefinery; as shown by Grossmann and Martin (2010) and Karuppiah et al. (2008), water and energy consumption can be significantly reduced through the implementation of such techniques.

A good number of synthesis routes can be defined in order to obtain one product; however, several of them can be technically or economically infeasible, and few options will meet all the requirements in terms of technical, economic, and environmental aspects. The correct and quick screening of alternatives is a challenge for synthesis and design areas, especially when information about process alternatives is not well known. One approach consists of comparing options for each individual processing step. For example, Eggeman and Elander (2005) developed a technoeconomic analysis for comparing pretreatment options; they assessed five methods and observed the global process performance. Different conversion flowsheets were analyzed by Morales-Rodriguez et al. (2011b) based on dynamic models; they showed that processes based on simultaneous saccharification and cofermentation (SSCF) flowsheets seem to have the highest potential. A more general screening can be done using optimization techniques (Bao et al., 2011; Pham and El-Halwagi, 2011). Another option would be to use a hierarchical approach, in which the overall problem is decomposed into individual tasks and implemented sequentially (Douglas, 1988; Dimian and Bildea, 2008). Moreover, integration technics can complement this methodology in order to achieve process design with efficient use of basic resources (Conde-Mejía et al., 2012, 2013). This chapter illustrates how the hierarchical approach can provide an excellent tool for the selection of processing routes in biorefinery facilities.

2.2 Problem Statement

Several feedstocks in SGB can be grouped as LCM, which are characterized by their cellulose, hemicellulose, and lignin content as principal components; Table 2.1 shows a typical composition for some agricultural residues. Bioethanol production from LCM has several options for implementation. We have done an extensive literature review on pretreatment and conversion configuration options (Conde-Mejía et al., 2012, 2013), and we consider here six pretreatment methods and six conversion configurations with positive potential (see Figure 2.1). In Table 2.2, the abbreviations for pretreatments and conversion configurations are given. This gives rise to 36 possible ways to produce bioethanol, from which the best alternatives are to be found.

The analysis is based on an LCM feed of 42 tonnes/h, with composition (wt%) of cellulose 40, hemicellulose 27, lignin 23, and others 10, based on dry biomass. This composition is similar to that of sugarcane bagasse.

Table 2.1 *Content of cellulose, hemicellulose, and lignin for some LCM on dry basis (wt%)*

Agricultural residues	Cellulose	Hemicellulose	Lignin
Corn cobs	33.7–41.2	31.9–36	6.1–15.9
Sugarcane bagasse	40	27–37.5	10–20
Wheat straw	32.9–50	24–35.5	8.9–17.3
Rice straw	36.2–47	19–24.5	9.9–24
Corn stalks	35–39.6	16.8–35	7–18.4
Barley straw	33.8–37.5	21.9–24.7	13.8–15.5
Soya stalks	34.5	24.8	19.8
Cotton stalks	38.4–42.6	20.9–34.4	21.45
Switch grass	32.0	25.2	18.1
Corn stover	40	25	17
Coastal bermuda grass	25	35	6

Adapted from Garrote et al. (1999), Saha (2003), and Balat et al. (2008).

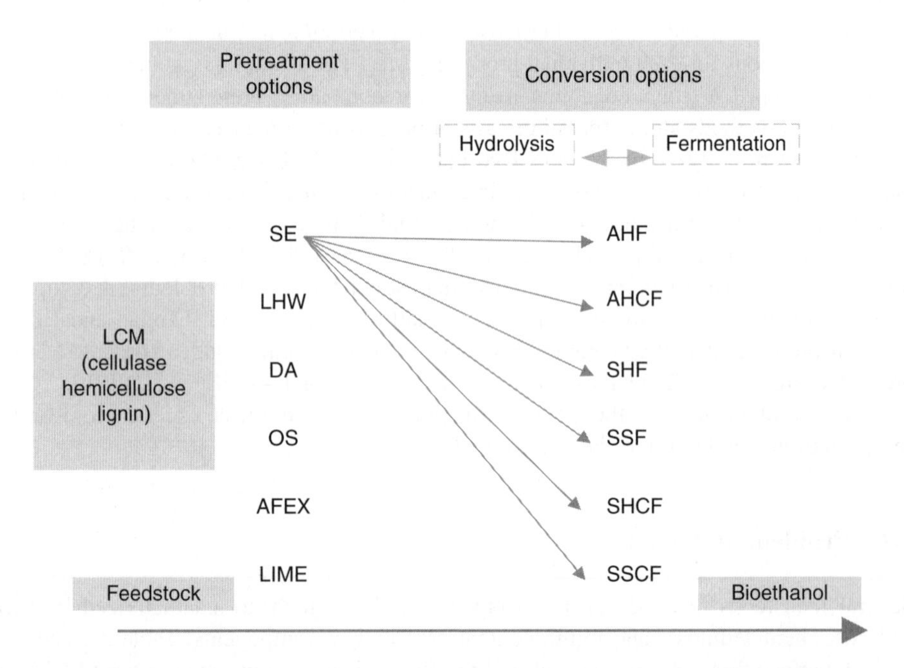

Figure 2.1 *Case study: bioethanol production based on biochemical platform*

2.3 General Methodology

The overall strategy for the hierarchical procedure is as follows. First, the six pretreatment methods are compared in terms of energy requirements. Then, the implementation of a direct recycle integration strategy is analyzed. The pretreatment options are then combined with conversion configurations and analyzed via process simulation. Next, a separation step based on conventional distillation is implemented. Finally, reactor fixed costs are estimated for some options to complete the analysis. Figure 2.2 summarizes the overall methodology.

Table 2.2 *Abbreviations for pretreatments and conversion configurations*

Abbreviation	Pretreatment
SE	Steam explosion
LHW	Liquid hot water
DA	Dilute acid using H_2SO_4
LIME	Alkali extraction with lime
AFEX	Ammonia fiber explosion
OS	Organosolvent
	Conversion configuration
AHF	Acid hydrolysis and fermentation
AHCF	Acid hydrolysis and cofermentation
SHF	Separated enzymatic hydrolysis and fermentation
SHCF	Separated enzymatic hydrolysis and cofermentation
SSF	Simultaneous saccharification and fermentation
SSCF	Simultaneous saccharification and cofermentation

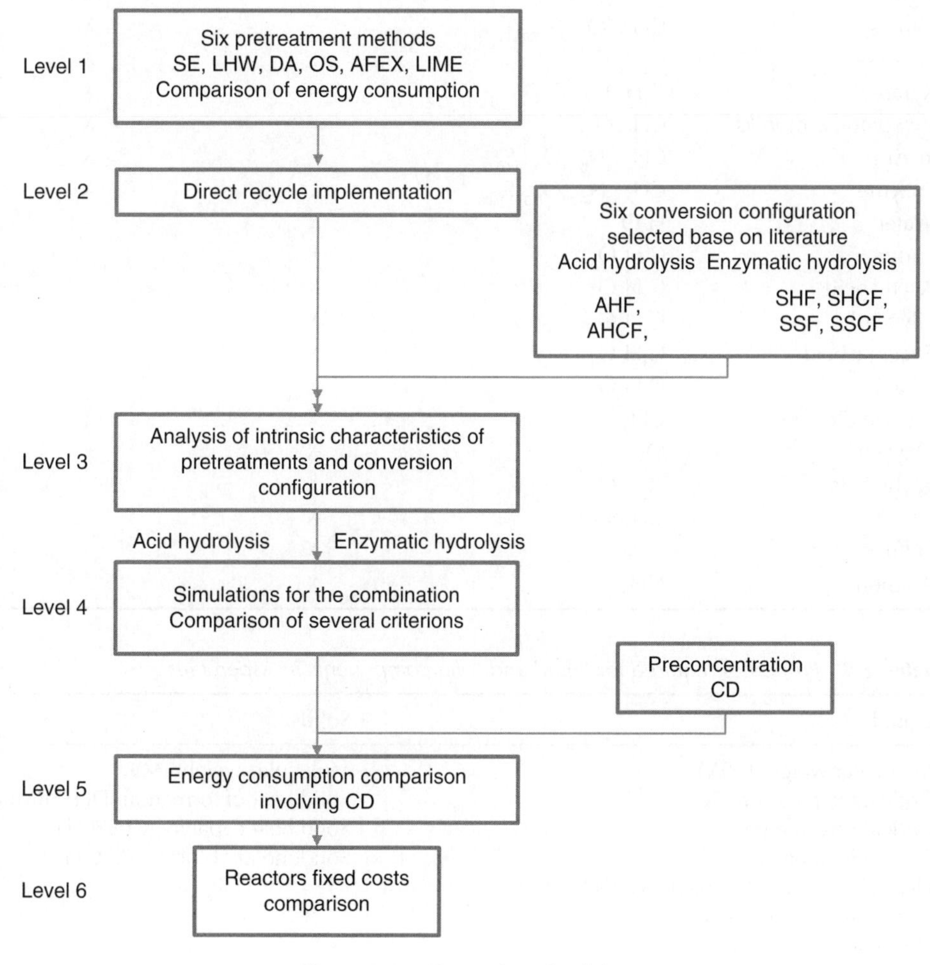

Figure 2.2 *General methodology*

2.4 Simulation of Flowsheets

The list of components involved in the simulations is presented in Table 2.3. Some components are not available in the Aspen Plus database, so it was necessary to include the properties presented in Table 2.4 for such components; such data were taken from Wooley and Putsche (1996).

The simulations were based on stoichiometric relations and yield data. Table 2.5 presents the data used for the simulations. These data were taken from different sources, as reported

Table 2.3 *Components for simulations in Aspen Plus*

Component	Formula	Available in Aspen database	Not available in Aspen database
Cellulose	$C_6H_{10}O_5$		X
Hemicellulose	$C_5H_8O_4$		X
Lignin	$C_{7.3}H_{13.9}O_{1.3}$		X
Others	$CH_{1.48}O_{0.19}S_{0.0013}$		X
Glucose	$C_6H_{12}O_6$		X
Xylose	$C_5H_{10}O_5$		X
Zymomonas mobilis	$CH_{1.8}O_{0.5}N_{0.2}$		X
Yeast	$CH_{1.64}N_{0.23}O_{0.39}S_{0.0035}$		X
Enzyme	$CH_{1.57}N_{0.29}O_{0.31}S_{0.007}$		X
Water	H_2O	X	
Furfural	$C_5H_4O_2$	X	
Acetic acid	$C_2H_4O_2$	X	
Glycerol	$C_3H_8O_3$	X	
Succinic acid	$C_4H_6O_4$	X	
Ethanol	C_2H_6O	X	
Carbon dioxide	CO_2	X	
Oxygen	O_2	X	
Lactic acid	$C_3H_6O_3$	X	
Urea	CH_4N_2O	X	
Sulfuric acid	H_2SO_4	X	
Ammonia	NH_3	X	

Table 2.4 *Properties required for liquid and solid components in Aspen Plus*

Liquid	Solids
Molecular weight (MW)	Molecular weight (MW)
Critical temperature (TC)	Solid heat of formation (DHSFRM)
Critical pressure (PC)	Solid heat capacity (CPSPO1)
Critical volume (VC)	Solid molar volume (VSPOLY)
Heat of formation for ideal gas (DHFORM)	
Vapor pressure (PLXANT)	
Heat capacity for ideal gas (CPIG)	
Heat of vaporization (DHVLWT)	
Acentric factor (OMEGA)	

Table 2.3 Reactions and yield date used in the simulations

Process	Reaction	SE	LHW	DA	OS
Pretreatment	Hemicellulose + H_2O → xylose	26.00	42.31	62.30	45.39
	Hemicellulose → furfural + 2 H_2O	2.97	2.97	2.97	2.97
	2 hemicellulose + 2 H_2O →5 acetic acid	1.25	1.25	1.25	1.25
	Cellulose + H_2O → glucose	4.00	5.62	6.26	1.01
		AH			
Acid hydrolysis	Hemicellulose + H_2O → xylose	17.80			
	Hemicellulose → furfural + 2 H_2O	71.50			
	2 hemicellulose + 2 H_2O → 5 acetic acid	10.60			
	Cellulose + H_2O → glucose	76.50			
	Xylose → furfural + 3 H_2O	80.00			
		EH_{SE}	**EH_{LHW}**	**EH_{DA}**	**EH_{AFEX}**
Enzymatic hydrolysis	Hemicellulose + H_2O → xylose	30.30	30.30	16.44	67.14
	2 hemicellulose + 2 H_2O → 5 acetic acid	10.60	10.60	10.60	10.60
	Cellulose + H_2O → glucose	90.00	90.00	91.12	96.01
		GF	**XF**	**CF**	
Fermentation	Glucose → 2 ethanol + 2 CO_2	94.00	92	92.00	
	Glucose + 2 H_2O → 2 glycerol + O_2	2.67	0.20	0.20	
	Glucose + 2 CO_2 → 2 succinic acid + O_2	0.29	0.80	0.80	
	Glucose → 3 acetic acid	1.19	2.20	2.20	
	Glucose +1.2 NH_3 → 6 yeast + 2.4 H_2O + 0.3 O_2	1.37			
	Glucose +1.2 NH_3 → 6 *Z. mobilis* + 2.4 H_2O + 0.3 O_2		2.70	2.70	
	3 xylose → 5 ethanol + 5 CO_2		85	75.00	
	3 xylose + 5 H_2O → 5 glycerol + 2.5 O_2		2.90	2.90	
	3 xylose + 5 CO_2 → 5 succinic acid + 2.5 O_2		0.90	0.90	
	2 xylose → 5 acetic acid		2.40	2.40	
	Xylose + NH_3 → 5 *Z. mobilis* + 2.4 H_2O + 0.3O_2		2.90	2.90	
	3 xylose → 5 lactic acid		1.40	1.40	

Source: Lawford et al. (1998), Aguilar et al. (2002), Laser et al. (2002), Xiang et al. (2003), Wyman et al. (2005a, 2005b), Dimian and Bildea (2008), Dias et al. (2009a), Brethauer and Wyman (2010), and Kazi et al. (2010).
AH, acid hydrolysis; CF, cofermentation; EH, enzymatic hydrolysis; GF, glucose fermentation; XF, xylose fermentation.

Table 2.6 *Operating conditions specified into Aspen Plus units*

Unit	Operation condition fixed		
	T (°C)	P (atm)	v_f
SE	220		1
LHW	190		0
DA	160		0
LIME	120		0
AFEX	90		0
OS	180		0
AH	220		0
EH	45	1	
XF	37	1	
GF	35	1	
CF	34	1	
SSF	35	1	
SSCF	34	1	

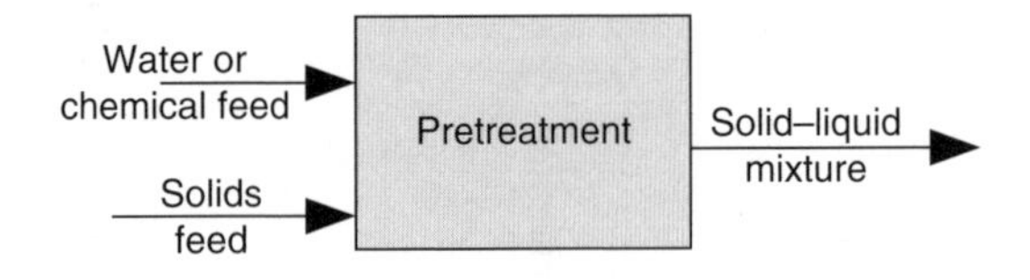

Figure 2.3 *Pretreatment flowsheet*

in the works by Conde-Mejía et al. (2012) and (2013). In Table 2.5, there is a distinction for EH depending on which pretreatment was used, for example, EHDA means that DA pretreatment was used before EH.

In order to implement the process steps into the Aspen Plus simulator, the stoichiometric reactor unit was used when the process involved a reaction system, and the flash unit was used for the LIME and ammonia fiber explosion (AFEX) pretreatments (in which there are no reactions). The operation conditions specified in these units, for each process, are shown in Table 2.6.

The flowsheets were sequentially implemented into the Aspen Plus process simulator. First, the pretreatments without recycle implementation were solved. Figure 2.3 shows a simplified scheme of the pretreatment step without direct recycle.

A direct recycle configuration was then added (Figure 2.4). Depending on the type of pretreatment, the recycle stream vapor can contain water or a mixture of water and chemicals. From the convergence methods available in Aspen, the Broyden method was selected to solve the direct recycle flowsheet. Several simulations were run in order to analyze the operation cost (energy cost + chemical cost) as a function of the vapor fraction sent to the recycle; the vapor fraction with the minimum cost was then identified.

Once the schemes of pretreatment with direct recycle were defined, the conversation step was added. The six conversion configurations are represented in Figures 2.5 and 2.6. Figure 2.5 shows the two options based on AH, while Figure 2.6 shows the four options

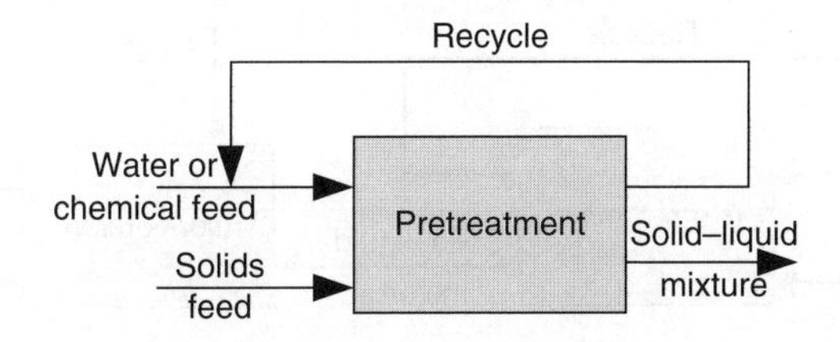

Figure 2.4 Pretreatment with direct recycle

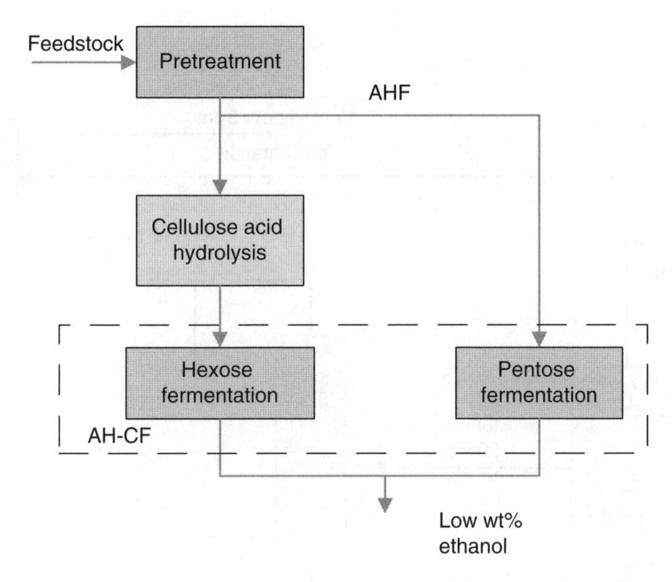

Figure 2.5 Conversion configurations based on acid hydrolysis

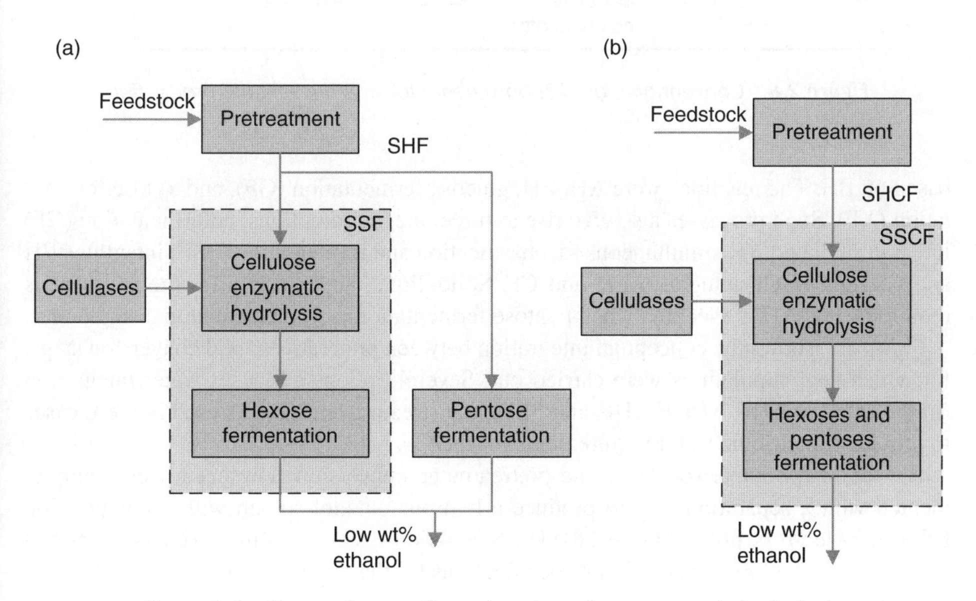

Figure 2.6 Conversion configurations based on enzymatic hydrolysis

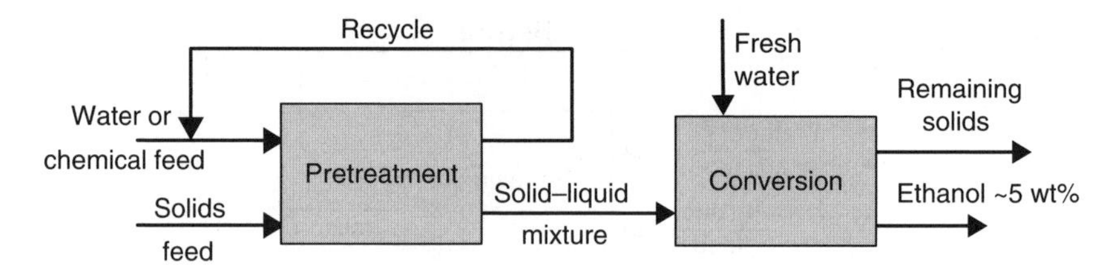

Figure 2.7 Integration between pretreatment and conversion steps

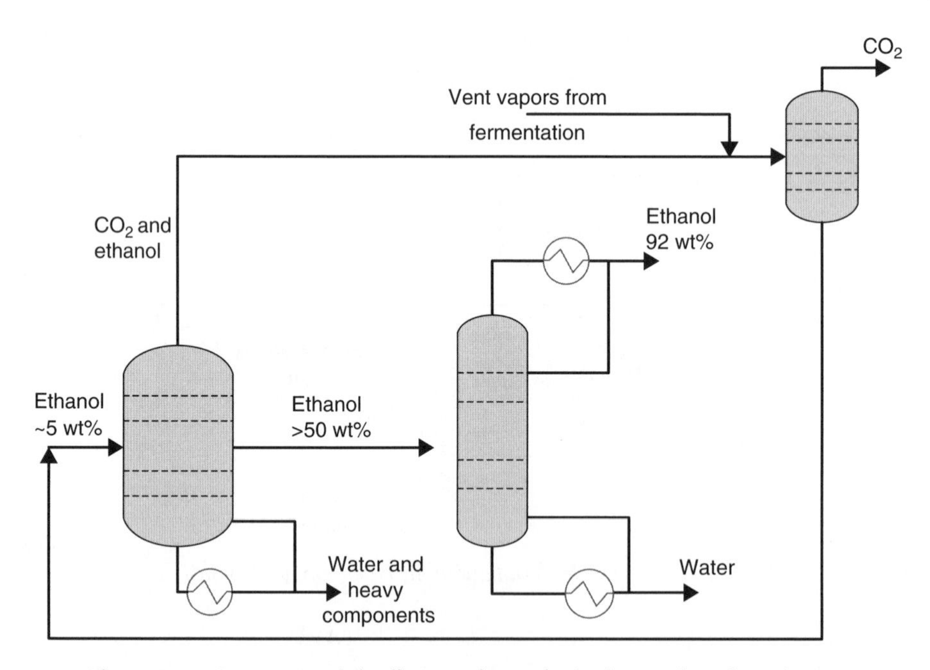

Figure 2.8 Conventional distillation scheme for hydrous ethanol production

based on EH. The reactions were AH, EH, glucose fermentation (GF), and xylose fermentation (XF). Such processes also give rise to three integrated options: cofermentation (CF) integrates GF and XF, simultaneous saccharification and fermentation (SSF) integrates EH and GF, and SSCF integrates EH and CF. Solid–liquid separation after pretreatment is required when AH is used and when pentose fermentation is implemented in a single unit.

Figure 2.7 shows the conceptual integration between pretreatment and conversion steps, for which new simulations were carried out. Several options allow for water integration (in the cases of AHF, AHCF, SHF, and SSF); the stream from XF was used in these cases to provide some of the water requirements for the GF.

For the next comparison level, the pretreatment–conversion flowsheets were complemented with a separation step to produce a hydrous ethanol stream with a composition below the azeotrope point (Figure 2.8). This separation system is similar to the one reported by Wooley et al. (1999). The column specifications for the separation scheme are presented in Table 2.7.

Table 2.7 Specification for separation system used in Aspen Plus

	Stripping with side draw stream					
	# Stages	Feed stage	*P* (atm)	Side draw	Wt% recovery (ETOH)	Wt% purity (ETOH)
DA-AHF	35	1	1	14	99.11	50.52
DA-AHCF	35	1	1	14	99.11	50.61
DA-SHF	22	1	1	11	99.11	52.27
DA-SHCF	22	1	1	10	99.11	55.65
DA-SSF	22	1	1	12	99.11	55.06
DA-SSCF	22	1	1	10	99.11	55.66
AFEX-SSCF	28	1	1	16	99.11	50.60
	Rectification column with partial condenser					
	# Stages	Feed stage	*P* (atm)	Reflux ratio	Wt% recovery (ETOH)	Wt% purity (ETOH)
DA-AHF	15	11	1	2.95	99.51	92.01
DA-AHCF	15	11	1	2.80	99.51	92.01
DA-SHF	15	12	1	2.60	99.51	92.01
DA-SHCF	15	11	1	2.44	99.51	92.01
DA-SSF	15	13	1	6.89	99.51	92.01
DA-SSCF	15	11	1	2.48	99.51	92.01
AFEX-SSCF	25	17	1	9.96	99.51	92.01

For this analysis, the inhibition problem due to the furfural and acetic acid by-products was considered. The composition of such components was observed downstream in pretreatment and conversion steps. Based on experimental reports (Larsson et al., 1999), the maximum composition values of furfural and acid acetic were fixed as 3.75 and 6 g/L. We used these composition values as a reference in order to ensure a low level of potential inhibition in the conversion steps.

2.5 Results and Discussion

Figure 2.9 gives the overall results and shows the effectiveness of the application of the design methodology. Initially, there were 36 possible ways to combine pretreatments and conversion configurations, and after the application of the hierarchical approach proposed in this work, two synthesis routes with high potential were identified. A detailed description of the application of the hierarchical method is given in the following, where the different steps, or levels, of the methodology are described.

2.5.1 Level 1

The first comparison level was based on the energy consumption for the pretreatment methods (Figure 2.10). The utility prices were \$6/MMBtu for heating, \$4/MMBtu for cooling, and 0.07 kW h for power. After this comparison, the OS option was discarded, because it showed the highest energy demand.

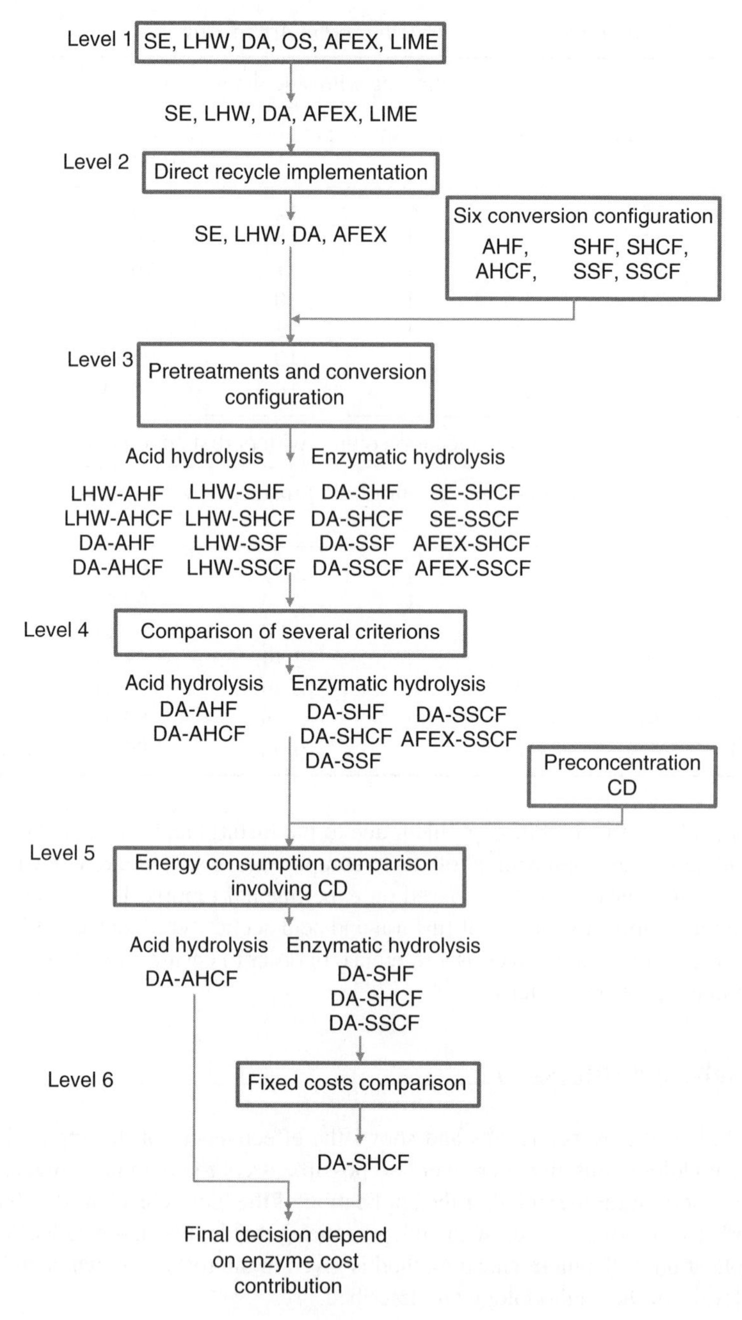

Figure 2.9 *A summary of results from the methodology application*

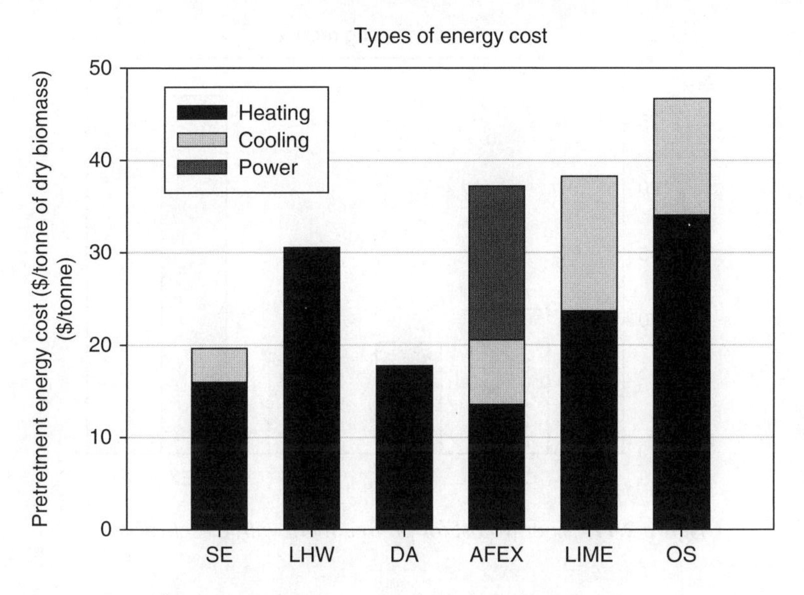

Figure 2.10 *Energy cost for pretreatment options*

2.5.2 Level 2

In the second comparison level, direct recycle was considered for the five remaining methods. We found that the LIME option did not present any opportunity for direct recycle implementation. Therefore, direct recycle was only implemented in the SE, DA, LHW, and AFEX cases. As can be observed in Figure 2.11, the energy requirements were considerably reduced. DA pretreatment showed the best performance when the direct recycle was implemented.

Water consumption was also reduced for SE, LHW, and DA methods, and ammonia consumption has a significant reduction in the AFEX method. Table 2.8 shows the comparison of water consumption with and without direct recycle.

After the pretreatment simulations, the composition of furfural and acetic acid was observed and compared with the values reported by Larsson et al. (1999). We found that for SE pretreatment there is a high inhibition potential when a solid–liquid separation is used after pretreatment. Therefore, it is recommended to avoid such separation alternative in that case.

2.5.3 Level 3

Once the SE, LHW, AFEX, and DA pretreatment methods were selected, 24 possible combinations with the six conversion configurations are established. However, based on the pretreatment characteristics, only 16 potential combinations were considered (Figure 2.12). SE pretreatment is not recommended for implementation with any configuration using solid–liquid separation because of the potential bioethanol inhibition due to high furfural and acetic acid compositions in the fermentation step. The AFEX pretreatment can be combined with separated enzymatic hydrolysis and cofermentation (SHCF) and

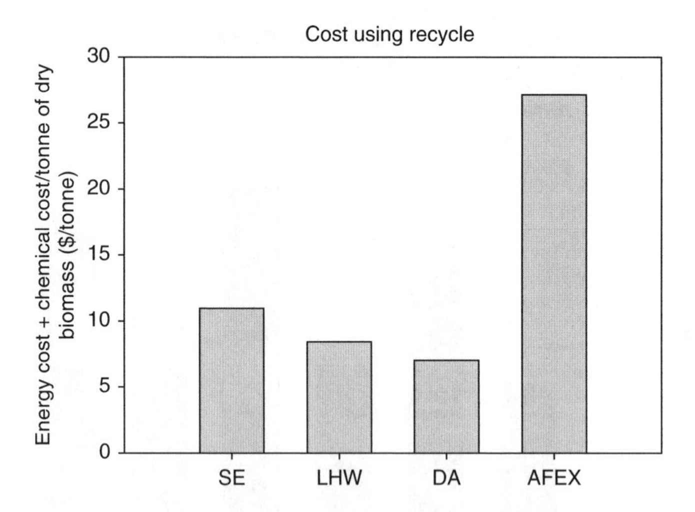

Figure 2.11 *Energy cost for pretreatment with direct recycle*

Table 2.8 *Effect of using direct recycle on water and ammonia consumption*

Freshwater and ammonia feed (tonne/tonne of LCM)			
	With recycle	Without recycle	
SE	0.9877	0.1060	Water
LHW	4.5767	3.5067	Water
DA	2.3316	1.7100	Water
AFEX	1	0.0109	Ammonia

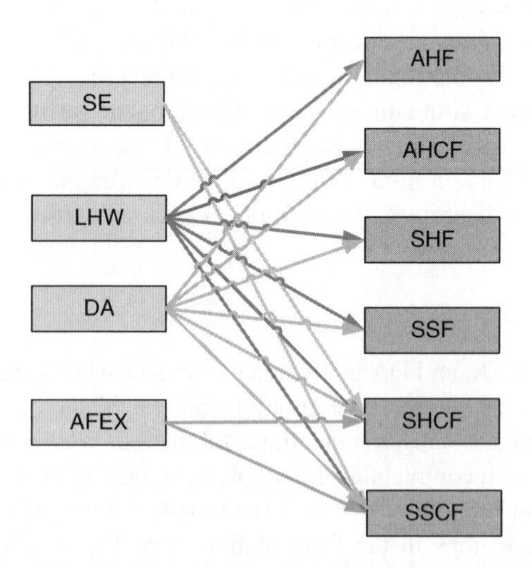

Figure 2.12 *Combinations among pretreatment and conversion configurations*

SSCF configurations; AFEX shows no significant production of xylose and glucose, so a solid–liquid separation is not required (see Table 2.5; no reaction is considered for the AFEX method).

2.5.4 Level 4

The sixteen pretreatment–conversion combinations were simulated in Aspen Plus. An initial aspect was to determine the dilution level for the reaction mixtures, for which the furfural and acetic acid compositions of 3.75 and 6 g/L, taken from Larsson et al. (1999), were considered. These parameters were used to iterate with the freshwater fed to the conversion step. Table 2.9 shows the final solid wt% and the acetic acid and furfural composition after the reacting steps. For example, in the SE-SHCF case, the acetic acid composition is 4.05 g/L after EH and 5.85 g/L after CF.

The results obtained for the sixteen combinations are summarized in Table 2.10. The unit energy cost was used to assess economic performance. A mass intensity index (MII), defined as [(mass of feedstock-mass of bioethanol produced)/mass of bioethanol produced], and a water intensity index (WII), defined as (mass of fresh water/mass of ethanol produced) (El-Halwagi, 2012), were calculated. These two indices show the efficient use of basic resources, feedstocks and water. The ethanol composition after conversion is also reported. It can be observed that the best values for the WII are associated to

Table 2.9 *Solid content and furfural or acetic acid composition obtained from the simulations*

Combinations	Solids wt%	[Significant component] (g/L)
SE-SHCF	12.0	$[\text{Acetic acid}]_{EH}=4.05$
		$[\text{Acetic acid}]_{CF}=5.83$
SE-SSCF	12.5	$[\text{Acetic acid}]_{SSCF}=5.85$
LWH-AHF	9.5	$[\text{Furfural}]_{F}=3.44$
LWH-AHCF	10.5	$[\text{Furfural}]_{CF}=3.70$
LWH-SHF	20.0	$[\text{Acetic acid}]_{EH}=5.66$
		$[\text{Acetic acid}]_{F}=4.04$
LWH-SHCF	12.5	$[\text{Acetic acid}]_{EH}=3.87$
		$[\text{Acetic acid}]_{CF}=5.81$
LWH-SSF	14.5	$[\text{Acetic acid}]_{SSF}=5.63$
LWH-SSCF	13.0	$[\text{Acetic acid}]_{SSCF}=5.84$
DA-AHF	10.0	$[\text{Furfural}]_{F}=3.70$
DA-AHCF	10.0	$[\text{Furfural}]_{CF}=3.70$
DA-SHF	20.0	$[\text{Acetic acid}]_{EH}=3.99$
		$[\text{Acetic acid}]_{F}=4.32$
DA-SHCF	14.0	$[\text{Acetic acid}]_{EH}=3.15$
		$[\text{Acetic acid}]_{CF}=5.89$
DA-SSF	17.0	$[\text{Acetic acid}]_{SSF}=5.86$
DA-SSCF	14.0	$[\text{Acetic acid}]_{SSCF}=5.85$
AFEX-SHCF	11.5	$[\text{Acetic acid}]_{EH}=4.27$
		$[\text{Acetic acid}]_{CF}=5.97$
AFEX-SSCF	13.0	$[\text{Acetic acid}]_{SSCF}=5.83$

Table 2.10 *Comparison of the 16 combination alternatives*

Combination	Based on one ton of dry biomass			Unit energy cost	Based on total mass of bioethanol produced		Bioethanol composition before separation
	Energy cost ($)	Bioethanol (gal)	Water consumption (kg)	$/gal (spent in energy)	Mass intensity index (MII)	Water intensity index (WII)	Wt % kg/kg of mixture
SE-SHCF	13.40	78.92	4973.72	0.1697	3.25	21.13	4.02
SE-SSCF	12.42	81.49	4708.11	0.1524	3.11	19.37	4.20
LWH-AHF	53.95	66.47	9340.28	0.8118	4.04	47.12	2.41
LWH-AHCF	49.44	68.19	8586.75	0.7251	3.92	42.22	2.64
LWH-SHF	13.45	72.38	5186.55	0.1858	3.63	24.03	3.75
LWH-SHCF	13.74	81.85	4345.53	0.1679	3.10	17.80	4.68
LWH-SSF	13.71	72.66	3506.88	0.1886	3.61	16.18	4.96
LWH-SSCF	13.92	81.43	4114.25	0.1709	3.12	16.94	4.89
DA-AHF	46.07	72.45	7147.48	0.6358	3.63	33.08	3.41
DA-AHCF	46.17	72.79	7147.48	0.6343	3.61	32.92	3.43
DA-SHF	9.72	79.07	3667.27	0.1229	3.24	15.55	5.13
DA-SHCF	9.95	85.05	3284.88	0.1170	2.94	12.95	6.07
DA-SSF	9.94	77.18	2408.60	0.1287	3.34	10.46	6.97
DA-SSCF	10.09	84.80	3284.88	0.1190	2.95	12.99	6.08
AFEX-SHCF	27.48	90.87	6673.80	0.3024	2.69	24.63	3.79
AFEX-SSCF	24.54	90.33	6047.71	0.2716	2.71	22.45	4.15

Shaded rows show options selected for the next level of the hierarchical procedure.

the highest compositions. In order to compare the combination options, a distinction is made among options based on AH and options based on EH, since enzyme costs were not considered in the EH-based cases. Four combinations with AH and twelve alternatives based on EH arise. A priority order was assigned to the parameters calculated here; the unit energy cost was considered to be the most important one, followed by the MII and the WII. From the AH-based options, the DA-AHF and DA-AHCF were selected as the best options because they have lower unit energy costs than LHW-AHF and LHW-AHCF; moreover, the MII and WII indices also show better values. In the EH-based cases, the four alternatives based on DA as pretreatment method have the best unit energy costs; hence, they were selected for further analysis. If the MII is compared among EH cases, the options with AFEX pretreatment show better performance; from the AFEX-SHCF and AFEX-SSCF options, AFEX-SSCF shows a lower unit energy cost; therefore, it was selected for further analysis.

2.5.5 Level 5

Up to this point, seven alternatives remain, two based on AH and five based on EH. For these seven options obtained from level 4, the flowsheets were complemented with a basic separation process, and energy requirements were estimated. The results from level 5 are presented in Figure 2.13. From a comparison of the five EH cases, the DA-SSF and AFEX-SSCF options are in clear disadvantage, while the DA-SHF, DA-SHCF, and DA-SSCF arrangements show similar performance. For the two cases based on AH, there is not a clear difference, but DA-AHCF shows lower energy consumption than DA-AHF, so

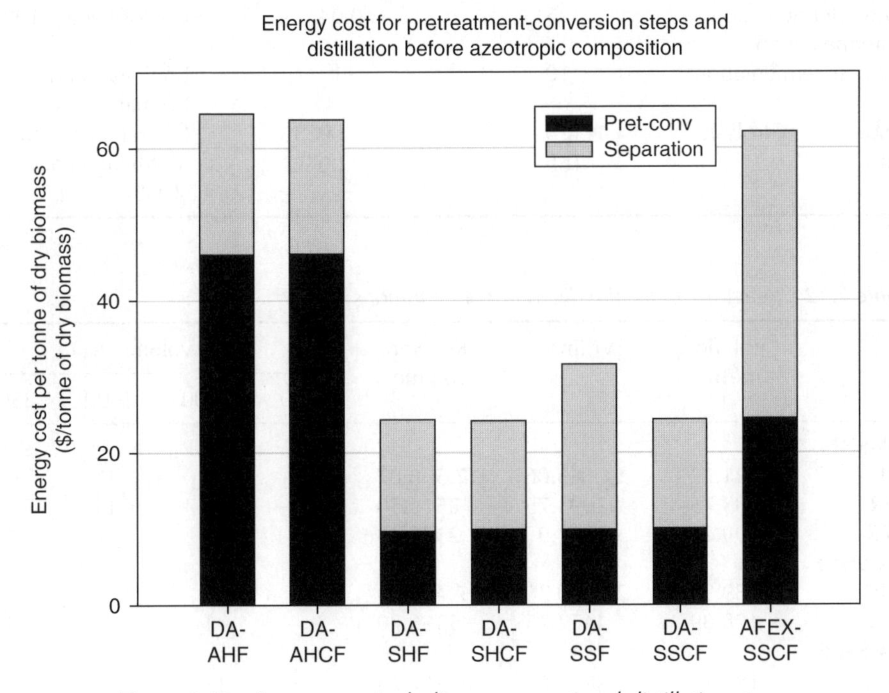

Figure 2.13 *Energy cost including a conventional distillation step*

DA-AHF was selected as the best alternative route based on AH. From this analysis step, the DA-SHF, DA-SHCF, and DA-SSCF options were selected for further analysis.

2.5.6 Level 6

The main difference in the three synthesis routes detected so far—DA-SHF, DA-SHCF, and DA-SSCF—is the hydrolysis and fermentation reactor distributions. Then, a comparison of annual fixed costs for reactor tanks is used in this level as a final factor. The reacting volume was estimated by relating the volume flow obtained from the simulations and the dilution rate reported in Table 2.11. The total reactor volume was then estimated as 20% above the reacting volume, and four different tank capacities were considered in order to achieve the total reactor volume (Table 2.12). The tank cost estimations were based on the data reported by Aden et al. (2002) with a linear depreciation in 10 years. Table 2.13 shows the annual fixed cost and the annual production for each option, from which unit costs per gallon of ethanol were estimated.

DA-SHCF seems to be the most promising synthesis route for bioethanol production based on EH, and DA-AHCF could be considered as the best option based on AH. The principal disadvantage of DA-AHCF is the high energy consumption required in the AH step. For the DA-SHCF case, the EH is a low-energy-consuming process; however, the enzyme

Table 2.11 *Residence time and dilution rate*

Conversion step	Residence time (h)	Dilution rate (h^{-1})	References
Xylose fermentation Cofermentation	25	0.04	Lawford et al. (1998)
Glucose fermentation	10	0.1	Brethauer and Wyman (2010)
Enzymatic hydrolysis	120	0.0083	Öhgren et al. (2007)
SSCF	168	0.0059	McMillan et al. (1999)

Table 2.12 *Reactor size for the different configurations*

	Vol. flow (m^3/h)	Vr (m^3)	Reactor volume	Tank volume (m^3)			
				3500	3000	2600	2500
DA-SHF							
EHR	121.72	14,665.06	17,598.07		5	1	
XFR	83.75	2093.75	2512.50			1	
GFR	199.29	1992.9	2391.48				1
DA-SHCF							
EHR	189.99	22,890.36	27,468.43	1	8		
CFR	178.89	4472.25	5366.70		1		1
DA-SSCF							
SSCFR	188.15	31,889.83	38,267.79		13		

Table 2.13 Annual costs and yearly production

Combination	Annual energy cost ($/year)	Annual reactor fixed cost ($/year)	Total cost ($/year)	Total annual (gal)	$/gal
DA-SHF	8,600,780	479,944	9,080,724	27,832,699	0.3263
DA-SHCF	8,539,056	682,395	9,221,452	29,995,775	0.3074
DA-SSCF	8,610,715	806,960	9,417,675	29,928,179	0.3147

cost contribution has to be considered. A current problem is that there is not an enzyme price available for industrial use. However, it is possible to estimate a maximum target value for the enzyme contribution using the difference between the unit energy costs for DA-AHCF and DA-SHCF combinations (considering only the energy cost for the conversion step) (Conde-Mejia et al., 2013). Such difference turns out to be $0.6343/gal − 0.1170/gal = $0.5173/gal. Higher values for enzyme contribution will make the DA-SHCF process unsuitable for industrial implementation.

2.6 Conclusions

A methodology was developed based on a hierarchical approach for the synthesis of biorefineries. The methodology was applied for bioethanol production based on a biochemical platform. It is shown that the hierarchical approach can be an excellent procedure to screen bioprocess alternatives. The analysis started with 36 possible alternatives for bioethanol production, from which the two options with the best potential for industrial implementation were identified. The DA showed the best performance as a pretreatment step, and its combination with enzymatic hydrolysis and cofermentation (SHCF) or acid hydrolysis and cofermentation (AHCF) configurations provided the most promising bioethanol production routes, depending on what kind of hydrolysis is selected. It is important to highlight that from an energy point of view, DA-SHCF shows the best performance; however, the final decision between DA-SHCF and DA-AHCF alternatives depends on enzyme costs.

In order to complement this work, purification alternatives for bioethanol production should be considered. Also, the effect of the application of process integration techniques on energy and water consumption for the overall process should be analyzed.

References

Aden, A. (2008). Biochemical production of ethanol from corn stover: 2007 state of technology model. NREL/TP-510-43205. Golden, CO: National Renewable Energy Laboratory.

Aden, A., M. Ruth, K. Ibsen, J. Jechura, K. Neeves, J. Sheehan, B. Wallace, L. Montague, A. Slayton and J. Lukas (2002). Lignocellulosic biomass to ethanol process design and economics utilizing co-current dilute acid prehydrolysis and enzymatic hydrolysis for corn stover. NREL/TP-510-32438. Golden, CO: National Renewable Energy Laboratory.

Aguilar, R., J. A. Ramırez, G. Garrote and M. Vazquez (2002). "Kinetic study of the acid hydrolysis of sugar cane bagasse." *Journal of Food Engineering* **55**: 309–318.

Álvarez, V. H., P. Alijó, D. Serrão, R. M. Filho, M. Aznar and S. Mattedi (2009). "Production of anhydrous ethanol by extractive distillation of diluted alcoholic solutions with ionic liquids." *Computer Aided Chemical Engineering* **27**: 1137–1142.

Anex, R. P., A. Aden, F. K. Kazi, J. Fortman, R. M. Swanson, M. M. Wright, J. A. Satrio, R. C. Brown, D. E. Daugaard, A. Platon, G. Kothandaraman, D. D. Hsu and A. Dutta (2010). "Techno-economic comparison of biomass-to-transportation fuels via pyrolysis, gasification, and biochemical pathways." *Fuel* **89**(Suppl. 1): S29–S35.

Avilés Martínez, A., J. Saucedo-Luna, J. G. Segovia-Hernandez, S. Hernandez, F. I. Gomez-Castro and A. J. Castro-Montoya (2011). "Dehydration of bioethanol by hybrid process liquid–liquid extraction/extractive distillation." *Industrial and Engineering Chemistry Research* **51**(17): 5847–5855.

Balat, M., H. Balat and C. Öz (2008). "Progress in bioethanol processing." *Progress in Energy and Combustion Science* **34**(5): 551–573.

Bao, B., D. K. S. Ng, D. H. S. Tay, A. Jiménez-Gutiérrez and M. M. El-Halwagi (2011). "A shortcut method for the preliminary synthesis of process-technology pathways: An optimization approach and application for the conceptual design of integrated biorefineries." *Computers and Chemical Engineering* **35**(8): 1374–1383.

Binod, P., M. Kuttiraja, M. Archana, K. U. Janu, R. Sindhu, R. K. Sukumaran and A. Pandey (2012). "High temperature pretreatment and hydrolysis of cotton stalk for producing sugars for bioethanol production." *Fuel* **92**(1): 340–345.

Bowling, I. M., J. M. a. Ponce-Ortega and M. M. El-Halwagi (2011). "Facility location and supply chain optimization for a biorefinery." *Industrial and Engineering Chemistry Research* **50**(10): 6276–6286.

Brethauer, S. and C. E. Wyman (2010). "Review: Continuous hydrolysis and fermentation for cellulosic ethanol production." *Bioresource Technology* **101**(13): 4862–4874.

Conde-Mejía, C., A. Jiménez-Gutiérrez and M. El-Halwagi (2012). "A comparison of pretreatment methods for bioethanol production from lignocellulosic materials." *Process Safety and Environmental Protection* **90**(3): 189–202.

Conde-Mejía, C., A. Jiménez-Gutiérrez and M. M. El-Halwagi (2013). "Assessment of combinations between pretreatment and conversion configurations for bioethanol production." *ACS Sustainable Chemistry and Engineering* **1**(8): 956–965.

Dias, M., A. Ensinas, S. Nebra, R. Maciel, C. Rossell and M. Wolf (2009a). "Production of bioethanol and other bio-based materials from sugarcane bagasse: Integration to conventional bioethanol production process." *Chemical Engineering Research and Design* **87**: 1206–1216.

Dias, M. O. S., T. L. Junqueira, R. Maciel Filho, M. R. W. Maciel and C. E. V. Rossell (2009b). "Anhydrous bioethanol production using bioglycerol—simulation of extractive distillation processes." *Computer Aided Chemical Engineering* **26**: 519–524.

Dimian, A. and C. S. Bildea (2008). *Chemical process design*. Weinheim: Wiley-VCH.

Douglas, J. M. (1988). *Conceptual design of chemical processes*. New York: McGraw-Hill.

Eggeman, T. and R. T. Elander. (2005). "Process and economic analysis of pretreatment technologies." *Bioresource Technology* **96**(18): 2019–2025.

El-Halwagi, M. (2012). *Sustainable design through process integration*. College Station, TX: Elsevier

Garrote, G., H. Domínguez and J. C. Parajó (1999). "Hydrothermal processing of lignocellulosic materials." *European Journal of Wood and Wood Products* **57**(3): 191–202.

Granda, C., M. Holtzapple, G. Luce, K. Searcy and D. Mamrosh (2009). "Carboxylate platform: The MixAlco process. Part 2: Process economics." *Applied Biochemistry and Biotechnology* **156**(1): 107–124.

Gregg, D. J. and J. N. Saddler (1996). "Factors affecting cellulose hydrolysis and the potential of enzyme recycle to enhance the efficiency of an integrated wood to ethanol process." *Biotechnology and Bioengineering* **51**(4): 375–383.

Gregg, D. J., A. Boussaid and J. N. Saddler (1998). "Techno-economic evaluations of a generic wood-to-ethanol process: Effect of increased cellulose yields and enzyme recycle." *Bioresource Technology* **63**(1): 7–12.

Grossmann, I. E. and M. Martín (2010). "Energy and water optimization in biofuel plants." *Chinese Journal of Chemical Engineering* **18**(6): 914–922.

Gupta, R., Y. P. Khasa and R. C. Kuhad (2011). "Evaluation of pretreatment methods in improving the enzymatic saccharification of cellulosic materials." *Carbohydrate Polymers* **84**(3): 1103–1109.

Holtzapple, M. and C. Granda (2009). "Carboxylate platform: The MixAlco process. Part 1: Comparison of three biomass conversion platforms." *Applied Biochemistry and Biotechnology* **156**(1): 95–106.

Huang, Y., R. W. Baker and L. M. Vane (2010). "Low-energy distillation-membrane separation process." *Industrial and Engineering Chemistry Research* **49**(8): 3760–3768.

Junqueira, T. L., M. O. S. Dias, R. M. Filho, M. R. W. Maciel and C. E. V. Rossell (2009). "Simulation of the azeotropic distillation for anhydrous bioethanol production: study on the formation of a second liquid phase." *Computer Aided Chemical Engineering* **27**: 1143–1148.

Karuppiah, R., A. Peschel and I. Grossmann (2008). "Energy optimization for the design of corn-based ethanol plants." *AIChE Journal* **54**: 1499–1525.

Kazi, F. K., J. A. Fortman, R. P. Anex, D. D. Hsu, A. Aden, A. Dutta and G. Kothandaraman (2010). "Techno–economic comparison of process technologies for biochemical ethanol production from corn stover." *Fuel* **89**(Suppl. 1): S20–S28.

Klinke, H. B., A. B. Thomsen and B. K. Ahring (2004). "Inhibition of ethanol-producing yeast and bacteria by degradation products produced during pre-treatment of biomass." *Applied Microbiology and Biotechnology* **66**(1): 10–26.

Krishnan, M., N. Ho and G. Tsao (1999). "Fermentation kinetics of ethanol production from glucose and xylose by recombinant Saccharomyces 1400(pLNH33)." *Applied Biochemistry and Biotechnology* **78**(1): 373–388.

Kumar, S., N. Singh and R. Prasad (2010). "Anhydrous ethanol: A renewable source of energy." *Renewable and Sustainable Energy Reviews* **14**(7): 1830–1844.

Larsson, S., E. Palmqvist, B. Hahn-Hägerdal, C. Tengborg, K. Stenberg, G. Zacchi and N.-O. Nilvebrant (1999). "The generation of fermentation inhibitors during dilute acid hydrolysis of softwood." *Enzyme and Microbial Technology* **24**: 151–159.

Laser, M., D. Schulman, S. G. Allen, J. Lichwa, M. J. Antal and L. R. Lynd (2002). "A comparison of liquid hot water and steam pretreatments of sugar cane bagasse for bioconversion to ethanol." *Bioresource Technology* **81**(1): 33–44.

Lawford, H., J. Rousseau, A. Mohagheghi and J. McMillan (1998). "Continuous culture studies of xylose-fermenting Zymomonas mobilis." *Applied Biochemistry and Biotechnology* **70–72**(1): 353–367.

Lloyd, T. A. and C. E. Wyman (2005). "Combined sugar yields for dilute sulfuric acid pretreatment of corn stover followed by enzymatic hydrolysis of the remaining solids." *Bioresource Technology* **96**(18): 1967–1977.

Lu, Y., B. Yang, D. Gregg, J. Saddler and S. Mansfield (2002). "Cellulase adsorption and an evaluation of enzyme recycle during hydrolysis of steam-exploded softwood residues." *Applied Biochemistry and Biotechnology* **98–100**(1): 641–654.

McMillan, J., M. Newman, D. Templeton and A. Mohagheghi (1999). "Simultaneous saccharification and cofermentation of dilute-acid pretreated yellow poplar hardwood to ethanol using xylose-fermenting *Zymomonas mobilis*." *Applied Biochemistry and Biotechnology* **79**(1): 649–665.

Morales-Rodriguez, R., K. V. Gernaey, A. S. Meyer and G. Sin (2011a). "A mathematical model for simultaneous saccharification and co-fermentation (SSCF) of C6 and C5 sugars." *Chinese Journal of Chemical Engineering* **19**(2): 185–191.

Morales-Rodriguez, R., A. S. Meyer, K. V. Gernaey and G. Sin (2011b). "Dynamic model-based evaluation of process configurations for integrated operation of hydrolysis and co-fermentation for bioethanol production from lignocellulose." *Bioresource Technology* **102**(2): 1174–1184.

Mosier, N., R. Hendrickson, N. Ho, M. Sedlak and M. R. Ladisch (2005). "Optimization of pH controlled liquid hot water pretreatment of corn stover." *Bioresource Technology* **96**(18): 1986–1993.

Naik, S. N., V. V. Goud, P. K. Rout and A. K. Dalai (2010). "Production of first and second generation biofuels: A comprehensive review." *Renewable and Sustainable Energy Reviews* **14**(2): 578–597.

Negro, M., P. Manzanares, I. Ballesteros, J. Oliva, A. Cabañas and M. Ballesteros (2003). "Hydrothermal pretreatment conditions to enhance ethanol production from poplar biomass." *Applied Biochemistry and Biotechnology* **105**(1–3): 87–100.

Neves, C. M. S S., J. F. O. Granjo, M. G. Freire, A. Robertson, N. M. C. Oliveira and J. A. P. Coutinho (2011). "Separation of ethanol-water mixtures by liquid–liquid extraction using phosphonium-based ionic liquids." *Green Chemistry* **13**: 1517–1526.

Öhgren, K., R. Bura, G. Lesnicki, J. Saddler and G. Zacchi (2007). "A comparison between simultaneous saccharification and fermentation and separate hydrolysis and fermentation using steam-pretreated corn stover." *Process Biochemistry* **42**(5): 834–839.

Palmqvist, E. and B. Hahn-Hägerdal (2000a). "Fermentation of lignocellulosic hydrolysates. I: Inhibition and detoxification." *Bioresource Technology* **74**(1): 17–24.

Palmqvist, E. and B. Hahn-Hägerdal (2000b). "Fermentation of lignocellulosic hydrolysates. II: Inhibitors and mechanisms of inhibition." *Bioresource Technology* **74**(1): 25–33.

Pham, V. and M. El-Halwagi (2011). "Process synthesis and optimization of biorefinery configurations." *AIChE Journal* **58**(4), 1212–1221.

Piccolo, C. and F. Bezzo (2007). Ethanol from lignocellulosic biomass: A comparison between conversion technologies. *Computer Aided Chemical Engineering* **24**(1): 1277–1282.

Piccolo, C. and F. Bezzo (2009). "A techno-economic comparison between two technologies for bioethanol production from lignocellulose." *Biomass and Bioenergy* **33**(3): 478–491.

Rinaldi, R. and F. Schüth (2009). "Acid hydrolysis of cellulose as the entry point into biorefinery schemes." *ChemSusChem* **2**(12): 1096–1107.

Saha, B. (2003). "Hemicellulose bioconversion." *Journal of Industrial Microbiology and Biotechnology* **30**(5): 279–291.

Silverstein, R. A., Y. Chen, R. R. Sharma-Shivappa, M. D. Boyette and J. Osborne (2007). "A comparison of chemical pretreatment methods for improving saccharification of cotton stalks." *Bioresource Technology* **98**(16): 3000–3011.

Steele, B., S. Raj, J. Nghiem and M. Stowers (2005). Enzyme Recovery and Recycling Following Hydrolysis of Ammonia Fiber Explosion-Treated Corn Stover. Twenty-Sixth Symposium on Biotechnology for Fuels and Chemicals. B. H. Davison, B. R. Evans, M. Finkelstein and J. D. McMillan, Humana Press: 901–910.

Stephen, J. D., S. Sokhansanj, X. Bi, T. Sowlati, T. Kloeck, L. Townley-Smith and M. A. Stumborg (2010). "Analysis of biomass feedstock availability and variability for the peace river region of Alberta, Canada." *Biosystems Engineering* **105**(1): 103–111.

Thomsen, M., A. Thygesen and A. Thomsen (2009). "Identification and characterization of fermentation inhibitors formed during hydrothermal treatment and following SSF of wheat straw." *Applied Microbiology and Biotechnology* **83**(3): 447–455.

Thorsell, S., F. M. Epplin, R. L. Huhnke and C. M. Taliaferro (2004). "Economics of a coordinated biorefinery feedstock harvest system: Lignocellulosic biomass harvest cost." *Biomass and Bioenergy* **27**(4): 327–337.

Vane, L. M. and F. R. Alvarez (2008). "Membrane-assisted vapor stripping: Energy efficient hybrid distillation–vapor permeation process for alcohol–water separation." *Journal of Chemical Technology and Biotechnology* **83**(9): 1275–1287.

Wingren, A., M. Galbe and G. Zacchi (2003). "Techno-economic evaluation of producing ethanol from softwood: Comparison of SSF and SHF and identification of bottlenecks." *Biotechnology Progress* **19**(4): 1109–1117.

Wooley, R. and V. Putsche (1996). Development of an ASPEN PLUS physical property database for biofuels components. NREL/MP-425-20685. Golden, CO: National Laboratory of the US Department of Energy, pp. 1–38.

Wooley, R., M. Ruth, J. Sheehan, K. Ibsen, H. Majdeski and A. Galvez (1999). Lignocellulosic biomass to ethanol process design and economics utilizing co-current dilute acid prehydrolysis and enzymatic hydrolysis current and futuristic scenarios NREL/TP-580-26157. Golden, CO: National Laboratory of the US Department of Energy.

Wyman, C. E., B. E. Dale, R. T. Elander, M. Holtzapple, M. R. Ladisch and Y. Y. Lee (2005a). "Comparative sugar recovery data from laboratory scale application of leading pretreatment technologies to corn stover." *Bioresource Technology* **96**(18): 2026–2032.

Wyman, C. E., B. E. Dale, R. T. Elander, M. Holtzapple, M. R. Ladisch and Y. Y. Lee (2005b). "Coordinated development of leading biomass pretreatment technologies." *Bioresource Technology* **96**(18): 1959–1966.

Xiang, Q., J. Kim and Y. Y. Lee (2003). "A comprehensive kinetic model for dilute-acid hydrolysis of cellulose." *Biotechnology for Fuels and Chemicals* **106**(1–3): 337–352.

Xu, Q., A. Singh and M. E. Himmel (2009). "Perspectives and new directions for the production of bioethanol using consolidated bioprocessing of lignocellulose." *Current Opinion in Biotechnology* **20**(3): 364–371.

3

A Systematic Approach for Synthesis of an Integrated Palm Oil-Based Biorefinery

Rex T. L. Ng[1] *and Denny K. S. Ng*[2]

[1]*Department of Chemical and Biological Engineering, University of Wisconsin–Madison, Madison, WI, USA*

[2]*Department of Chemical and Environmental Engineering/Centre of Sustainable Palm Oil Research (CESPOR), The University of Nottingham, Malaysia Campus, Selangor, Malaysia*

3.1 Introduction

Oil palm, scientifically known as *Elaeis guineensis* Jacg., is a unique crop that produces two distinct types of oils: crude palm oil (CPO) from mesocarp and crude palm kernel oil (CPKO) from kernels. According to Malaysia Palm Oil Council (2006), both CPO and CPKO possess excellent cooking properties and are widely used for both food (i.e. cooking oil, margarine, shortening, cocoa butter, etc.) and non-food applications (i.e. soap, cosmetics, detergents, etc.). According to statistics reported by the United States Department of Agriculture (USDA, 2014), CPO is the world's largest vegetable oil produced. There was a significant increase in CPO production from 2010/2011 to 2013/2014 as illustrated in Figure 3.1. As shown, both Indonesia and Malaysia were the world's largest producers accounting for 52% and 39%, respectively, out of 55.56 million tonnes of CPO production in 2013/2014.

Figure 3.2 illustrates a schematic flow diagram of palm oil milling processes. Upon harvesting, oil palm fruits (fresh fruit bunches (FFBs)) are collected from plantations and transported

Process Design Strategies for Biomass Conversion Systems, First Edition. Edited by Denny K. S. Ng, Raymond R. Tan, Dominic C. Y. Foo, and Mahmoud M. El-Halwagi.
© 2016 John Wiley & Sons, Ltd. Published 2016 by John Wiley & Sons, Ltd.

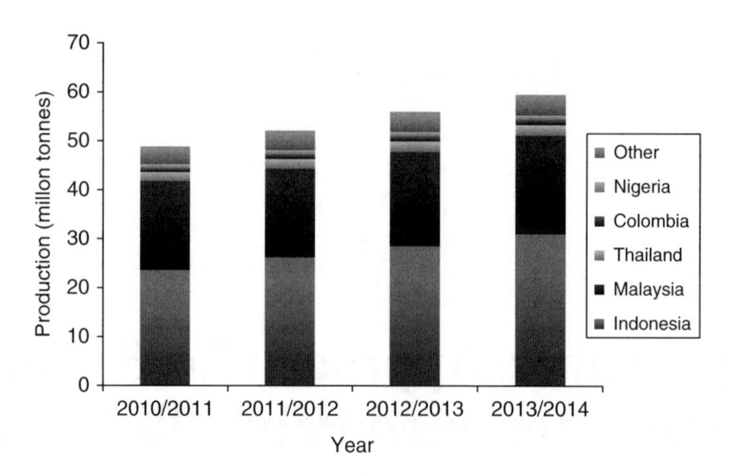

Figure 3.1 *Global palm oil production 2008/2009 to 2012/2013 (USDA, 2014)*

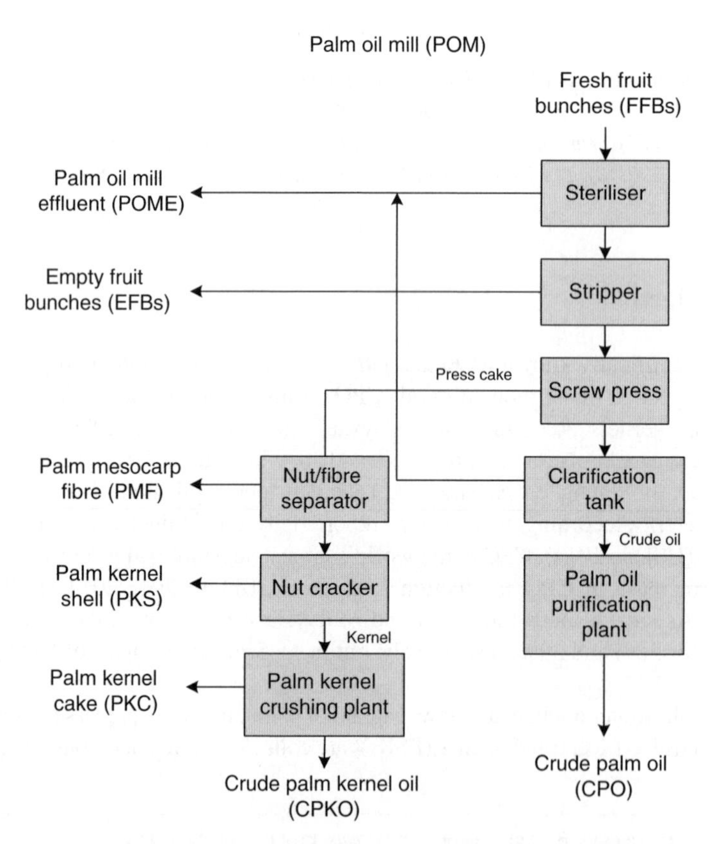

Figure 3.2 *Schematic diagram of palm oil mill. Reproduced with permission from Ng and Ng (2013b), © 2013, American Chemical Society*

instantly to the palm oil mill (POM). The screening of FFB is first performed in the POM to filter damaged oil palm fruits before they are sent to the steriliser. Screened FFB is loaded into the steriliser to facilitate the removal of fruitlets from FFB in a stripper. Sterilised FFB is fed to the stripper where fruitlets are separated from bunches. Empty fruit bunches (EFBs) are generated as by-product after stripping process. Meanwhile, the loose fruitlets are being pressed via screw press to extract crude oil. Crude oil is then pumped to a clarification tank and purified to produce CPO. Steam condensate, wastewater and sludge from separator and clarification tank are discharged as palm oil mill effluent (POME).

On the other hand, press cake which is produced from screw press is sent to the nut/fibre separator to remove palm mesocarp fibre (PMF) from the nut. The nut is further cracked in nut cracker and palm kernel shell (PKS) is removed. The remaining palm kernels are sent to the kernel crushing plant for CPKO production, and palm kernel cake (PKC) is generated as by-product after crushing process. CPO and CPKO can be further processed in the palm oil refinery (POR) to meet various specifications of downstream products (i.e. cocoa butter equivalent, emulsifiers, margarine, etc.). All by-products generated in the POM (e.g. EFBs, PMF, PKS, POME, PKC) are shown in Figure 3.2. According to Husain et al. (2002), 1 tonne (1000 kg) of oil palm fruits generates 210 kg of CPO, 60–70 kg of palm kernel, 230 kg of EFB, 140–150 kg of PMF, 60–70 kg of PKS, 30–40 kg of PKC and 600 kg of POME.

Conventionally, palm-based biomasses are either incinerated or dumped in the plantation as mulching material through natural decomposition (Ng et al., 2012a). The governments of Indonesia and Malaysia, as the major CPO and CPKO producers in the world, have introduced various biomass strategies and biomass energy-related policies to promote the utilisation of palm-based biomass for the production of bioenergy and value-added products. For instance, Presidential Regulation No. 5/2006 on National Energy Policy was introduced by Indonesia; National Biomass Strategy 2020 and 1Malaysia Biomass Alternative Strategy (1MBAS) were initiated by Malaysia. Different research works have been conducted to further convert palm-based biomasses into value-added products via physical, biological and thermochemical platforms. Table 3.1 shows the summary of palm-based biomass conversion technologies. Note that some of the value-added products act as intermediates which can be further upgraded for other applications. For instance, palm briquette and palm pellet can be used as replacement of coal in industrial boiler; biogas can be utilised in

Table 3.1 *Summary of palm-based biomass conversion technologies*

Category	Technology	Value-added product
Physical	Densification	Palm briquette, palm pellet, hybrid plywood, MDF
Biological	Anaerobic digestion	Biogas
	Fermentation	Bioethanol, biopolymer, compost
Thermochemical	Gasification	Syngas
	Pyrolysis	Bio-oil
	Torrefaction	Torrefied biomass
	Liquefaction	Bio-oil

Reproduced with permission from Ng and Ng (2013a), © 2013, Elsevier.

electricity generation via gas turbine or internal combustion engine; bio-oil and syngas can be further upgraded to produce bio-based chemicals and transportation fuel.

In addition to rapid development of different palm-based biomass conversion technologies, an integrated palm oil-based biorefinery (POB) has been proposed to integrate multiple platforms to produce value-added products (Kasivisvanathan et al., 2012; Ng et al., 2012b). Process, heat and power integration were considered simultaneously in synthesising an integrated POB via modular optimisation (Ng et al., 2012b). Kasivisvanathan et al. (2012) adopted fuzzy optimisation approach for retrofitting POM into an integrated POB with consideration of both economic and environmental perspectives. Inherent safety and occupational health impacts were included in synthesising a sustainable integrated POB (Ng et al., 2013a) with the consideration of economic, environmental and social aspects simultaneously. Ng and Ng (2013b) then introduced the concept of palm oil processing complex (POPC) consisting of POB, POM, POR and combined heat and power (CHP) to promote the interaction (e.g. mass and energy integration) of all processing facilities in palm oil industry.

Most recently, the concept of industrial symbiosis (IS) was included in promoting process integration within the processing plants with processing facilities owned by different owners (Chertow, 2004). An eco-industrial park (EIP) is an industrial cluster developed based on IS concept that shares a common infrastructure that involves energy, water and materials (Chertow, 2007). In practice, every participant in EIPs has their own individual goal and may conflict each other. The conflicts of interest of each participant in an IS system were considered (Ng et al., 2013b, 2014). Ng et al. (2013b) presented the concept of IS in synthesising an integrated POPC, and all processing facilities in an integrated POPC were predefined to join the IS scheme. Ng et al. (2014) introduced disjunctive fuzzy optimisation approach to enhance the flexibility of potential participants' decision in the IS of palm oil industry. Disjunctive fuzzy optimisation is a combination of disjunctive constraints and multi-objective optimisation that eases the decision of companies in participating IS system systematically. This approach utilises disjunctive constraints to deactivate the processing plants if their interest (e.g. economic performance) unsatisfied the agreed upon bounds.

This chapter presents a systematic approach for the synthesis and optimisation of an integrated POB with CHP. Optimum material and energy allocation network is determined via proposed approach. In this chapter, problem statement is stated and it is followed by problem formulations. An industrial case study is then solved to illustrate the proposed approach.

3.2 Problem Statement

A generic superstructure of the POB integrated with the CHP is shown in Figure 3.3. Given palm-based biomass $i \in I$ generated from the palm oil mill (POM) can be allocated to POB and CHP for value added product production and heat and power generation. In the POB, palm-based biomass i can be sent to technology $j \in J$ to produce intermediate $k \in K$. Intermediate k is then converted to produce product $q \in Q$ via technology $j' \in J'$. Furthermore, palm-based biomass $i \in I$ can be converted to primary energy $e \in E$ or secondary energy $e' \in E'$ via technologies $g \in G$ and $g' \in G'$ in CHP, respectively. The main objective of this work is to maximise the economic performance (EP) of the entire integrated POB with CHP.

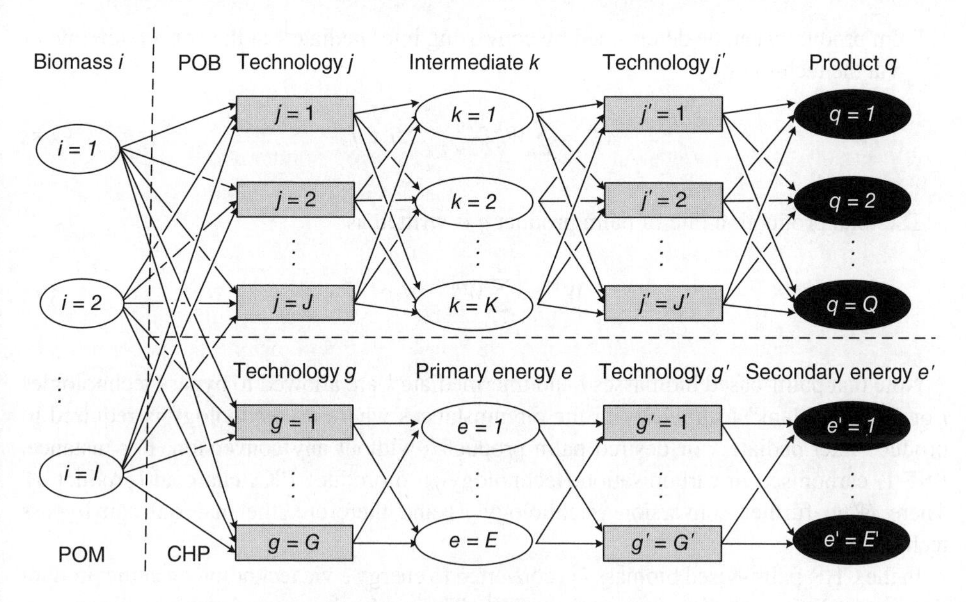

Figure 3.3 *Generic superstructure*

3.3 Problem Formulation

Palm-based biomass i with flowrate W_i^{BIO} is split into the potential technology j with the flowrate of W_{ij}^{I} and the potential technology g in the CHP with the flowrate of W_{ig}^{I}:

$$W_i^{\mathrm{BIO}} = \sum_{j=1}^{J} W_{ij}^{\mathrm{I}} + \sum_{g=1}^{G} W_{ig}^{\mathrm{I}} \quad \forall i \tag{3.1}$$

Palm-based biomass i is converted into intermediate k via technology j at the production rate of W_{jk}^{I}, with the conversion of $\mathrm{X}_{ijk}^{\mathrm{I}}$:

$$W_{jk}^{\mathrm{I}} = \sum_{i=1}^{I} W_{ij}^{\mathrm{I}} \mathrm{X}_{ijk}^{\mathrm{I}} \quad \forall j \forall k \tag{3.2}$$

The production rate of intermediate k is given as

$$W_k^{\mathrm{INT}} = \sum_{j=1}^{J} W_{jk}^{\mathrm{I}} \quad \forall k \tag{3.3}$$

Next, the intermediate k can be distributed to potential technology j' for further process to produce palm product q. The splitting constraint of intermediate k is written as

$$W_k^{\mathrm{INT}} = \sum_{j'=1}^{J'\sqrt{a^2+b^2}} W_{kj'}^{\mathrm{II}} \quad \forall k \tag{3.4}$$

Palm product q can be determined by converting intermediate k at the conversion rate of $X^{\mathrm{II}}_{kj'q}$ via the technology j':

$$W^{\mathrm{II}}_{j'q} = \sum_{k=1}^{K} W^{\mathrm{II}}_{kj'} X^{\mathrm{II}}_{kj'q} \quad \forall j' \forall q \tag{3.5}$$

The total production rate of palm product q is written as

$$W^{\mathrm{PR}}_{q} = \sum_{j'=1}^{J'} W^{\mathrm{II}}_{j'q} \quad \forall q \tag{3.6}$$

Note that palm-based biomasses i and intermediate k are allowed to bypass technologies j or j' via a 'blank' technology in the circumstances where no technology is required to produce intermediate k or desired palm product q without any conversion. For instance, PKS is carbonised in carbonisation (technology j) to produce PKS charcoal (product q). There is no further conversion (technology j') and therefore, the material can bypass technology j'.

In the CHP, palm-based biomass i is converted to energy e via technology g at the production rate of E^{Gen}_{e}, with given conversion of Y^{I}_{ige}. The production rate of energy e is given as

$$E^{\mathrm{I}}_{ge} = \sum_{i=1}^{I} W^{\mathrm{I}}_{ig} Y^{\mathrm{I}}_{ige} \quad \forall g \forall e \tag{3.7}$$

$$E^{\mathrm{Gen}}_{e} = \sum_{g=1}^{G} E^{\mathrm{I}}_{ge} \quad \forall e \tag{3.8}$$

Primary energy e such as biogas is further upgraded via technology g' for production of other types of energy (e.g. electricity). E^{Gen}_{e} is split and further converted to secondary energy e' via technology g' with conversion of $Y^{\mathrm{II}}_{eg'e'}$. The splitting constraint of energy e is written as

$$E^{\mathrm{Gen}}_{e} = \sum_{g'=1}^{G'} E^{\mathrm{Gen}}_{eg'} \quad \forall e \tag{3.9}$$

The total production rate of secondary energy e' through technology g' is written as

$$E^{\mathrm{II}}_{g'e'} = \sum_{e=1}^{E} E^{\mathrm{Gen}}_{eg'} Y^{\mathrm{II}}_{eg'e'} \quad \forall g' \forall e' \tag{3.10}$$

$$E^{\mathrm{Gen}}_{e'} = \sum_{g'=1}^{G'} E^{\mathrm{II}}_{g'e'} \quad \forall e' \tag{3.11}$$

The total energy consumption $E^{\mathrm{Con}}_{e'}$ in the POB is determined based on the energy consumed in technologies j and j'. The total energy consumption is determined as

$$E^{\mathrm{Con}}_{e'} = \sum_{k=1}^{K} \sum_{j=1}^{J} \left(W^{\mathrm{I}}_{jk} Y^{\mathrm{I}}_{e'jk} \right) + \sum_{q=1}^{Q} \sum_{j'=1}^{J'} \left(W^{\mathrm{II}}_{j'q} Y^{\mathrm{II}}_{e'j'q} \right) \quad \forall e' \tag{3.12}$$

In a sustainable integrated POB, the excess energy $E_{e'}^{\text{Exp}}$ can be sold and exported to any third-party plants if the total energy generation exceeds the total energy consumption ($E_{e'}^{\text{Gen}} > E_{e'}^{\text{Con}}$). In contrast, external import of energy $E_{e'}^{\text{Imp}}$ is needed in case where the total energy generated is insufficient to fulfil the total consumption ($E_{e'}^{\text{Gen}} < E_{e'}^{\text{Con}}$). Therefore, the energy correlation can be written as

$$E_{e'}^{\text{Con}} = E_{e'}^{\text{Gen}} + E_{e'}^{\text{Imp}} - E_{e'}^{\text{Exp}} \quad \forall e' \tag{3.13}$$

In order to reduce the complexity of model, all energy correlations in this study focus on secondary energy e'. In case where primary energy e (e.g. high-pressure steam) is required in the POB, primary energy e is allowed to bypass technology g' via a 'blank' process where there is no conversion in the technology g'. Thus, primary energy e is equal to secondary energy e'.

The gross profit (GP) of an integrated POB is determined as given in the equation below:

$$GP = \text{AOT}\left(\begin{array}{l} \sum_{q=1}^{Q} W_q^{\text{PR}} \text{C}_q^{\text{PR}} + \sum_{e'=1}^{E'} E_{e'}^{\text{Exp}} \text{C}_{e'}^{\text{Exp}} - \sum_{e'=1}^{E'} E_{e'}^{\text{Imp}} \text{C}_{e'}^{\text{Imp}} \\ -\sum_{i=1}^{I} W_i^{\text{BIO}} \text{C}_i^{\text{BIO}} - \sum_{k=1}^{K}\sum_{j=1}^{J} W_{jk}^{\text{I}} \text{C}_{jk}^{\text{Proc}} - \sum_{q=1}^{Q}\sum_{j'=1}^{J'} W_{j'q}^{\text{II}} \text{C}_{j'q}^{\text{Proc}} \\ -\sum_{e=1}^{E}\sum_{g=1}^{G} E_{ge}^{\text{I}} \text{C}_{ge}^{\text{Proc}} - \sum_{e'=1}^{E'}\sum_{g'=1}^{G'} E_{g'e'}^{\text{II}} \text{C}_{g'e'}^{\text{Proc}} \end{array}\right) \tag{3.14}$$

where AOT is the annual operating time, C_q^{PR} is the selling price of palm product q, $\text{C}_{e'}^{\text{Exp}}$ is the cost of energy e' export, $\text{C}_{e'}^{\text{Imp}}$ the cost of energy e' import and C_i^{BIO} is the cost of palm-based biomass i. $\text{C}_{jk}^{\text{Proc}}$, $\text{C}_{j'q}^{\text{Proc}}$, $\text{C}_{ge}^{\text{Proc}}$ and $\text{C}_{g'e'}^{\text{Proc}}$ are the overall expenses of technology j per unit flowrate of k produced, technology j' per unit flowrate of q produced, technology g per unit energy e generated and technology g' per unit energy e' generated, respectively. Note that overall expenses cover costs of start-up, working capital, maintenance, manpower, installation, etc. of each technology.

In addition, payback period (PP) is calculated to determine the length of time required to recover the total investment cost. PP is expressed as the total cost of investment over GP and it is shown in the following equation:

$$\text{PP} = \frac{\sum_{k=1}^{K}\sum_{j=1}^{J} W_{jk}^{\text{I}} \text{C}_{jk}^{\text{Cap}} + \sum_{q=1}^{Q}\sum_{j'=1}^{J'} W_{j'q}^{\text{II}} \text{C}_{j'q}^{\text{Cap}} + \sum_{e=1}^{E}\sum_{g=1}^{G} E_{ge}^{\text{I}} \text{C}_{ge}^{\text{Cap}} + \sum_{e'=1}^{E'}\sum_{g'=1}^{G'} E_{g'e'}^{\text{II}} \text{C}_{g'e'}^{\text{Cap}} + \text{C}^{\text{Cap-Fixed}}}{\text{GP}} \tag{3.15}$$

where $\text{C}_{jk}^{\text{Cap}}$, $\text{C}_{j'q}^{\text{Cap}}$, $\text{C}_{ge}^{\text{Cap}}$ and $\text{C}_{g'e'}^{\text{Cap}}$ are the technology j per unit flowrate of k produced, technology j' per unit flowrate of q produced, technology g per unit energy e generated and technology g' per unit energy e' generated, respectively, while $\text{C}^{\text{Cap-Fixed}}$ is the miscellaneous fixed capital (industrial land, vehicles, building, etc.) needed to be invested in an integrated POB.

In this work, the economic performance (EP) is evaluated based the net present value (NPV), which indicates the profit or loss of an integrated palm oil-based biorefinery over its operational lifespan (Ng and Ng, 2013b). The NPV is expressed in the following equation:

$$\mathrm{NPV} = \sum_{t}^{t_{\max}} \frac{\left[\mathrm{GP} \times (1 - \mathrm{TAX}) + \mathrm{DEP} \times \mathrm{TAX} - \mathrm{HEDGE} + \mathrm{GOV}\right]}{(1 + \mathrm{ROR})^{t}} \tag{3.16}$$

where GP is the gross profit and TAX and DEP are the marginal tax rate and depreciation rate, respectively. HEDGE and GOV are expenses associated with hedging against catastrophic market actions and net benefits realised through governmental incentives or penalties, respectively. t_{max} is the operating lifespan and ROR is the expected rate of return.

In this chapter, the optimisation objective is set to maximise the NPV subject to Equations 3.1–3.16. The optimisation model is solved via LINGO v13.0 in Dell Inspiron 15 3000 Series with Intel® Core™ i5-4210U CPU and 8GB DDR3 RAM within few seconds.

3.4 Case Study

The superstructure of the POM and POB integrated with CHP is presented in Figure 3.4. In this case study, three palm-based biomasses (EFB, PMF and PKS) generated from the POM are chosen as feedstocks fed into both POB and CHP. In POB, there are two options

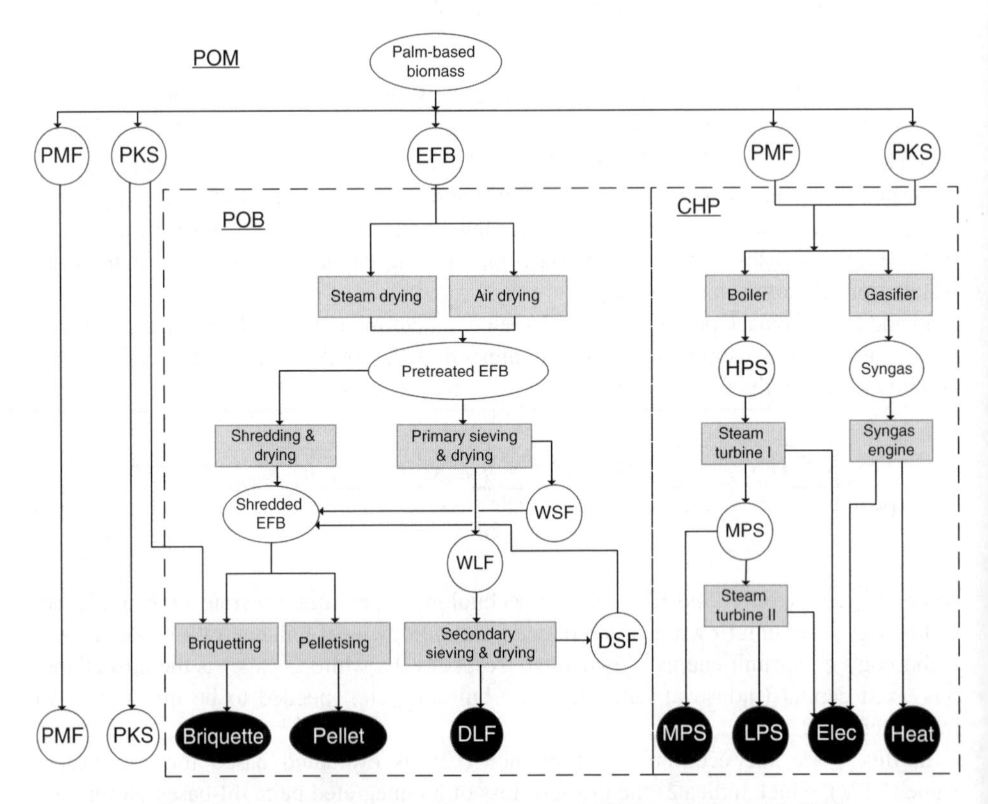

Figure 3.4 *Superstructure of an integrated POB and POM with CHP*

of pretreatment (e.g. air drying and steam drying) that are taken into consideration. Pretreated EFB can be sent to two different conversion technologies (e.g. hammer mill and primary sieving and drying) in order to generate briquette, pellet and dried long fibre (DLF). As shown in Figure 3.4, by-products such as wet short fibre (WSF) and dry short fibre (DSF) can be recycled as feedstocks that are fed into other technologies (e.g. pelletising and briquetting processes). In CHP, pretreated EFB, PMF and PKS can be directly combusted in the boiler to generate high-pressure steam (HPS). HPS is then sent to steam turbine for the production of electricity and medium-pressure steam (MPS) or low-pressure steam (LPS). In addition, PKS from the POM and DSF from the POB can be fed into the gasifier to generate syngas and burnt in gas engine for electricity generation. Heat generated from gasification system is recovered and to be sent to air drying system in the pretreatment of EFB.

In the POM, biomass i can be converted into MPS at 12 bar, 250°C via technology h and sent to the upgrading technology h' for production of electricity and LPS at 4 bar, 145°C. In the CHP, biomass i and intermediate k can be fed into the boiler and generated HPS at 40 bar, 400°C via technology g. HPS is then sent to the upgrading technology g' for production of electricity and MPS at 12 bar, 250°C or LPS at 4 bar, 145°C. The price of raw material, palm product and energy, mass conversion factor, energy consumptions (electricity and steam) and economic data (including fixed capital investment and overall expenses) for each technology are shown in Tables 3.2, 3.3, 3.4 and 3.5, respectively. Please note that the information in Tables 3.2, 3.3, 3.4 and 3.5 is extracted from the supporting information in the previous work (Ng et al., 2014).

It is assumed that the POM has an operating capacity of 60 t/h of fresh fruit bunches (FFBs) and all processing facilities are designed based on an operating lifespan of 15 years with annual operating time (AOT) of 8000 h. It is further assumed that the main factor of economic performance is based on GP; thus, terms of TAX, DEP, HEDGE and GOV in Equation 3.16

Table 3.2 *Price of palm-based biomass, product and energy*

Item	Price (USD)
Biomass, *i*	
EFB	6/t
PKS	50/t
PMF	22/t
DSF	10/t
WSF	10/t
Product, *q*	
Pellet	140/t
DLF	210/t
Briquette	120/t
Energy, *e′*	
Electricity import from external grid	0.14/kW h
Electricity import/export from CHP	0.09/kW h
HPS import/export from CHP	26/t
MPS import/export from CHP	17/t
LPS import/export from CHP	12/t

Reproduced with permission from Ng et al. (2014), © 2013, Elsevier.

Table 3.3 *Conversion factor for technology pathway*

Input	Output	Conversion value
POB		
Pretreated EFB	Shredded EFB	0.611 shredded EFB/pretreated EFB
Shredded EFB + PKS	Briquette	0.921 briquette/(shredded EFB + PKS)[a]
Shredded EFB + PKS	Pellet	0.921 pellet/(shredded EFB + PKS)
EFB	Pretreated EFB	0.636 pretreated EFB/(EFB)
Pretreated EFB	WSF	0.237 WSF/pretreated EFB
Pretreated EFB	WLF	0.480 WLF/pretreated EFB
WLF	DSF	0.071 DSF/WLF
WLF	DLF	0.929 DLF/WLF
CHP		
DSF + PKS	Syngas	0.723 syngas/PKS
		0.650 syngas/DSF
Syngas	Electricity	1165 kW electricity/t/h syngas
Syngas	Heat	3496 kW heat/t/h syngas
HPS	MPS	0.947 MPS/HPS
MPS	LPS	1.000 LPS/MPS
HPS to steam turbine I	Electricity	60.26 electricity/MPS generated
MPS to steam turbine I	Electricity	43.64 electricity/LPS generated

Reproduced with permission from Ng et al. (2014), © 2013, Elsevier.
[a] Ratio of shredded EFB : PKS = 80 : 20.

Table 3.4 *Energy consumption of each technology*

Description	Product	Value
Steam/hot air consumption		
Palm oil mill	FFB	0.300 t/h LPS/t/h FFB
Air pretreatment		160.00 kW hot air/t/h FFB
	Pretreated EFB	120.00 kW hot air/t/h pretreated EFB
		1.000 t/h MPS/t/h pretreated EFB
Primary sieving	WLF	0.929 t/h MPS/t/h WLF
		325.15 kW hot air/t/h WLF
Secondary sieving	DLF	0 t/h MPS/t/h DLF
		0 kW hot air/t/h DLF
Shredding	Shredded EFB	0.725 t/h MPS/t/h shredded EFB
		253.63 t/h kW hot air/t/h shredded EFB
Pellet production	Pellet	0.220 t/h MPS/t/h pellet
		77.00 kW hot air/t/h pellet
Briquette production	Briquette	0.220 t/h MPS/t/h briquette
		77.00 kW hot air/t/h briquette
Electricity consumption		
Palm oil mill	FFB	13.50 kW/t/h FFB
Air pretreatment	Pretreated EFB	50.00 kW/t/h pretreated EFB
Steam pretreatment	Pretreated EFB	0 kW/t/h pretreated EFB
Primary sieving	WLF	148.64 kW/t/h WLF
Secondary sieving	DLF	50.00 kW/t/h DLF
Shredding	Shredded EFB	70.60 kW/t/h shredded EFB
Pellet production	Pellet	106.5 kW/t/h pellet
Briquette production	Briquette	63.00 kW/t/h briquette

Reproduced with permission from Ng et al. (2014), © 2013, Elsevier.

Table 3.5 *Economic data of each technology*

Item	Price (USD)
Fixed capital investment	
Boiler in POM	USD 400,000/(1000 kW electricity)
Steam turbine in POM	USD 80,000/(1000 kW electricity)
Shredding machine	USD 592,000/(1 t/h shredded EFB)
Pelletising machine	USD 82,000/(1 t/h pellet)
Briquette machine	USD 80,000/(1 t/h briquette)
Air treatment	USD 18,000/(1 t/h EFB)
Steam treatment	USD 18,000/(1 t/h EFB)
Primary and secondary sieving	USD 460,000/(1 t/h DLF)
Boiler in CHP	USD 400,000/(1000 kW electricity)
Steam turbine I	USD 80,000/(1000 kW electricity)
Steam turbines I and II	USD 140,000/(1000 kW electricity)
Gasifier	USD 3,900,000/(1000 kW electricity)
Fixed capital cost of POB (structural, land, etc.)	USD 400,000
General expenses	
Boiler in POM	USD 20.00/(1000 kW electricity)
Steam turbine in POM	USD 10.00/(1000 kW electricity)
Shredding machine	USD 13.50/(1 t/h shredded EFB)
Pelletising machine	USD 6.50/(1 t/h pellet)
Briquette machine	USD 4.50/(1 t/h briquette)
Air treatment	USD 4.00/(1 t/h EFB)
Steam treatment	USD 3.00/(1 t/h EFB)
Primary and secondary sieving	USD 31.50/(1 t/h DLF)
Boiler in CHP	USD 20/(1000 kW electricity)
Steam turbine I	USD 10/(1000 kW electricity)
Steam turbines I and II	USD 15/(1000 kW electricity)
Gasifier	USD 35/(1000 kW electricity)

Reproduced with permission from Ng et al. (2014), © 2013, Elsevier.

are neglected. Besides, payback period of an integrated POPC is assumed to be less than or equal to three years. The linear programming model is solved by maximising NPV subject to Equations 3.1–3.16 and data in Tables 3.2, 3.3, 3.4 and 3.5. The economic analysis of this case study is tabulated in Table 3.6. Maximum NPV of the entire palm oil-based biorefinery is targeted as USD 9.98 million over 15-year lifespan with annual GP of USD 1.46 million and its payback period is around 2.57 years. The optimal network configuration is shown in Figure 3.5.

In this case study, 60 t/h of FFB is sent to POM, producing 14.04 t/h EFB, 4.38 PKS and 7.80 t/h PMF. DLF and pellet production pathways are selected to produce palm products in the POB, while boiler and steam turbine pathways are selected as steam and electricity generation in the CHP. As shown, all EFBs are pretreated in steam drying before they are sent to DLF production and boiler combustion. A total of 3.04 t/h DLF, 1.61 t/h WSF and 0.23 t/h DSF are produced throughout the DLF production. All by-products (WSF and DSF) are sent to pelletising processes to produce 1.49 t/h pellet. There is 2.12 t/h pretreated EFB sent to CHP and mixed with all PMF and PKS before being further combusted in the boiler.

Table 3.6 *Detailed economic analysis of case study*

Description	Value
Net present value, NPV (USD)	9,976,282
Gross profit, GP (USD/year)	1,457,005
Total revenue (USD/year)	7,262,697
Total expenses (USD/year)	2,006,972
Total raw material (USD/year)	3,798,720
Total capital investment (USD)	3,764,337
Payback period, PP (year)	2.57

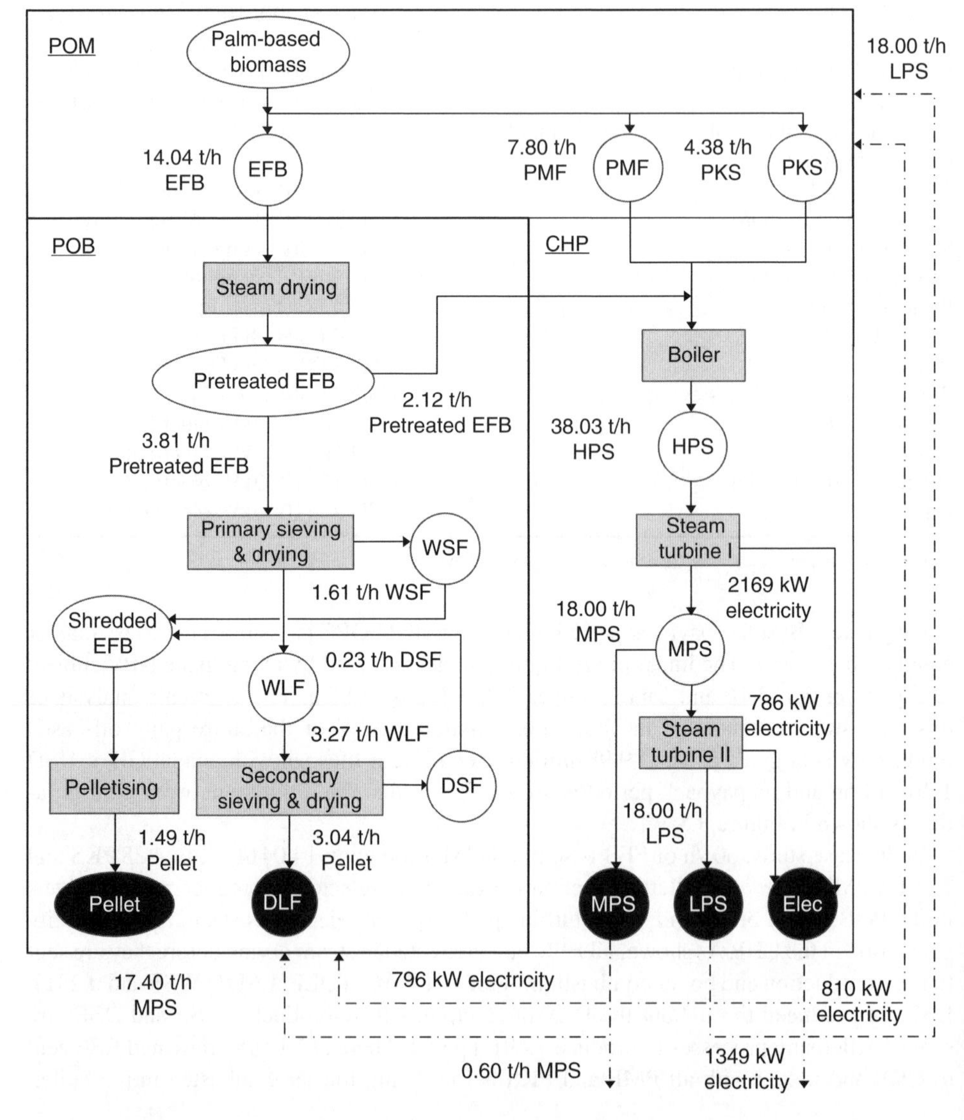

Figure 3.5 *Optimal configuration of an integrated POB and POM with CHP*

It is noted that 38.03 t/h HPS is produced from combustion in the boiler and 2955 kW electricity is generated via both steam turbines I and II. A total of 1606 kW electricity is supplied to the POM and the POB whereas the excess of 1349 kW of electricity can be exported to the grid. After CHP exported steams based on requirement of POM and POB, there is an excess of 0.60 t/h MPS that can be exported to external facilities.

3.5 Conclusions

Biomass utilisation in palm oil industry is gaining significant attention with the increasing volume of global palm oil production. In this chapter, a systematic approach for the synthesis of an integrated POB and POM with CHP is presented. It is expected that the concept of an integrated POB will be able to convert the palm oil industry into a greener industry and as an independent electricity supply to external facilities. The proposed approach can be easily revised and reformulated to handle different case studies from different industries for systematic allocation of biomass (e.g. rubber seed, rice husk).

References

Chertow, M.R. (2004) Industrial symbiosis. *Encyclopedia of Energy*, **3**, 407–415.

Chertow, M.R. (2007) "Uncovering" industrial symbiosis. *Journal of Industrial Ecology*, **11**(1), 11–30.

Husain, Z., Zainac, Z., Abdullah, Z. (2002) Briquetting of palm fibre and shell from the processing of palm nuts to palm oil. *Biomass and Bioenergy*, **22**: 505–509.

Kasivisvanathan, H., Ng, R.T.L., Tay, D.H.S., Ng, D.K.S. (2012) Fuzzy optimisation for retrofitting a palm oil mill into a sustainable palm oil-based biorefinery. *Chemical Engineering Journal*, **200–202**: 694–709.

Malaysian Palm Oil Council (2006) *Palm oil & palm kernel oil applications*. Petaling Jaya: Malaysian Palm Oil Council, i–v.

Ng, D.K.S., Ng, R.T.L. (2013a) Applications of process system engineering in palm-based biomass processing industry. *Current Opinion in Chemical Engineering*, **2**(4), 448–454.

Ng, R.T.L., Ng, D.K.S. (2013b) Systematic approach for synthesis of integrated palm oil processing complex. Part 1: Single owner. *Industrial & Engineering Chemistry Research*, **52**(30), 10206–10220.

Ng, R.T.L., Hassim, M.H., Ng, D.K.S. (2013a) Process synthesis and optimisation of a sustainable integrated biorefinery. *AIChE Journal*, **59**(11), 4212–4227.

Ng, R.T.L., Ng, D.K.S., Tan, R.R. (2013b) Systematic approach for synthesis of integrated palm oil processing complex. Part 2: Multiple owners. *Industrial & Engineering Chemistry Research*, **52**(30), 10221–10235.

Ng, R.T.L., Ng, D.K.S., Tan, R.R., El-Halwagi, M.M. (2014) Disjunctive fuzzy optimisation for planning and synthesis of bioenergy-based industrial symbiosis system. *Journal of Environmental and Chemical Engineering*, **2** (1), 652–664.

Ng, R.T.L., Tay, D.H.S., Ng, D.K.S. (2012a) Simultaneous process synthesis, heat and power integration in a sustainable integrated biorefinery. *Energy & Fuels*, **26**(12):7316–7330.

Ng, W.P.Q., Lam, H.L., Ng, F.Y., Kamal, M., Lim, J.H.E. (2012b) Waste-to-wealth: green potential from palm biomass in Malaysia. *Journal of Cleaner Production*, **34**, 57–65.

United States Department of Agriculture (2013–2014) Oilseeds: World Markets and Trade. http://www.fas.usda.gov/psdonline/circulars/oilseeds.pdf (accessed 28 November, 2014).

[illegible]

Conclusions

[illegible]

References

[illegible]

4

Design Strategies for Integration of Biorefinery Concepts at Existing Industrial Process Sites: Case Study of a Biorefinery Producing Ethylene from Lignocellulosic Feedstock as an Intermediate Platform for a Chemical Cluster

Roman Hackl and Simon Harvey

Department of Energy and Environment, Chalmers University of Technology, Gothenburg, Sweden

4.1 Introduction

4.1.1 Biorefinery Concepts

According to the definition provided by the International Energy Agency (IEA), biorefineries are based upon 'the sustainable processing of biomass into a spectrum of marketable products and energy'. Possible end products of biorefineries include heat, power, fuels, chemicals and materials (De Jong et al., 2009).

Figure 4.1 shows an overview of biorefinery concepts and potential products. The raw materials that can be used as feedstock for a biorefinery include crops and crop residues,

Process Design Strategies for Biomass Conversion Systems, First Edition. Edited by Denny K. S. Ng, Raymond R. Tan, Dominic C. Y. Foo, and Mahmoud M. El-Halwagi.
© 2016 John Wiley & Sons, Ltd. Published 2016 by John Wiley & Sons, Ltd.

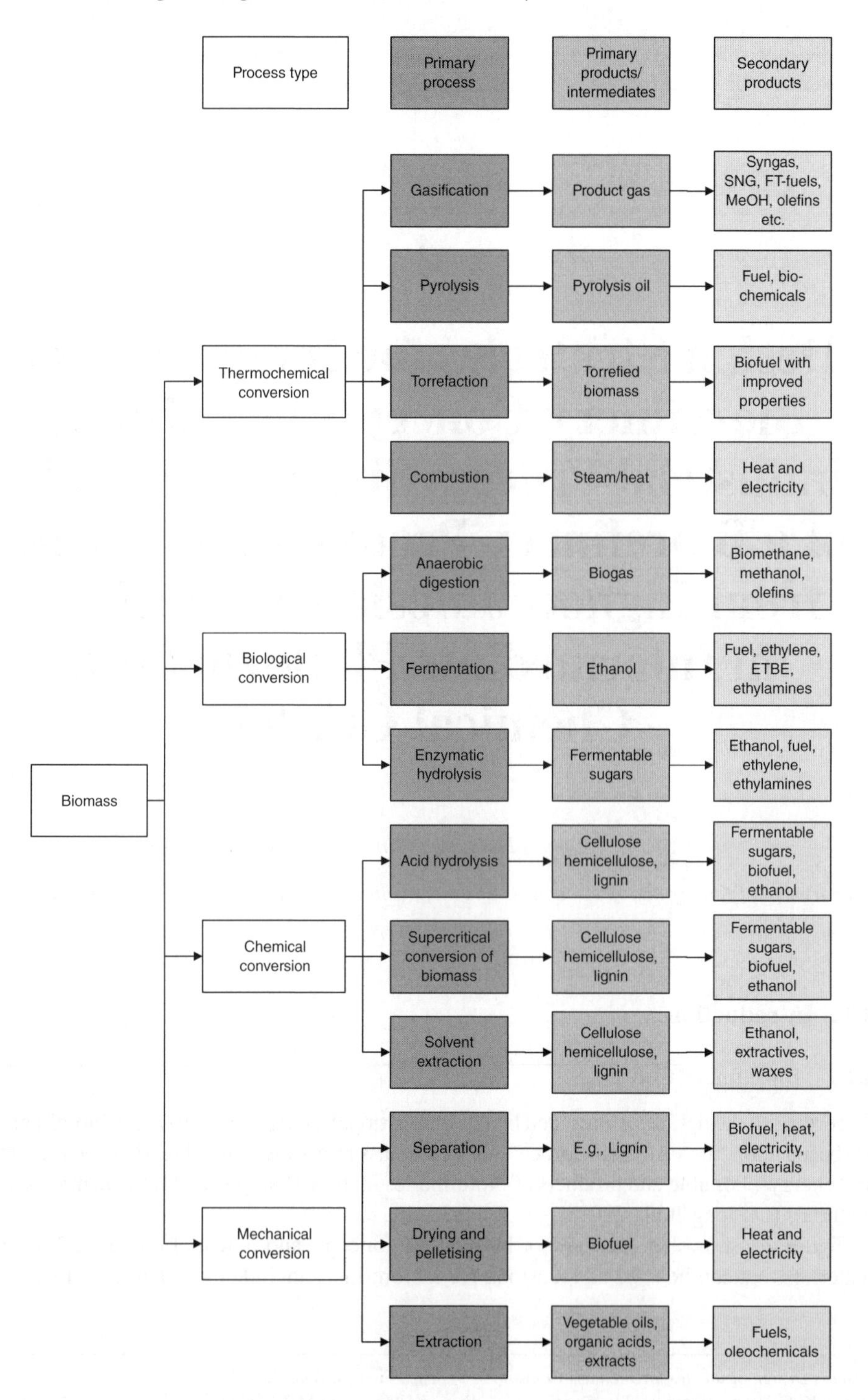

Figure 4.1 *Overview of biomass conversion processes and potential products. Adapted from UOP LLC (2004), Bludowsky and Agar (2009), De Jong et al. (2009), Sadaka and Negi (2009) and Naik (2010)*

lignocellulosic material, municipal solid waste and algae. There are four main groups of processes commonly applied in a biorefinery to decompose the incoming biomass feedstock into their constituent building blocks (see primary products/intermediates in Figure 4.1): thermochemical, biochemical, mechanical and chemical processes. These (intermediate) products can then be converted into value-added products (see secondary products in Figure 4.1) by a combination of conversion and separation processes (Cherubini, 2010).

4.1.2 Advantages of Co-locating Biorefinery Operations at an Industrial Cluster Site

In order to become economically competitive, a major challenge for biorefineries is to maximise the conversion of biomass feedstock into high-value products (Bludowsky & Agar, 2009). Integrating a biorefinery in an industrial cluster can be advantageous from an energy point of view as the cluster can serve as both a source of excess heat to the biorefinery process and a sink for excess heat from the biorefinery, thus decreasing the amount of biomass or fossil fuel necessary to satisfy the combined heat demand of both processes. Biomass not used for process heating can be either converted into products, thus increasing the conversion efficiency, or sold as feedstock or fuel to external customers.

Another advantage of integration into an industrial cluster is that the existing infrastructure (boilers, utility systems, air separation plant, etc.) is already in place. Compared to a stand-alone biorefinery unit, this can have positive impact on process economics since, for example, parts of the infrastructure for electricity, steam, waste water, safety, etc. can be shared (Kimm, 2008).

The chemical cluster used as a case study in this chapter is located in Stenungsund on the west coast of Sweden and is Sweden's largest agglomeration of its kind. The companies involved and their main products are shown in Figure 4.2. A common vision called 'Sustainable Chemistry 2030' was recently announced aiming at increasing the use of renewable feedstock and energy flows, reducing fuel usage within the cluster through site-wide inter-process heat exchanging and decreasing overall global CO_2 emissions. A steam cracker plant at the heart of the cluster converts not only naphtha but also other hydrocarbon feedstocks into intermediate products (mainly ethylene and propylene) that are used as feedstock in downstream plants within the cluster. Off-gas products from the cracker plant are used as boiler fuel throughout the cluster. Figure 4.2 shows an overview of the plants within the chemical cluster in Stenungsund and the current material and energy flows between them. Collaboration between the companies is currently limited to material exchange.

Biorefineries can produce a wide range of products including both final and intermediate products which can be used as feedstock to other chemical processes. Thereby, it is possible to integrate a suitable biorefinery in an existing chemical cluster which produces the starting materials for the existing chemical processes. One option is ethylene which can be used as a feedstock to several processes at the chemical cluster.

4.1.3 Ethylene Production from Biomass Feedstock

The chemical cluster studied consumes a large amount of ethylene (polymer grade; purity 99.95 mol% (Kochar et al., 1981)) of which 200 kt/year is imported and about 500 kt/year is produced in a conventional steam cracking plant fed mainly with

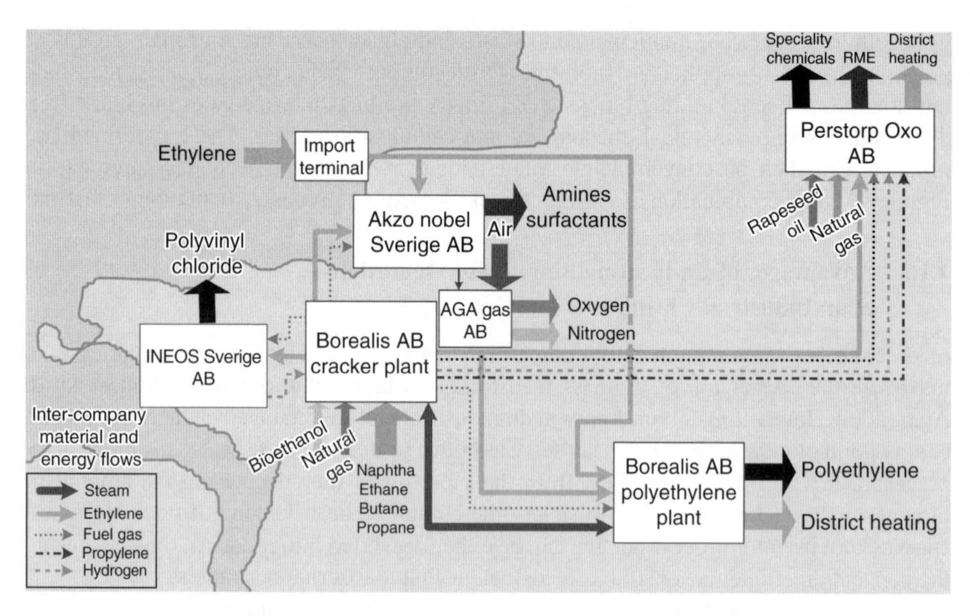

Figure 4.2 Material and energy flows across the chemical cluster in Stenungsund (Jönsson et al., 2012)

naphtha. In this study, it is assumed that the imported ethylene is replaced by ethylene from lignocellulosic feedstock. This assumption can be seen as part of the clusters' long-term strategy. In their vision, the companies within the cluster aim for increasing the amount of renewable feedstock materials. This can be achieved by gradually replacing fossil feedstocks with biomass. As a first step, the fossil feedstock that can be easily replaced in large volumes is the imported ethylene to the cluster. Thereby, the steam cracker, currently producing the major amount of olefins in the cluster, can still be in operation, while imported ethylene is replaced by renewable ethylene. If proven successful, also the olefins produced by the cracker can be replaced renewable olefins.

Figure 4.3 illustrates different possible conversion routes for the production of ethylene from biomass. One way to produce ethylene from renewable feedstock is catalytic dehydration of bioethanol. In the short term, it is likely that bioethanol dehydration for the production of ethylene will be established in regions with cheap access to bioethanol, for example, Brazil, where ethanol usage is on the same level as usage of fossil-based fuels (on an energy basis) in the transportation sector. In Europe and the United States, this trend is expected to occur after the commercial introduction of lignocellulosic ethanol (Jones et al., 2010).

This chapter investigates the heat savings potential that can be achieved by integrating an ethanol production plant based on fermentation of lignocellulosic feedstock and an ethanol dehydration plant producing ethylene with an existing chemical cluster.

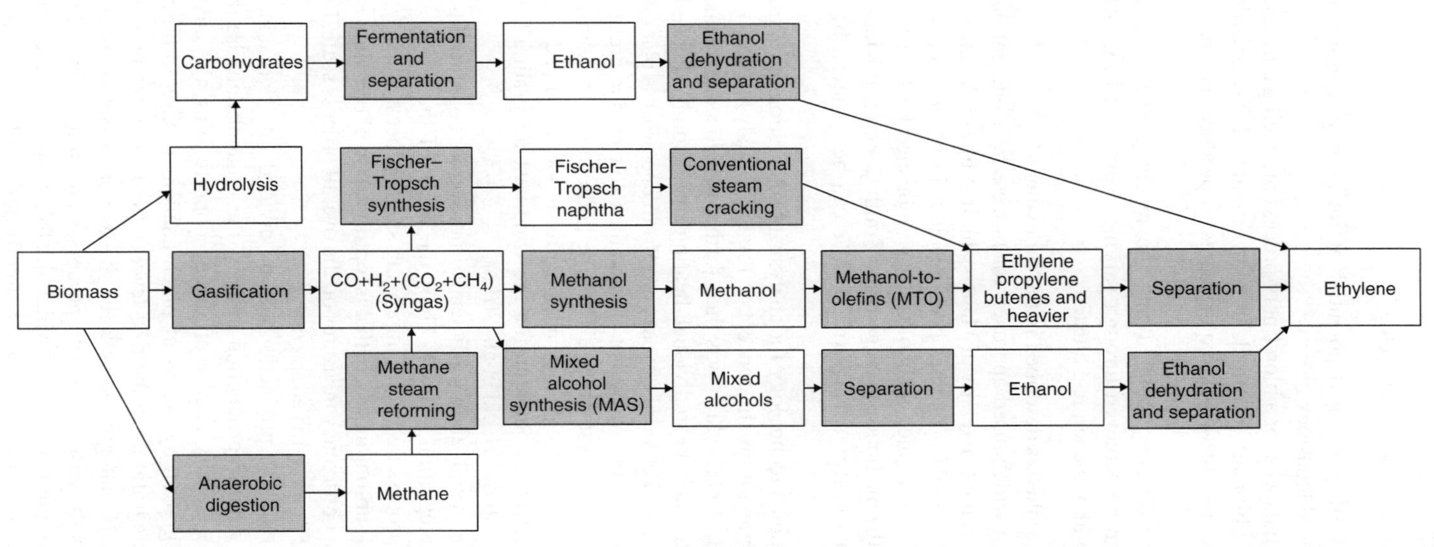

Figure 4.3 *Overview of biomass-to-ethylene production routes: different biomass conversion technologies to produce ethylene from biomass. Adapted from Bludowsky and Agar (2009), De Jong et al. (2009) and Naik (2010)*

4.1.4 Design Strategy

This chapter aims to investigate the following:

- Process integration (utility savings) potential of integrating a biorefinery producing ethylene from bio-based feedstock
- Energy and material balances for a biorefinery process producing ethylene via fermentation of lignocellulosic biomass feedstock and downstream ethanol dehydration
- Estimation of the biomass feedstock requirements to produce the specified amount of ethylene
- Estimation of energy savings potential and the resulting overall energy efficiency improvement that can be achieved through varying degrees of heat integration of the biorefinery process with the chemical cluster

The design strategy applied is described in the following.

Pinch technology is a widely used approach for process heat integration. It was developed by Bodo Linnhoff at the University of Manchester Institute of Science and Technology (UMIST) in the end of the 1970s and has been developed further since then. An updated version of the user guide on pinch technology was published by Kemp (2007). A more detailed description of the methodology is given by Smith (2005) and Klemeš et al. (2010). Studies have shown that energy savings of 20–40% can be achieved by pinch analysis (Heck et al., 2009).

To be able to apply pinch technology for analysis of integration opportunities in single processes to site-wide integration, the concept of Total Site Analysis (TSA) was introduced by Dhole and Linnhoff (1993) which was then further developed by Raissi (1994). The concept is used to integrate the individual heating and cooling demands of different processes at a total site. In other words, excess heat from one process plant is transferred to a common utility (e.g. steam, hot water, hot oil) and then delivered to processes with a heat deficit by the common utility system. The TSA method enables to determine targets for site-wide heat recovery, which can be used as guideline to identify and design specific heat recovery systems. The cluster's potential for site-wide heat integration has been investigated in a previous TSA study (Hackl et al., 2011).

In order to investigate overall energy efficiency improvement opportunities in a systematic manner, process integration is used. Thus, the focus of the study is on a holistic approach of the whole process instead of improving single process steps.

This approach avoids sub-optimisation, that is, optimisation of one single process step in a way that hinders further efficiency measures on the overall process scale. In this work, the approach illustrated in Figure 4.4 is adopted.

The lignocellulosic ethylene production can be divided into two steps of processes: (i) the lignocellulosic ethanol production and (ii) the ethanol dehydration to ethylene production. In order to identify energy efficiency opportunities, the analysis was performed using heat integration by means of pinch analysis at three different levels with increasing degree of integration. On the *first level* (Case I), the two processing steps are investigated separately. No integration between the two processes is accordingly considered. In practice, this is the case if the both processes are situated in different locations. This case represents the reference case to which the following cases are compared. At the *second level* (Case II), material (i.e. ethanol is directly delivered in gaseous phase from the

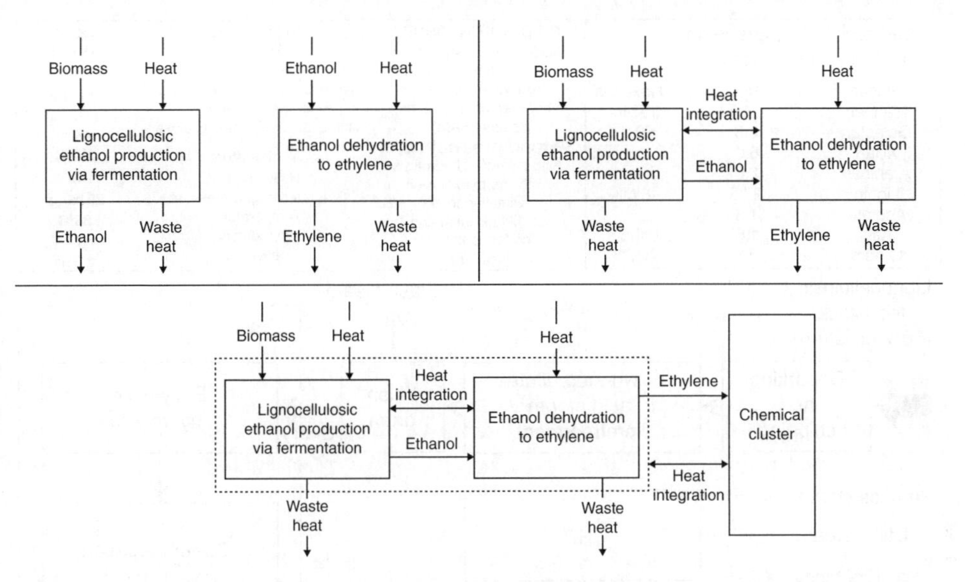

Figure 4.4 *Illustration of the heat integration approach. Upper left: base case with no integration between ethanol and ethylene process. Upper right: heat and material integration between the two processes. Lower: heat and material integration and design of a utility system to enable site-wide heat integration*

lignocellulosic ethanol production to the ethanol dehydration process) and heat integration of the two processes is investigated. In practice, this requires co-location of the two processes. At the *third level* (Case III), the integration potential of the combined lignocellulosic ethylene process with the existing chemical cluster through a common utility system is estimated using results obtained in a previous TSA study (Hackl et al., 2011). The energy targets determined by pinch analysis assume possible direct heat exchange between all the process streams across the site. This is very often not the case as most heat within processes is transferred by the utility system. In order to be able to transfer heat within the biorefinery and between the existing clusters, the new plant has to be integrated with the cluster's utility system. At the moment, the cluster has no common utility system. In the previous TSA study, several measures to increase the cluster's energy efficiency by site-wide energy collaboration were identified, and a common utility system was suggested. The suggested utility system is assumed to be implemented in order to estimate the site-wide heat integration potential between the ethylene production plant and the chemical cluster.

For each case, the potential for cogeneration of electricity in a combined heat and power (CHP) plant with the specifications given in Figure 4.5 is estimated. Waste heat rejected from the processes which cannot be utilised for process heating is potentially available for district heating. Currently, the demand for district heating of the system in Stenungsund is met. Therefore, this option is not investigated in this study. If delivering district heating is an option in other locations, investigating opportunities for selling excess heat to heat sinks outside the cluster should be part of the design strategy.

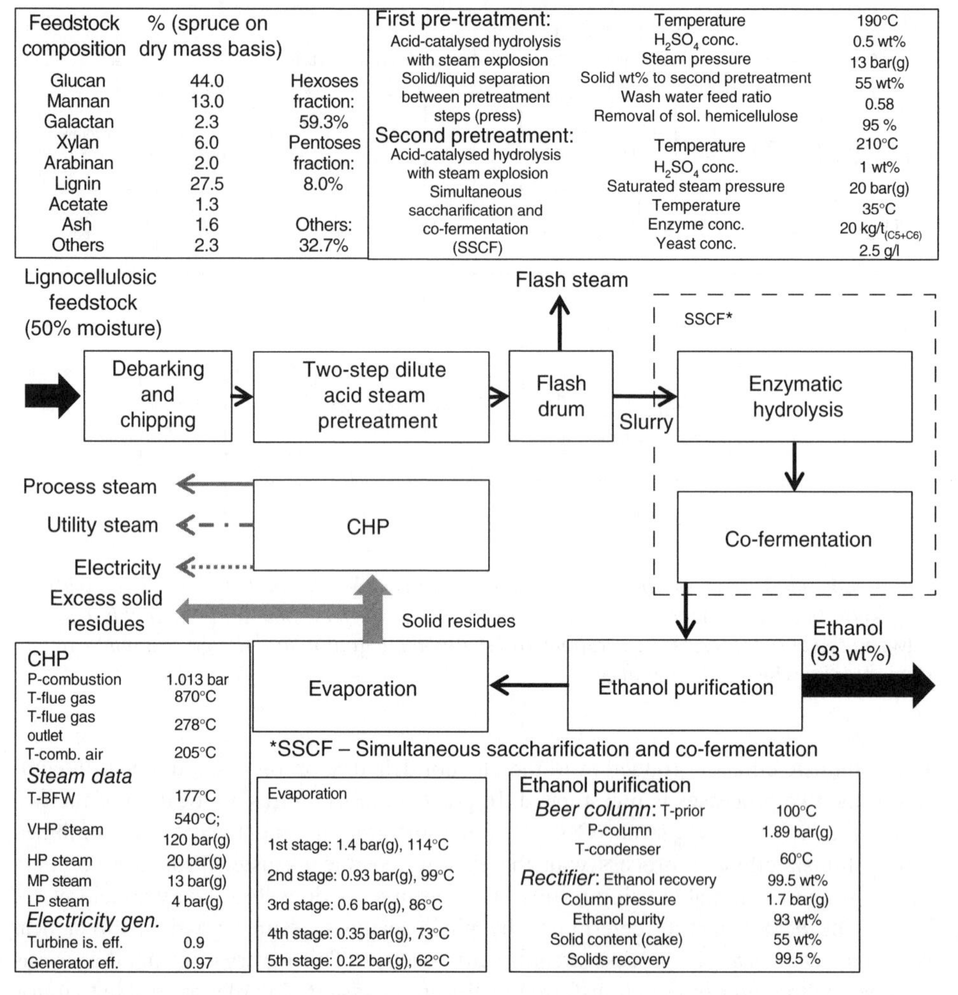

Figure 4.5 *Overview of bioethanol production process from lignocellulosic feedstock. Data included shows the process parameters used for process simulation (Stenberg et al., 1998; Wooley et al., 1999; Aden et al., 2002; Galbe & Zacchi, 2002; Söderström et al., 2003; National Renewable Energy Laboratory, 2004; Söderström et al., 2004; Sassner et al., 2008; Fornell & Berntsson, 2009)*

The different process integration levels are compared by quantifying the minimum process energy requirements ($Q_{\text{heating,min}}$ and $Q_{\text{cooling,min}}$) of a plant with an annual ethylene production of 200 kt.

4.2 Methodology

The first step in this work is establishing a simulation model (Aspen Plus v. 7.1) of the process based on literature data given in the references. The model is then run to obtain material and energy balances of the processes. Thereafter, pinch analysis is used to systematically identify heat integration opportunities (Kemp, 2007). Process simulation and parts of the heat integration study described in this chapter are based on the work by Arvidsson and Lundin (2011).

4.2.1 Process Simulation

Process simulation is used to establish process mass and energy balances as well as relevant process stream data. In this work, process simulation is conducted based on the process description given in Sections 4.2.1.1 and 4.2.1.2. Key input data for process simulation is presented in Figures 4.5 and 4.6. The process simulation software Aspen Plus version 7.2 (Aspen Technology, 2009) was used in order to model the lignocellulosic ethanol and the ethanol dehydration process. Relevant process data is calculated for base case process conditions. Process data can also be generated for different operating conditions in order to investigate the impact of such changes on heat integration opportunities.

For the ethanol production process (including the CHP), the non-random two liquid (NRTL) property method with Henry components was used, which was also recommended by the Aspen Plus guidelines, as it is suitable for, among others, liquid-phase reactions and azeotropic alcohol separation. Some compounds involved in the ethanol production do not exist in the conventional Aspen Plus database. Therefore, physical properties of these components were taken from a database developed by the National Renewable Energy Laboratory (NREL) for biofuel components (Wooley et al., 1999; Aspentech, 2011).

For simulation of the ethanol dehydration plant, the Peng–Robinson property method was applied for all non-electrolyte systems. This method is suitable for handling light hydrocarbons, which are very typical in an ethylene production facility. For simulations involving electrolytes (caustic tower), the electrolyte NRTL (ELECNRTL) property method was applied as this property method is more suitable (Aspentech, 2011).

4.2.1.1 Lignocellulosic ethanol production

Production of ethanol from lignocellulosic materials such as wood is still at the research stage. Today, only pilot plants are in operation (Jones et al., 2010). A variety of process configurations are suggested in the literature.

Figure 4.5 gives an overview of the process steps, process conditions and specifications used for process simulation in Aspen Plus.

In Sweden, softwood is expected to be the main source of bioethanol from lignocellulosic feedstock. Figure 4.5 shows the composition of spruce.

Debarking and chipping: The first step in processing softwood is debarking and chipping. Bark can be used for energy purposes on-site or sold as a solid fuel. To which degree this process step is necessary strongly depends on the type of feedstock used. The design of this first step differs when forest residues are utilised. In this study, mainly different process integration design strategies are compared, and therefore, the influence of the process chosen in this step is of minor importance.

Pretreatment: The size-reduced lignocellulosic feedstock is pretreated to disintegrate the complex wood matrix consisting of cellulose, hemicelluloses and lignin. Thereby, the yield of the downstream hydrolysis stage can be increased from 20 up to 90%. In order to increase the hydrolysis yield, a two-step pretreatment is applied, since the optimal conditions for hemicellulose and cellulose sugar recovery differ. Two-step pretreatment shows a higher

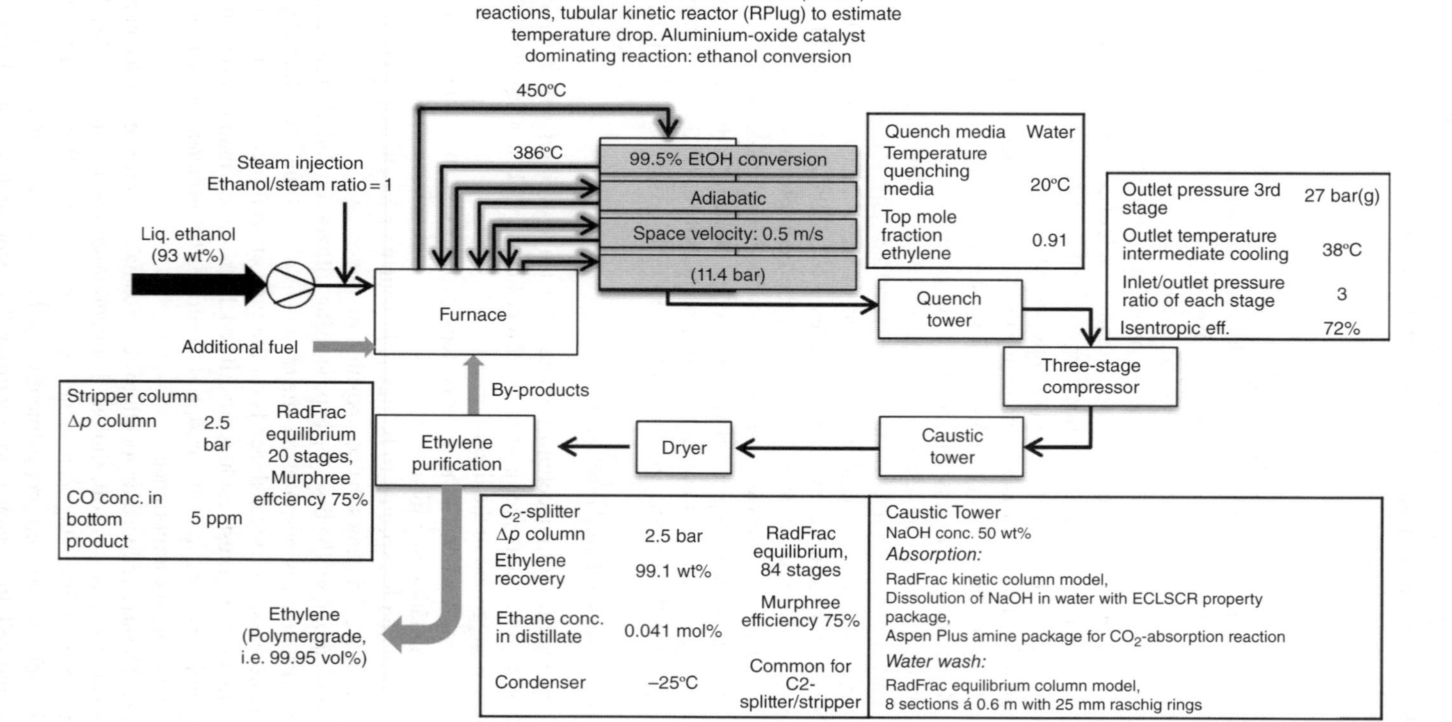

Figure 4.6 *Overview of the ethanol dehydration to ethylene process configuration. Data included shows the process parameters used for process simulation (Stauffer & Kranich, 1962; Barrocas et al., 1980; Kochar et al., 1981; Chematur, 2010; Huang, 2010)*

ethanol yield and lower enzyme consumption, but comes at the cost of higher investment and energy costs due to the addition of a second pretreatment stage. In the first step, acid-catalysed hydrolysis is performed using sulphuric acid (H_2SO_4) and steam explosion under low severity conditions (Wooley et al., 1999). The woodchips are impregnated with dilute acid and then treated with steam. Steam explosion is applied in order to obtain a high fraction of solubilised hemicelluloses, while the lignin stays in the solid fraction. The steam condenses in the woodchips and rapidly evaporates when the pressure is decreased. Thereby, the wood structure 'explodes' (Carvalheiro et al., 2008) and 95% of the hemicellulose sugars can be recovered. After this step, the resulting slurry is separated. The solid fraction is further treated in the second, more severe pretreatment step (higher acid concentration and steam pressure), while the liquid fraction is bypassed. Advantages of the two-step pretreatment method are increased utilisation of and less enzyme consumption in the subsequent hydrolysis step. On the other hand, the investment costs increase as two different reactors are needed and the energy costs are higher. Despite the higher investment and energy costs, the two-step pretreatment shows lower ethanol production costs (Söderström et al., 2004).

Hydrolysis and fermentation: After the pretreatment steam is recovered in a flash drum before the slurry enters the simultaneous saccharification and co-fermentation (SSCF) reactor. The ethanol concentration of the slurry leaving the reactor is 4–5 wt%.

Purification: For ethanol purification, a two-step process is applied. First, the ethanol–water mixture is separated from the solid and non-volatile substances in a stripper column (beer column). In the following rectification column, the ethanol is concentrated to 93 wt%. Further concentration is not necessary as the downstream ethanol dehydration process only requires an ethanol concentration of 92–96 wt%. Thereby, energy-demanding azeotropic distillation can be avoided (Hamelinck et al., 2005). The solid and non-volatile substances at the bottom of the beer column are further treated. The solid fraction (mainly lignin) is separated in a solid–liquid separator, while the liquid fraction is sent to an evaporation unit in order to recovery the dissolved, non-volatile substances. These can then be used as fuel to cover the plant's energy demand or be sold. In the evaporator unit, steam is used for heating in the first stage. By lowering the pressure in the following stages, the evaporated water at one stage can be utilised for heating the subsequent stage. After evaporation, the syrup is mixed with the previously separated solid fraction and then utilised as fuel in the plant's CHP production unit.

CHP production: The solid residue from the beer column and the concentrated syrup from the evaporation plant can be used to generate heat to the process and electricity in a CHP plant. If necessary, bark from the debarking unit can also be used. The CHP plant (back-pressure steam turbine unit with steam extraction at appropriate pressure levels to supply utility and process steam) was simulated in Aspen Plus (for input data, see Figure 4.5). Utility steam is used for process heating in, for example, distillation column reboilers and can be therefore replaced by another hot utility at the same temperature. Process steam is used in the process directly as, for example, in the pretreatment step where steam is used for steam explosion to separate to wood components. As steam in this case is an essential part of the process, it cannot be replaced by another utility.

4.2.1.2 *Ethanol dehydration to ethylene*

The production of ethylene by dehydration of ethanol is a proven technology and was demonstrated and implemented on large scale (Winter, 1976). Braskem started a full-scale plant in Brazil in 2010 (Braskem, 2012). The process consists of a dehydration reactor and several subsequent purification steps in order to obtain polymer-grade ethylene (composition: 99.95 wt% ethylene, 0.05 wt% ethane, 5 ppm CO and 10 ppm CO_2 (Kochar et al., 1981)). Figure 4.6 illustrates the ethanol dehydration process investigated in this study and lists the input data used for process simulation.

Ethanol is first pressurised to reactor operating pressure and then evaporated. After that, steam is injected. Steam injection has a positive impact on conversion and selectivity in temperature regions above 375°C (Kochar et al., 1981). Furthermore, the amount of catalyst needed is lower and its lifetime is increased (Barrocas et al., 1980; Morschbacker, 2009). A multi-bed adiabatic reactor system consisting of four reactors in series with a SynDol catalyst bed in each reactor is assumed. The reactor tubes are located inside a furnace in order to heat and subsequently reheat the reactants to reaction temperature. A mixture of combustible by-product gases from the ethylene dehydration process and natural gas (additional fuel) is used to heat the furnace.

After the reactor, water, unconverted ethanol and other impurities are removed from the effluents in a direct contact quench tower (Winter, 1976). The reaction products enter the tower in the bottom. Water is used as a quenching media and enters the column at the top. Part of the bottom product is cooled and recirculated to the top of the column. The rest is sent to waste water treatment.

The gas stream leaving the quench tower mainly consists of ethylene. The gas is compressed in a multistage compressor with intercooling and then fed to the caustic tower, where CO_2 is absorbed with a sodium hydroxide (NaOH) solution. NaOH in the gas is removed in a water wash section in the tower. The remaining water in the gas stream is removed in a molecular sieve dryer before the gas enters the final purification stage.

In the ethylene column (C_2 splitter), heavier impurities such as ethane, ethanol, diethyl ether and acetaldehyde are removed by cryogenic distillation. In the stripper column, lighter impurities such as CO, CH_4 and H_2 are removed. The two columns have a common condenser (Chematur, 2010) which operates at −25°C. The lighter impurities are vented from the condenser to the atmosphere. After the stripper column, the ethylene has reached polymer-grade purity.

4.2.2 Performance Indicator for Heat Integration Opportunities

In order to compare the different process integration alternatives, the overall energy efficiency (η_{overall}) of the processes is calculated using Equation 4.1. Using the given equation enables for the investigation of the performance of the bio-based process assumed in this study integrated with an existing energy system based on a primary energy basis. Electricity ($\dot{W}_{\text{el}}$) and other energy carriers and feedstocks as by-products and/or input to the processes are accounted for:

$$\eta_{\text{sys}} = \frac{\dot{m}_{\text{ethylene}} \cdot \text{HHV}_{\text{ethylene}} + \dot{m}_{\text{excess solid residues}} \cdot \text{HHV}_{\text{excess solid residues}} + \dfrac{\dot{W}_{\text{el}}^{-}}{\eta_{\text{el,ref}}}}{\dot{m}_{\text{biomass,in}} \cdot \text{HHV}_{\text{biomass}} + \dot{m}_{\text{yeast}} \cdot \text{HHV}_{\text{yeast}} + \dot{Q}_{\text{fuel to ethylene reactors}} + \dfrac{\dot{W}_{\text{el}}^{+}}{\eta_{\text{el,ref}}}} \tag{4.1}$$

Table 4.1 Main results of process simulation of the lignocellulosic ethanol production process

Ethanol production	43.9 t/h
Biomass feedstock	147 t-dry mass/h[a]
Ethanol yield	277.6 kg/t dry mass
Available solid residues (from beer column and evaporation plant)	32.64 MJ/kg-EtOH[b]
Bark available	6.84 MJ/kg-EtOH
Process heat demand	
Pretreatment 1 (MP)	3.71 MJ/kg-EtOH
Pretreatment 2 (HP)	0.80 MJ/kg-EtOH
Ethanol purification	9.32 MJ/kg-EtOH
Solid residue evaporation	3.69 MJ/kg-EtOH
Ethylene production	23.4 t/h[c]
Ethylene yield	570 kg/t-ethanol
Available by-products for energy purposes	1.15 MJ/kg-ethylene[d]
Process heat demand	
Preheat ethanol feed	0.66 MJ/kg-ethylene
Direct steam injection	3.84 MJ/kg-ethylene
Ethylene reactors (high *T*)	4.86 MJ/kg-ethylene
C2 splitter	0.94 MJ/kg-ethylene
Stripper	0.21 MJ/kg-ethylene

[a] HHV 18.3 MJ/kg (Aspen Plus simulation).
[b] Based on a higher heating value (HHV) of 10.6 MJ/kg solid residue.
[c] HHV 47.2 MJ/kg (Aspen Plus simulation).
[d] Based on HHV of 43.83 MJ/kg (Schnelle & Brown, 2002).

Only net energy flows are considered. This means that $\dot{W}_{el}$ can only be part of the nominator or the denominator. $\dot{W}_{el}^{-}$ denotes for the net electricity generated, while $\dot{W}_{el}^{+}$ is the net electricity consumed by the processes. Distribution losses when exporting/importing electricity are not accounted for (Heyne, 2013). $\eta_{el,ref}$ is the conversion efficiency of the reference energy system in order to estimate the primary energy demand of the processes or the amount of primary energy replaced in the reference system. Estimating the efficiencies of the reference system in a future scenario is very complicated. A tool developed at the division of Heat and Power Technology at Chalmers was used for this estimation (Harvey & Axelsson, 2010). In this case, the marginal technology for electricity generation is assumed to be coal combustion with an efficiency of 0.46[1] by 2020 (Stephen et al., 2011) based on higher heating value (HHV) of coal. HHV of the different energy carriers is used for all combustible flows in the system. HHVs are given in the footnotes of Table 4.1. The HHV of yeast was estimated to be 21.6 MJ/kg in Aspen Plus.

[1] The reference gives the efficiency based on lower heating value (LHV) (0.48); therefore, the efficiency was adjusted for the difference between HHV (23.968 MJ/kg) and LHV (22.732 MJ/kg) of coal (Oakridge National Laboratory, 2010).

4.3 Results

4.3.1 Process Simulation

The obtained results from the Aspen Plus simulations of the ethanol production and the ethanol dehydration plant are presented in Table 4.1, which summarises the results obtained for a production capacity of 200 kt/year of polymer-grade ethylene. It can be seen that the most energy-intensive part of the ethanol production process is the ethanol purification, accounting for approximately 53% of the total process steam consumption. The ethanol yield found in this study corresponds to approximately 73% of the theoretical yield assuming complete conversion of hexoses and pentoses and is similar to that obtained by Cardona and Sánchez; however, the present study shows comparatively lower energy consumption (Cardona Alzate & Sánchez Toro, 2006). This is a result of the lower ethanol concentration requirement for downstream processing in the ethylene production (i.e. 93 wt% compared to >99.5 wt% assumed by Cardona and Sánchez), which consequently avoids the energy-demanding azeotropic distillation.

The assumed running time of both the ethanol and the ethanol dehydration plants is 8500 h/year. 73% of the total heat used in the ethanol dehydration process is consumed by the ethylene reactors which are by far the most energy-intensive part of the process. The reactors operate at high temperature (390–450°C) and therefore have to be heated by direct firing or flue gases.

If only the energy content of the main ethylene product is compared to the energy content of the biomass feedstock, a first law conversion efficiency of approximately 41% is achieved. This low value indicates the importance of maximising the amount of by-product utilisation in order to increase the processes' overall energy efficiency.

Detailed stream data obtained from process simulation for the ethanol production process and the ethylene production process are provided in Tables 4.A.1 and 4.A.2.

4.3.2 Integration of Separate Ethanol and Ethylene Production Processes

Direct steam injection in the pretreatment steps in ethanol production process (51.2 MW) and direct steam to the ethylene reactor (25.1 MW) is not included in the heat integration analysis. This is due to the fact that this steam usage is a process requirement and cannot be replaced by heat exchange with other process streams. These amounts of steam must be added to cover the total steam demand of the processes.

On the first integration level, the lignocellulosic ethanol production and ethanol dehydration process are investigated separately. Figure 4.7 shows the GCC of the ethanol production process. Assuming a global ΔT_{min} of 10 K, the minimum heating and cooling demands for the ethanol production process are 112 and 148 MW. The pinch temperature of the process is located at 96°C. Moreover, it can be seen that a large fraction of the heating demand occurs at temperature levels of 117°C (i.e. the ethanol purification) and above. Large sources of excess heat are the condensers in the purification section and the cooling demand of the hydrolysis and fermentation processes. Using excess solid residues for firing a CHP unit sized to cover the heat demand of the ethanol production plant, it was found that an excess of electricity of 24 MW_{el} can be generated (process electricity demand = 32 MW_{el}, electricity produced in CHP unit = 56 MW_{el}). Thereafter, it was estimated that 86 MW of excess solid residues are available for export from the plant.

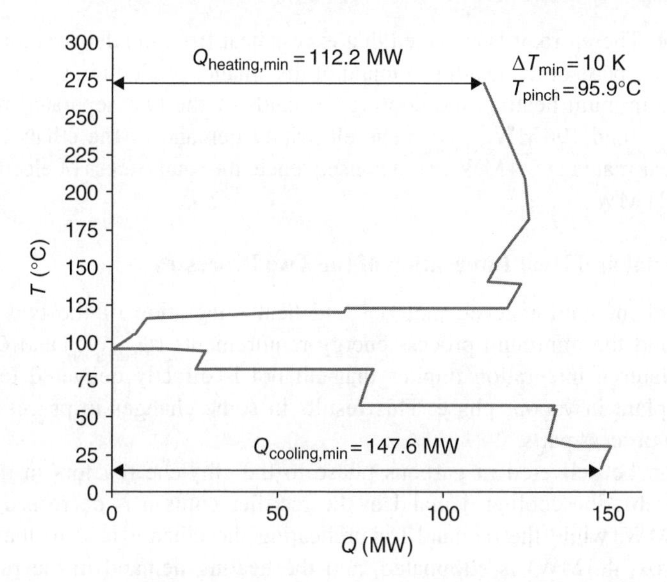

Figure 4.7 *GCC of the ethanol production process from lignocellulosic biomass; direct stream injection of 51 MW in the pretreatment steps is considered a process requirement and therefore not included*

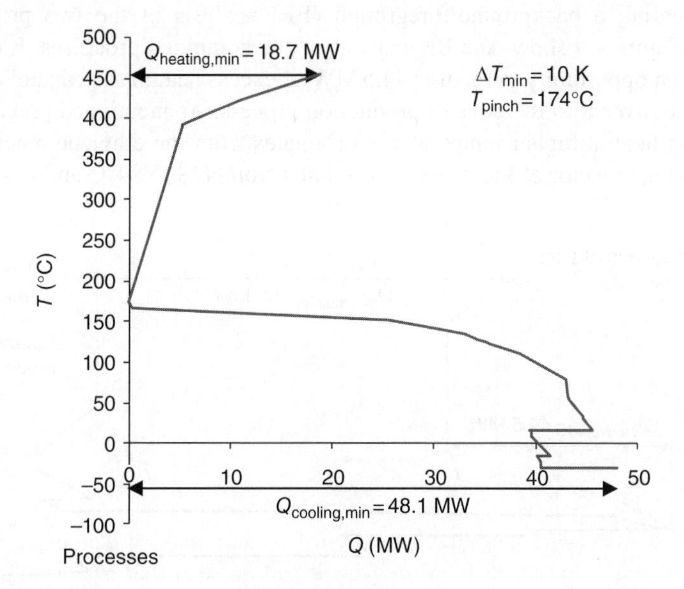

Figure 4.8 *GCC of the ethanol dehydration process; direct steam injection of 25 MW steam to the ethylene reactor is considered a process requirement and therefore not included*

The GCC of the ethanol dehydration process is shown in Figure 4.8. The minimum heating and cooling demands are 19 and 48 MW. The processes' pinch temperature is 174°C and therefore considerably higher than the pinch temperature of the ethanol production process (96°C). The GCC below the pinch point is relatively flat and indicates large amounts of excess heat. Therefore, it is expected that excess heat from the dehydration process can be used to cover parts of the heating demand of the ethanol production process.

The total minimum heating and cooling demands of the two separated processes are 131 $MW_{heating,min}$ and 196 $MW_{cooling,min}$. The electricity demand of the ethanol dehydration process was estimated at 3.4 MW_{el}. As a consequence, the total excess of electricity of both processes is 21 MW_{el}.

4.3.3 Material and Heat Integration of the Two Processes

At the second integration level, material and heat integration of the two processes is considered, and the minimum process energy requirements ($Q_{heating,min}$ and $Q_{cooling,min}$) are estimated. Material integration implies that ethanol is directly delivered to the ethanol dehydration plant in vapour phase. This results in some changes in process and energy flows in both process parts.

Ethanol can be delivered in gaseous phase to the ethylene reactors in the combined process. Thereby, the cooling demand in the rectifier column is decreased by approximately 14.3 MW, while the demand for preheating the ethanol feed to the dehydration reactor (approx. 4.3 MW) is eliminated, and the heating demand in the furnace of the ethylene plant is decreased by approximately 8.7 MW. Detailed stream data of the combined process is given in Table 4.A.3.

In order to illustrate heat integration opportunities and thus also to estimate the utility savings potential, a background/foreground (BF) analysis of the two processes was performed. Figure 4.9 shows the BF analysis of the combined processes. It can be seen that there is an opportunity to recover 44.5 MW of excess heat in the ethanol dehydration process and deliver it to the ethanol production process. As mentioned previously, most of the excess heat at higher temperatures originates from the ethylene reactor effluent. The hot ethylene reactor effluent stream is cooled from 428 to 84°C and has a relatively

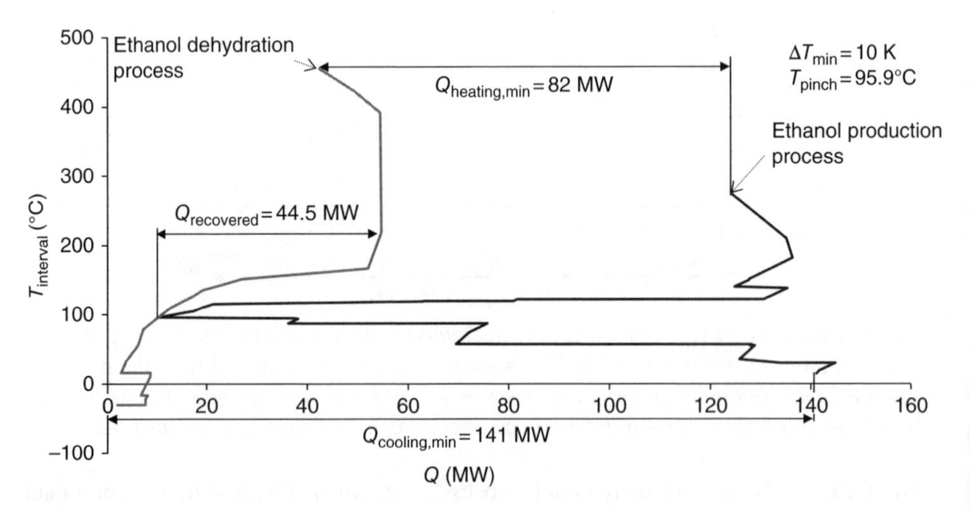

Figure 4.9 *Background/foreground analysis of the ethanol production and ethanol dehydration process; direct delivery of ethanol between the processes is accounted for in the stream data*

flat *T*–*Q* profile below 117°C due to condensation. It is therefore possible to utilise a large share of the heat in the stream. As a result, it can be shown that there is an opportunity to reduce the total minimum heating demand from 131 to 82 MW by material and heat integration of the two processes (i.e. a reduction of 49 MW compared to the case of two separate processes). This corresponds to hot utility savings potential of up to 37%. Similarly, it can be seen that there is an opportunity to reduce the total minimum cooling demand from 196 to 141 MW (corresponding to a reduction of 55 MW) by material and heat integration of the two processes. This corresponds to a cold utility savings potential of 28%.

The cogeneration potential is affected by the reduction of hot utility demand. As the minimum hot utility demand is reduced, the steam production in the CHP plant is reduced which consequently reduces the electricity production (which is estimated at 46 MW_{el}). The electricity demand of the combined processes is estimated at 38 MW_{el}.[2] Accordingly, material and heat integration of the two processes results in an excess electricity production of 8 MW_{el}. 123 MW of excess solid residues is available after fuelling the cogeneration unit.

4.3.4 Integration Opportunities with the Existing Chemical Cluster

The GCC shown in Figure 4.9 represents the minimum heating and cooling demand of the biomass-to-ethylene production plant. In this case, it is assumed that direct heat exchange between process streams is possible across the whole plant. In practice, most process heating and cooling are performed via the utility system. In order to design a utility system for the process which enables both high amount of heat recovery within the process and heat integration within the existing chemical cluster, TSA was used. The process stream data including which utilities are used for stream heating and cooling is given in Table 4.A.3.

In a previous study (Hackl et al., 2011), a common utility system for heat recovery and process heating was suggested consisting of four steam levels (85 bar(g), 40 bar(g), 8.8 bar(g) and 2 bar(g)) and a hot water circuit. By applying these utility levels to the new biomass-to-ethylene process, the following Total Site Profile (TSP) curves were obtained (see Figure 4.10). The TSP curves show the amount and level of utility used for process cooling (left side of the graph). It shows that a total of 119 MW_{heat} can be recovered from the process at different utility levels. The hot utility consumption is shown as well (right side of the graph), resulting in a total heating demand of the process of 222 MW.

By overlapping the TSP curves, the Total Site Composites (TSC) can be created, as shown in Figure 4.11. The TSC curves show the amount of external hot and cold utility, $Q_{heating}$ = 103 MW and $Q_{cooling}$ = 164 MW. 78.4 MW of the external heating demand is covered by 2 bar(g) steam and 1.2 MW by 8.8 bar(g) steam.

This steam is supplied by a CHP plant fired with solid residues from the ethanol process resulting in an excess of 13.3 MW_{el} (process electricity demand = 38 MW, electricity produced in CHP turbine = 51.3 MW). 102.5 MW of solid residues can be exported from the plant.

The increased heating and cooling demand compared to results from the pinch analysis study (see Section 3.3) is due to the fact that pinch analysis assumes direct heat exchange between process streams with a constant ΔT_{min} of 10 K, whereas heat recovery through the utility system requires a higher temperature difference (ΔT_{min} between source profile and

[2] Sum of the electricity demand of the two separate plants + 2.3 MW_{el} for compression of gaseous ethanol feed to the ethylene reactors.

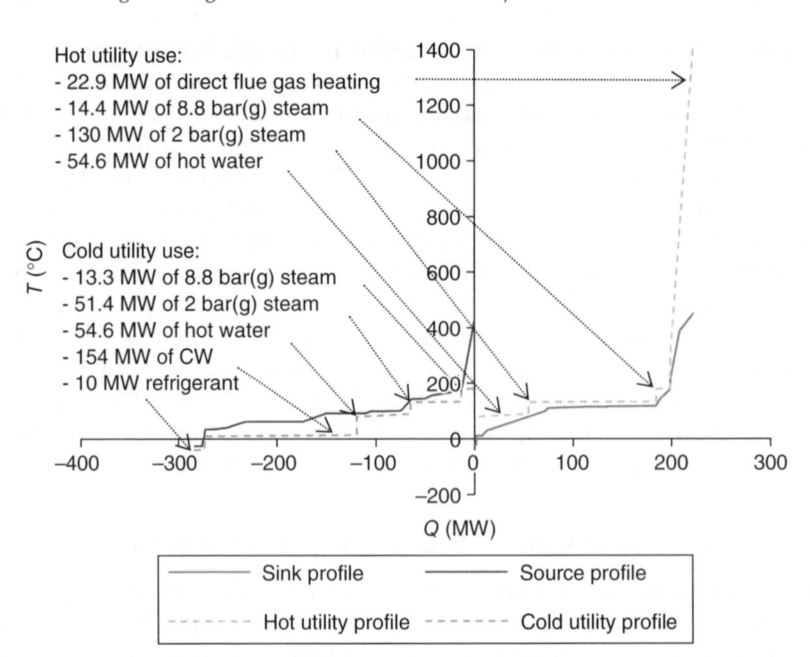

Figure 4.10 *Total site profile curves of the biorefinery, introducing a utility system with two steam levels (8.8 bar(g), 2 bar(g)): a hot water circuit and direct flue gas heating*

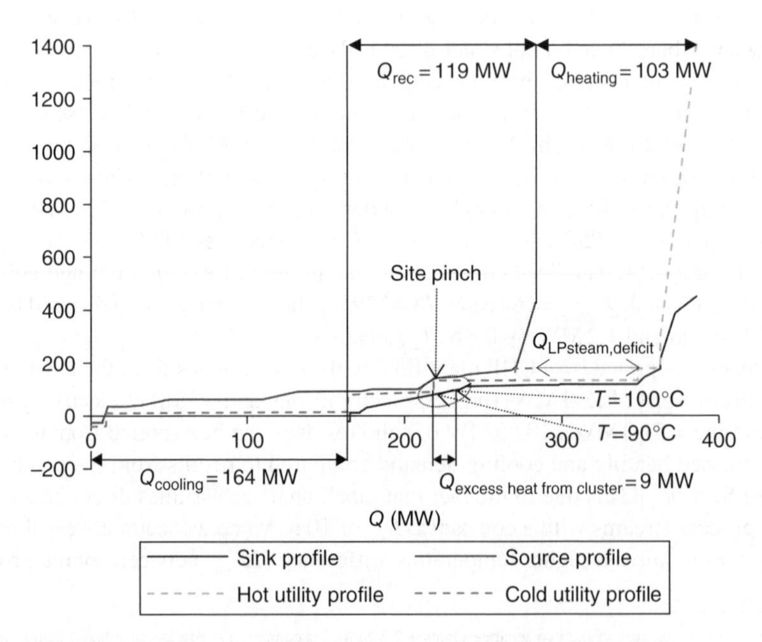

Figure 4.11 *Total site composite curves of the biomass-to-ethylene plant*

cold utility, plus ΔT_{min} between hot utility and sink profile). Therefore, it is not possible to achieve the same amount of heat recovery as by direct heat exchange. Another reason for the lower heat recovery lies in the nature of the utility steam system. Because of the constant temperature of condensing steam and evaporating water used for process heating and cooling, a site pinch is created (see Figure 4.11), which hinders increased heat recovery.

The dotted circle close to the site pinch in Figure 4.11 indicates a large gap between the hot utility curve and the sink profile. This means that a utility with a lower temperature could be used for process heating. Figure 4.12 shows the temperature level and the amount of excess process heat available after maximum heat integration within the cluster via an improved utility system. The GCC contains the hot process and the utility demands for process heating of the clusters' common improved utility system. The graph enables to estimate the amount of heat available from the processes after the maximum amount of process heat is recovered via the common improved utility system. It can be seen that 23 MW of excess heat is available at temperatures above 110°C. As indicated in Figure 4.11, there is a heating demand of 9 MW at a temperature between 90 and 100°C in the biorefinery process. This means that parts of the excess heat from the cluster can be used for heating streams in the biorefinery. This results in savings of 9 MW of 2 bar(g) steam from the CHP plant, reducing the external hot utility demand of the ethylene process to 94 MW.

As indicated in Figure 4.11, the biorefinery has a deficit of LP steam of approximately 78 MW. This can be an important aspect for integration of this process with other petrochemical sites with an excess of LP steam as it can be utilised in the biorefinery and thereby improve EE. Moreover, there are hot and cold utility savings if ethylene is directly delivered to the chemical cluster. These savings are not considered in this study.

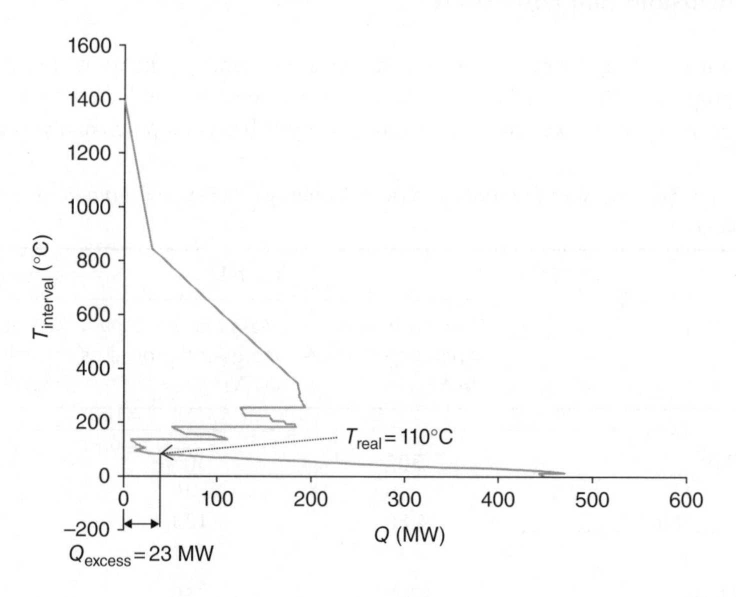

Figure 4.12 *GCC representing the transfer of heat from hot process streams to an improved utility system in the chemical cluster. Used to determine the amount of additional excess heat possible to deliver to the biorefinery (Hackl et al., 2011)*

4.3.5 Performance Indicator for Heat Integration Opportunities

Table 4.2 shows the flows of energy to and from the biomass-to-ethylene production processes considering the different levels of process integration. The level of heat integration has a strong influence on the amount of excess solid residues and electricity that can be exported from the processes and therefore influences their overall energy efficiency. It can be seen that in Case I the largest net amount of electricity can be exported from the site, while the amount of excess solid residues is at a minimum.

This is due to the fact that in this case more heat is needed for process heating which results in a larger potential for cogenerating electricity, while at the same time a larger share of the solid residues has to be combusted to supply heat to the processes. On an overall scale, this alternative shows the lowest overall energy efficiency. Case II shows the highest overall energy efficiency and is therefore regarded the most energy efficient.

Anyhow Case I and Case II assume the possibility of direct heat exchange throughout the respective site. Case III assumes heat exchange via a common utility system which enables for heat exchange with the existing chemical cluster. Assuming heat transfer via a utility system increases the necessary temperature difference for heat recovery which is why the overall energy efficiency in Case III is somewhat lower than for Case II, but it gives a more realistic potential.

Moreover, as the graphs in Figures 4.7, 4.8, 4.9, 4.10 and 4.11 indicate, there is excess process heat available at temperatures suitable to, for example, district heating. Finding use for the available excess process heat could further improve the overall energy efficiency of the processes.

4.4 Conclusions and Discussion

In this chapter, a design strategy for integrating a biorefinery process within a chemical cluster is proposed. The procedure is illustrated in a case study conducted on the chemical cluster in Stenungsund, Sweden. The design strategy is based on pinch analysis tools. In a

Table 4.2 *Energy inputs and outputs and overall energy efficiencies at different levels of process integration*

	Case I	Case II	Case III
	Base case with separate processes (MW)	Mass and heat integrated processes (MW)	Integration with existing site (MW)
Outputs			
$\dot{m}_{ethylene} \cdot HHV_{ethylene}$	307	307	307
$\dot{W}_{el}^{-}$	56	46	51
$\dot{m}_{excess\ solid\ residues} \cdot HHV_{excess\ solid\ residues}$	86	123	103
Inputs			
$\dot{m}_{biomass,in} \cdot HHV_{biomass}$	749	749	749
$\dot{m}_{yeast} \cdot HHV_{yeast}$	9	9	9
$\dot{W}_{el}^{+}$	35	38	38
$\dot{Q}_{NG}$	16	8	8
$\eta_{overall}$ (%)	56.4	58.3	57.1

first step, the two processes are investigated separately. In a second step, material and heat integration of the two processing steps is investigated, and finally, the integration potential of the combined processes with the existing chemical cluster through a common utility system is estimated.

The case study was based on a biorefinery producing 200 kt-ethylene/year from lignocellulosic biomass. Ethylene production was assumed to be conducted in two steps. Ethanol is produced first from lignocellulosic biomass and thereafter dehydrated to ethylene.

In order to identify opportunities to decrease the heat consumption of the processes, pinch analysis was used. The analysis of the separate ethanol production and dehydration processes indicates a minimum heating demand of 112 $MW_{heating,min}$ and 19 $MW_{heating,min}$. The cooling demand obtained was 148 MW for ethanol production and 48 MW for ethanol dehydration. This case represents the base case to which all further integration options are compared to.

Material and heat integration of the two processes potentially allows 49 MW of heat to be saved, corresponding to approximately 37% of the combined minimum heating demands (112 MW + 19 MW = 131 MW). The cooling demand could be reduced by 28% compared to the base case.

Pinch analysis assumes direct heat exchange between all process streams within a plant. In practice, this is rarely the case, and most of the heat between process streams is transferred by the utility system. Therefore, the process was further investigated using TSA and a utility system for the process was suggested. The utility system even allows for integration with an existing chemical cluster. Process integration within the combined ethanol production and dehydration processes via a common utility system indicates that the processes' combined heating and cooling demand can be reduced from 131 $MW_{heating}$ and 196 $MW_{cooling}$ to 103 $MW_{heating}$ and 164 $MW_{cooling}$. By integration with the existing cluster, an additional 9 MW of hot utility can be replaced by excess process heat from the cluster.

Heat and electricity to the processes can be produced by combustion of by-products obtained in both processes. The analysis of the cogeneration potential indicates that depending on the level of heat integration, an excess of electricity between 21 and 8 MW can be generated. The overall energy efficiency in the studied cases was between 56.4 and 58.3%.

The pinch temperature of the combined processes is approximately 96°C (see Figure 4.9) which makes it possible to deliver district heating. In the current situation, the district heating system of the Stenungsund region is saturated, but in the future, there are plans to expand the system to the neighbouring city of Gothenburg, making district heating a very interesting option.

Acknowledgements

This work was carried out under the auspices on the Energy Systems Programme, which is funded primarily by the Swedish Energy Agency. Additional funding was provided by the Swedish Energy Agency's programme for energy efficiency in industry and by participating industrial partners from the chemical cluster in Stenungsund. The authors would particularly like to thank Reine Spetz from Borealis AB, Carl Johan Franzén at the Industrial Biotechnology group at Chalmers as well as Björn Lundin and Maria Arvidsson for their exceptional work during his MSc thesis.

Appendix

Table 4.A.1 Stream data for the lignocellulosic ethanol production plant

Description	T_{start} (°C)	T_{target} (°C)	Q (kW)
FG cooler	278	155	24,018
Condensing flash steam	145.5	143.5	11,122
Condensing flash steam	100.9	99.6	27,898
Condensing flash steam	99.5	99	1,087
Cooler first pretreatment slurry (liquid frac) to SSCF	100.1	35	26,342
Cooler flash steam to WWT	99.6	37	4,589
Cooler evaporator condensate (stage #2)	109.4	37	6,155
Condenser rectifier column	87.5	87.4	53,828
Condenser flash steam	62.2	62.1	58,615
Condenser beer column	60.1	60	944
Cooling demand enzymatic hydrolysis	40.1	35	6,686
Cooling demand fermentation	35.1	35	10,805
Cooler rectifier bottoms to WWT	114.8	37	3,629
Cooler recirculation water	79.1	37	5,455
Air preheater combustion	10	205	28,087
Preheater condensate return	147	177	8,209
Feed water preheater combustion	14.3	177	19,106
Reboiler beer column	117.2	117.3	48,836
Reboiler rectifier column	114.7	114.8	17,882
Evaporator (stage #1 evaporation)	110.4	114.5	40,489
Evaporator (stage #1 heating to bubble point)	98.6	110.4	1,283.6
Preheater beer column	30.1	100	41,845

Table 4.A.2 Stream data of the ethylene dehydration plant

Description	T_{start} (°C)	T_{target} (°C)	Q (kW)
Reactor effluent #1	428	172	13,251.4
Reactor effluent #2	172	156.77	25,128.0
Reactor effluent #3	156.7	140.4	7,699.8
Reactor effluent #4	140.4	114.9	5,707.1
Reactor effluent #5	114.9	84	4,151.6
Midcooler #1 compressor	132.9	38	1,844.7
Midcooler #2 compressor	128	38	1,287.9
Midcooler #3 compressor	129.4	38	1,294.0
Cooler quench tower	61.2	20	1,939.4
Cooler before dryer	56.5	15	532.1
Precooler C2 splitter	15	−17.2	2,337.3
Condenser C2 splitter	−25	−25.1	7,578.7
Furnace #1	169	450	21,244.2
Furnace #2	385.6	450	3,562.0
Furnace #3	386.9	450	3,476.0
Furnace #4	389	450	3,345.0
Reboiler C2 splitter	11.2	11.3	6,091.2
Reboiler stripper	−21.4	−21.3	1,379.5
Preheat ethanol feed	25	135	4,274.4

Table 4.A.3 Stream data with hot and cold utility of the combined ethylene production process

Description	T_{start} (°C)	T_{target} (°C)	Q (kW)	Utility
FG cooler	278	186	15,670	Direct heat exchange with 'air preheater combustion'
FG cooler	186	142	7,452	BFW 2 bar(g)
Condensing flash steam	145.5	143.5	11,122	BFW 2 bar(g)
Condensing flash steam	100.9	99.6	27,898	Water circuit
Condensing flash steam	99.5	99	1,087	Water circuit
Cooler first pretreatment slurry (liquid frac) to SSCF	100.1	90	4,087	Water circuit
Cooler first pretreatment slurry (liquid frac) to SSCF	90	35	22,255	CW
Cooler flash steam to WWT	99.6	90	704	Water circuit
Cooler flash steam to WWT	90	37	3,885	CW
Cooler evaporator condensate (stage #2)	109.4	90	1,649	Water circuit
Cooler evaporator condensate (stage #2)	90	37	4,506	CW
Condenser rectifier column	92.1	92	30,553	CW
Condenser rectifier column	92.1	92	8,994	Water circuit
Condenser flash steam	62.2	62.1	58,615	CW
Condenser beer column	60.1	60	944	CW
Cooling demand enzymatic hydrolysis	40.1	35	6,686	CW
Cooling demand fermentation	35.1	35	10,805	CW
Cooler rectifier bottoms to WWT	114.8	90	1,157	Water circuit
Cooler rectifier bottoms to WWT	90	37	2,472	CW
Cooler recirculation water	79.1	37	5,455	CW
Reactor effluent #1	428	172	13,251	BFW 8.8 bar(g)
Reactor effluent #2	172	156.77	25,128	BFW 2 bar(g)
Reactor effluent #3	156.7	140.4	7,700	BFW 2 bar(g)
Reactor effluent #4	140.4	114.9	5,707	Water circuit
Reactor effluent #5	114.9	90	3,345	Water circuit
Reactor effluent #5	90	84	806	CW
Midcooler #1 compressor	132.9	38	1,845	CW
Midcooler #2 compressor	128	38	1,288	CW
Midcooler #3 compressor	129.4	38	1,294	CW
Cooler quench tower	61.2	20	1,939	CW
Cooler before dryer	56.5	15	532.1	CW
Precooler C2 splitter	15	−17.2	2,337	C3/−41°C
Condenser C2 splitter	−25	−25.1	7,579	C3/−41°C
Air preheater combustion	10	80	8,774	Water circuit
Air preheater combustion	80	205	15,670	Direct heat exchange with 'FG cooler'
Preheater condensate return	147	177	4,805	8.8 bar(g)
Feed water preheater combustion	120	177	9,611	8.8 bar(g)
Feed water preheater combustion	80	120	9,314	2 bar(g)
Feed water preheater combustion	14.3	80	8,510	Water circuit

(*continued overleaf*)

Table 4.A.3 (*continued*)

Description	T_{start} (°C)	T_{target} (°C)	Q (kW)	Utility
Reboiler beer column	117.2	117.3	48,836	Steam 2 bar(g)
Reboiler rectifier column	114.7	114.8	17,882	Steam 2 bar(g)
Evaporator (stage #1 evaporation)	110.4	114.5	40,489	2 bar(g)
Evaporator (stage #1 heating to bubble point)	98.6	110.4	1,284	2 bar(g)
Preheater beer column	30.1	80	29,873	Water circuit
Preheater beer column	80	100	11,973	2 bar(g)
Furnace #1	213.6	450	12,554	Flue
Furnace #2	385.6	450	3,562	Flue
Furnace #3	386.9	450	3,476	Flue
Furnace #4	389	450	3,345	Flue
Reboiler C2 splitter	11.2	11.3	6,091	Water circuit
Reboiler stripper	–21.4	–21.3	1,380	Water circuit

Nomenclature

BF	Background/foreground analysis
BFW	Boiler feed water
CHP	Combined heat and power
COP	Coefficient of performance
CW	Cooling water
EtOH	Ethanol
FG	Flue gas
GCC	Grand composite curve
HHV	Higher heating value
LP	Low pressure (steam)
LHV	Lower heating value
$Q_{cooling,min}$	Minimum cooling demand
$Q_{heating,min}$	Minimum heating demand
SHF	Separate hydrolysis and fermentation
SSCF	Simultaneous saccharification and co-fermentation
TSA	Total site analysis
ΔT_{min}	Minimum temperature difference

References

Aden, A., Ruth, M., Ibsen, K. et al. (2002). *Lignocellulosic Biomass to Ethanol Process Design and Economics Utilizing Co-current Dilute Acid Prehydrolysis and Enzymatic Hydrolysis for Corn Stover*. Golden, CO: National Renewable Energy Laboratory.

Arvidsson, M. and Lundin, B. (2011). Process Integration Study of a Biorefinery Producing Ethylene from Lignocellulosic Feedstock for a Chemical Cluster. Master's thesis. Chalmers University of Technology, Göteborg. Retrieved from publications.lib.chalmers.se/records/fulltext/140886.pdf. Accessed 25 July 2015.

Aspen Technology. (2009). *AspenPlus—Getting started building and running a process model.* (p. 101) Burlington, MA: Aspen Technology, Inc. Retrieved from http://www.che.eng.kmutt.ac.th/cheps/AspenPlusProcModelV7_1-Start.pdf. Accessed 25 July 2015.

Aspentech. (2011). *Aspen Plus*. Burlington, USA: Aspentech. Retrieved from www.aspentech.com/core/aspen-plus.aspx. Accessed 25 July 2015.

Barrocas, H. V. V., Silva, J. B., Assis, R. C. (1980). Process for preparing ethene. US Patent 4,232,179 A, filed 9 August 1978 and issued 4 November 1980.

Bludowsky, T. and Agar, D. W. (2009). Thermally integrated bio-syngas-production for biorefineries. *Chemical Engineering Research and Design*, **87**(9), 1328–1339.

Braskem. (2012). Green Products—Braskem. In *Green Polyethylene (Green PE)*. Retrieved from www.braskem.com.br/site.aspx/green-products-USA. Accessed 23 March 2012.

Cardona Alzate, C. A. and Sánchez Toro, O. J. (2006). Energy consumption analysis of integrated flowsheets for production of fuel ethanol from lignocellulosic biomass. *Energy*, **31**(13), 2447–2459.

Carvalheiro, F., Duarte, L. C., Gírio, F. M. (2008). Hemicellulose biorefineries: A review on biomass pretreatments. *Journal of Scientific Industrial Research*, **67**, 849–864.

Chematur (2010). Ethylene from Ethanol—Chematur. Retrieved from www.chematur.se/sok/download/Ethylene_rev_0904.pdf. Accessed 28 May 2010.

Cherubini, F. (2010). The biorefinery concept: Using biomass instead of oil for producing energy and chemicals. *Energy Conversion and Management*, **51**(7), 1412–1421.

De Jong, E., van Ree, R., Kwant, I. K. (2009). *Biorefineries: Adding Value to the Sustainable Utilisation of Biomass [Internet]*. Amsterdam: IEA Bioenergy. Retrieved from www.ieabioenergy.com/LibItem.aspx?id=6420. Accessed 25 July 2015.

Dhole, V. R. and Linnhoff, B. (1993). Total site targets for fuel, co-generation, emissions, and cooling. *Computers & Chemical Engineering*, **17**(1), 101–109.

Fornell, R. and Berntsson, T. (2009). Techno-economic analysis of energy efficiency measures in a pulp mill converted to an ethanol production plant. *Nordic Pulp and Paper Research Journal*, **24**(2), 183–192. Retrieved from http://publications.lib.chalmers.se/cpl/record/index.xsql?pubid=97740. Accessed 25 July 2015.

Galbe, M. and Zacchi, G. (2002). A review of the production of ethanol from softwood. *Applied Microbiology and Biotechnology*, **59**(6), 618–628.

Hackl, R., Andersson, E., Harvey, S. (2011). Targeting for energy efficiency and improved energy collaboration between different companies using total site analysis (TSA). *Energy*, **36**(8), 4609–4615.

Hamelinck, C. N., Hooijdonk, G. van, Faaij, A. P. (2005). Ethanol from lignocellulosic biomass: Techno-economic performance in short-, middle- and long-term. *Biomass and Bioenergy*, **28**(4), 384–410.

Harvey, S. and Axelsson, E. (2010). *Scenarios for Assessing Profitability and Carbon Balances of Energy Investments in Industry*. Chalmers University of Technology. Retrieved from http://publications.lib.chalmers.se/records/fulltext/98347.pdf. Accessed 25 July 2015.

Heck, L., Poth, N., Soni, V. (2009). Pinch Technology. In *Ullmann's Encyclopedia of Industrial Chemistry*, 435–447. Weinheim: Wiley-VCH Verlag GmbH & Co. KGaA. Retrieved from onlinelibrary.wiley.com/book/10.1002/14356007. Accessed 25 July 2015.

Heyne, S. (2013). Bio-SNG from Thermal Gasification—Process Synthesis, Integration and Performance. Doctoral thesis. Chalmers University of Technology, Göteborg. Retrieved from http://publications.lib.chalmers.se/publication/175377. Accessed 25 July 2015.

Huang, H. (2010). Microbial ethanol, its polymer polyethylene, and applications. *Microbiology Monographs*, **14**, 389–404.

Jones, M. J., Kresge, C. T., Maughon, B. R. (2010). Alternative feedstocks for olefin productions what role will ethanol play? *Oil Gas: European Magazine*, **36**(1), 34–39.

Jönsson, J., Hackl, R., Harvey, S. et al. (2012). Sustainable Chemistry in 2030: Exploring Different Technology Pathways for a Swedish Chemical Cluster. In *Proceedings of the 2012 eceee Industrial Summer Study*. Arnhem, the Netherlands: eceee. Retrieved from proceedings.eceee.org/vispanel.php?event=2. Accessed 25 July 2015.

Kemp, I. C. (2007). *Pinch Analysis and Process Integration* (2nd ed.). Oxford: Butterworth-Heinemann.
Kimm, N. K. (2008). *Economic and Environmental Aspects of Integration in Chemical Production Sites*. Karlsruhe: University of Karlsruhe. Retrieved from digbib.ubka.uni-karlsruhe.de/volltexte/documents/403969. Accessed 25 July 2015.
Klemes, J., Friedler, F., Bulatov, I., Varbanov, P. (2010). *Sustainability in the Process Industry: Integration and Optimization* (1st ed.). New York: McGraw-Hill Professional.
Kochar, N. K., Merims, R., Padia, A. S. (1981). Ethylene from ethanol. *Chemical Engineering Progress*, **77**(6), 66–70.
Morschbacker, A. (2009). Bio-ethanol based ethylene. *Polymer Reviews*, **49**(2), 79.
Naik, S. N. (2010). Production of first and second generation biofuels: A comprehensive review. *Renewable and Sustainable Energy Reviews*, **14**(2), 578–597.
National Renewable Energy Laboratory. (2004). Corn Stover to Ethanol. Aspen Plus Backup file. Available at http://www.nrel.gov/extranet/biorefinery/aspen_models/
Oakridge National Laboratory. (2010). Lower and Higher Heating Values of Gas, Liquid and Solid Fuels. In: Lower and Higher Heating Values of Gas, Liquid and Solid Fuels, Retrieved from http://cta.ornl.gov/bedb/appendix_a/Lower_and_Higher_Heating_Values_of_Gas_Liquid_and_Solid_Fuels.xls. Accessed October 10, 2013.
Raissi, K. (1994). Total Site Integration, Ph.D. thesis. University of Manchester Institute of Science and Technology, Manchester.
Sadaka, S. and Negi, S. (2009). Improvements of biomass physical and thermochemical characteristics via torrefaction process. *Environmental Progress and Sustainable Energy*, **28**(3), 427–434.
Sassner, P., Galbe, M., Zacchi, G. (2008). Techno-economic evaluation of bioethanol production from three different lignocellulosic materials. *Biomass and Bioenergy*, **32**(5), 422–430.
Schnelle, K. B. and Brown, C. A. (2002). *Air Pollution Control Technology Handbook*. Boca Raton, FL: CRC Press.
Smith, R. (2005). *Chemical Process Design and Integration* (1st ed.). Chichester: John Wiley & Sons, Ltd.
Söderström, J., Pilcher, L., Galbe, M., Zacchi, G. (2003). Two-step steam pretreatment of softwood by dilute H2SO4 impregnation for ethanol production. *Biomass and Bioenergy*, **24**(6), 475–486.
Söderström, J., Galbe, M., Zacchi, G., Wingren, A. (2004). Process considerations and economic evaluation of two-step steam pretreatment for production of fuel ethanol from softwood. *Biotechnology Progress*, **20**(5), 1421–1429.
Stauffer, J. E. and Kranich, W. L. (1962). Kinetics of the catalytic dehydration of primary alcohols. *Industrial & Engineering Chemistry Fundamentals*, **1**(2), 107–111.
Stenberg, K., Tengborg, C., Galbe, M. et al. (1998). Recycling of process streams in ethanol production from softwoods based on enzymatic hydrolysis. *Applied Biochemistry and Biotechnology*, **70–72**(1), 697–708.
Stephen, J. D., Mabee, W. E., Saddler, J. N. (2011). Will second-generation ethanol be able to compete with first-generation ethanol? Opportunities for cost reduction. *Biofuels, Bioproducts and Biorefining*, **6**(2), 159–176.
UOP LLC. (2004). *UOP/HYDRO MTO Process—Methanol to Olefins Conversion*. UOP LLC. Retrieved from http://www.uop.com/objects/26%20MTO%20process.pdf. Accessed 25 July 2015.
Winter, O. (1976). Make ethylene from ethanol. *Hydrocarbon Processing*, **55**(11), 125–133.
Wooley, R., Ruth, M., Sheehan, J. et al. (1999). *Economics Utilizing Co-current Dilute Acid Prehydrolysis and Enzymatic Hydrolysis Current and Futuristic Scenarios*. National Renewable Energy Laboratory. Retrieved from http://www.nrel.gov/docs/fy99osti/26157.pdf. Accessed 25 July 2015.

5

Synthesis of Biomass-Based Tri-generation Systems with Variations in Biomass Supply and Energy Demand

Viknesh Andiappan[1], *Denny K. S. Ng*[1], *and Santanu Bandyopadhyay*[2]

[1]*Department of Chemical and Environmental Engineering/Centre of Sustainable Palm Oil Research (CESPOR), The University of Nottingham, Malaysia Campus, Selangor, Malaysia*
[2]*Department of Energy Science and Engineering, Indian Institute of Technology Bombay, Mumbai, India*

5.1 Introduction

Tri-generation systems produce heat, power and cooling simultaneously from a single fuel source. In this system, high-pressure steam is produced through combustion of fuel in boilers. The pressure of the produced steam is then reduced in steam turbines to generate power. In addition, steam is extracted from different steam headers to provide heating energy based on process requirements. Besides generating heat and power, a tri-generation system also produces cooling energy with either mechanical chillers or thermally fed absorption chillers (Chicco & Mancarella, 2009). Since heat, power and cooling energy are produced simultaneously from a single fuel source (Wu & Wang, 2006), the application of tri-generation is highly fuel efficient (ranging from 70 to 90%). Therefore, installing such system on-site would reduce the need to import power from large distances. This would lead to significant reductions in energy expenses and environmental impact (Stojkov et al., 2011).

Process Design Strategies for Biomass Conversion Systems, First Edition. Edited by
Denny K. S. Ng, Raymond R. Tan, Dominic C. Y. Foo, and Mahmoud M. El-Halwagi.
© 2016 John Wiley & Sons, Ltd. Published 2016 by John Wiley & Sons, Ltd.

Additionally, a tri-generation system can also improve the local power quality and reliability of the grid power. However, to maximize these benefits, it is important that the tri-generation system addresses fundamental aspects in the conceptual synthesis of energy systems. These aspects include synthesis optimization (e.g. system configuration and technology selection), design optimization (e.g. capacity, number of units) and operational optimization (e.g. seasonal variations in flow rates). Typically, these aspects are addressed in a hierarchically structured manner. In other words, synthesis, design and operation optimization are performed in three subsequent stages.

Several research works have been performed with respect to the hierarchical approach. For instance, Papaulias and Grossmann (1983) presented one of the earliest superstructure-based approaches to synthesize a cogeneration system which is required to provide fixed demands of power for drivers and steam at various at levels. Next, Maréchal and Kalitventzeff (1998) proposed a three-step superstructure-based approach to synthesize an optimal cogeneration system. In the first step, a generic cogeneration superstructure was used to identify the technologies required for meeting process energy requirements. Based on the first step results, the available technologies that fit such requirements are selected in the second step. Next, the optimal process configuration is targeted. Meanwhile, Lozano et al. (2009a) proposed an approach to determine the optimal configuration of a tri-generation system installed in the tertiary sector based on a superstructure approach that considered the type, number and capacity of equipment. Later, Lozano et al. (2010) continued this work by developing a cost optimization approach for the design of a tri-generation system. Carvalho et al. (2011) developed a model to determine the optimal configuration from a tri-generation superstructure to meet specific demands of a hospital subject to environmental constraints.

On the other hand, there has been a vast amount of works presented on operational optimization, mainly for existing or predefined energy systems. Some of these works have purely focused on studying the economic performance of energy systems. For instance, Cardona et al. (2006) optimized the operation of a tri-generation system installed in an airport in Italy based on economic performance. Ziher and Poredos (2006) focused on optimizing the economics and cooling production of a tri-generation system in a hospital. Arcuri et al. (2007) presented the optimal operation of a tri-generation system to maximize annual economic returns. Li et al. (2008) optimized the operation of tri-generation system in China to achieve the minimum cost. Sugiartha et al. (2009) optimized the operation of a tri-generation system for a supermarket located in United Kingdom. Meanwhile, Lozano et al. (2009b) proposed an approach to analyse the operational strategy of a simple tri-generation system. In addition, a thermo-economic analysis was included to analyse the marginal production costs (Lozano et al., 2009b).

Apart from economic performance, environmental impact has also been a focus of operation optimization studies. For example, Fumo et al. (2009) minimized primary energy consumption and carbon dioxide (CO_2) emissions separately in the operation of tri-generation systems in the America. Meanwhile, Cho et al. (2009) presented an optimization approach for the operation of tri-generation systems in different climate conditions based on primary energy consumption and CO_2 emissions. Wang et al. (2010) maximized primary energy savings and minimized pollutant emissions in the operation of a tri-generation system for a hotel in China. Similarly, Kavvadias et al. (2010) optimized the economic performance for operational strategy of a tri-generation system based on various operation parameters such as energy tariffs. Carvalho et al. (2012b) presented a simple optimization model for the operation of tri-generation system based on environmental loads.

Apart from economic and environmental considerations, several operation optimization studies have considered operational uncertainties. Typically, the operations of an energy system are susceptible to potential uncertainties. Sources for such uncertainties can be categorized into short-term and long-term uncertainties. Short-term uncertainties typically include operational variations, equipment failure, etc. (Subrahmanyam et al., 1994). Meanwhile, long-term uncertainties may refer to fluctuations or variations (Hastings & McManus, 2004) in energy demand, raw material supply, energy prices, etc. (Shah, 1998). If such uncertainties are not considered, the performance of operations may deviate significantly from the optimal one. In this respect, several operational optimization studies have taken uncertainties into account. For instance, Gamou et al. (2002) proposed an optimal unit sizing method for cogeneration system using energy demands as continuous variables. Li et al. (2010) proposed a model to optimize a tri-generation system under uncertainty in energy demands using Monte Carlo method. Lozano et al. (2011) presented a model which analyses the allocation of cost for the operation of a tri-generation system based on variations in energy supply services, fuel prices and energy prices. Meanwhile, Carpaneto et al. (2011a, 2011b) presented a framework which identifies (large-scale as well as small-scale) uncertainties within cogeneration operations and selects the best planning solution. Rezvan et al. (2013) presented a stochastic approach to determine the capacity of a tri-generation system based on uncertainties in energy demand for a hospital in Iran. In the previous work (Rezvan et al., 2013), the uncertainty in energy demand was analysed using probabilistic theory. Mitra et al. (2013) presented a mathematical model to determine the optimal scheduling of industrial cogeneration systems under time-dependent electricity prices.

Despite the vast number of research works presented on the hierarchical approach, such approach often leads to a suboptimal design (Herder & Weijnen, 2000). This is evident as key decisions concerning the basic system flow sheet are made in the initial stages of design and are not supposedly changed in the later stages. As such, uncertainties are typically considered only at the later stages of the conceptual synthesis, namely, in operational optimization. At this stage, the optimal system configuration (which was defined in the earlier synthesis and design optimization) might not be sufficiently flexible to cope with variations introduced. This would lead to an under-designed system and induce a need to reconsider some of the first-stage decisions. However, changing an early design decision would imply a large amount of rework and a lot of extra cost for completion of the design. If the objective is to establish a completely optimized energy system, the three optimization stages (e.g. synthesis, design and operation) cannot be considered independently (Voll, 2014). To address this issue, Herder and Weijnen (2000) proposed an approach which requires relevant uncertainties to be considered in parallel with each stage of the conceptual synthesis. This approach is otherwise known as concurrent engineering. In this respect, Yoshida et al. (2007) proposed a method of determining the optimal system structure and operational strategy for tri-generation system for a hospital. As shown in the previous work (Yoshida et al., 2007), a sensitivity analysis was performed on variations related to energy prices and decline in equipment costs. Meanwhile, Aguilar et al. (2007a, 2007b) presented a systematic methodology which is able to simultaneously synthesize, design and optimize the capital investment of a cogeneration system subject to variable design conditions. Dimopoulos et al. (2008) presented an approach using an evolutionary programming to solve the synthesis, design and operation optimization of a marine cogeneration system. Buoro et al. (2010) presented an optimization model to determine the optimal synthesis and operation of an urban tri-generation system based on total annual costs of operations. The optimization specifies the type, number, location and optimal operating strategy of the tri-generation technologies based on total annual cost for owning, maintaining and operating the entire system. Buoro et al. (2011) then extended this by proposing a model which determines the optimal tri-generation system based on varying amortization periods. Later, Buoro et al. (2012) presented a model which obtains the optimal synthesis, design and operation of a cogeneration system for standard and domotic homes, whereby a sensitivity analysis was performed based on different economic constraints. Carvalho et al. (2012a) presented a multi-criteria approach for the synthesis and operation of tri-generation systems considering environmental and economic aspects. In the presented approach (Carvalho et al., 2012a), optimal solutions are obtained and analysed via a Pareto front. This was then extended by incorporating a sensitivity analysis on electricity prices towards the optimal configuration (Carvalho et al., 2013). Similarly, Buoro et al. (2013) developed a multi-objective optimization approach to obtain the optimal structure, capacity and operational strategy of a cogeneration system in an industrial area. Mehleri et al. (2013) presented an optimization model to determine the optimal synthesis and operation for a residential cogeneration system based on variations in site energy loads, local climate data and utility tariff structure. Recently, Yokoyama et al. (2015) developed an optimization model to determine the types, capacities and number of equipment in consideration of their operational strategies corresponding to seasonal and hourly variations in energy demands for a cogeneration system.

Despite the usefulness of the aforementioned works, most of the contributions have focused on fossil fuel-based energy systems. The use of fossil fuel (e.g. oil, natural gas, coal, etc.), however, is regarded as a key contributor to greenhouse gas (GHG) emissions. As such, biomass has been identified as a sustainable alternative to replace fossil fuel in energy systems (Bridgwater, 2003). In spite of this, the economic viability of biomass-based energy systems remains unclear in cases where seasonal variations in biomass supply and energy demand exist. In this respect, it is essential to develop a systematic approach to synthesize a biomass-based energy system considering seasonal biomass supply and energy demand. Thus, this work presents a systematic multi-period optimization approach for the synthesis of a biomass-based tri-generation system (BTS) based on seasonal variations in raw material supply and energy demand. Multi-period optimization is an approach that has been applied in several design and operation optimization studies in energy systems (Aaltola, 2002; Chen et al., 2011; Iyer & Grossmann, 1998; Yunt et al., 2008). This approach provides a unique solution that is feasible for a given set of scenarios. Such solution is desirable as the economic feasibility has been assessed for all potential scenarios. To illustrate the proposed approach, a palm-based biomass case study is presented.

5.2 Problem Statement

As various established technologies are available in the market, the synthesis of an optimal BTS is a highly complex problem. Generic representation of the problem is shown in Figure 5.1. The synthesis problem addressed is stated as follows: Biomass i with flow rate

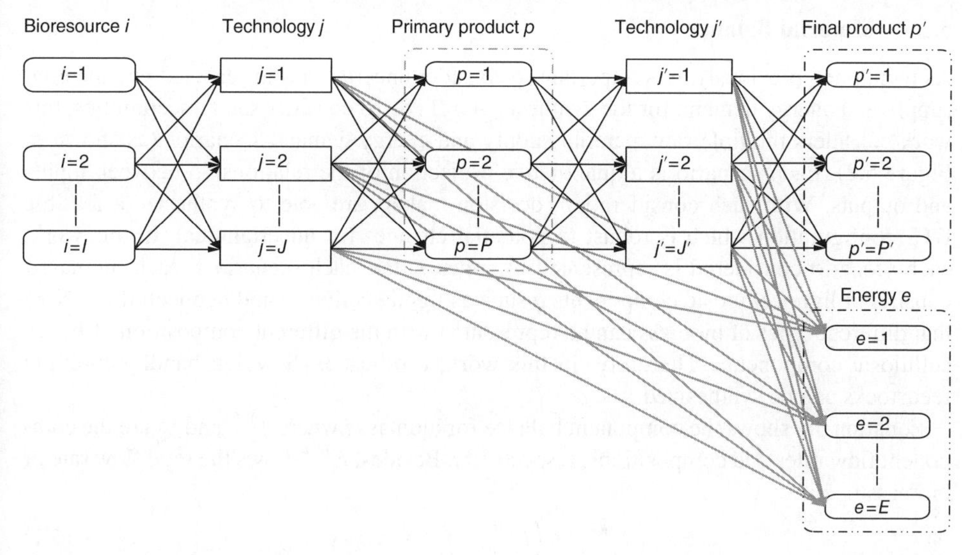

Figure 5.1 *Generic representation of superstructure for scenarios. Reproduced with permission from reference Andiappan et al. (2014), © 2014, American Chemical Society*

F_i^{BIO} and its composition $q \in Q$ (e.g. lignin, cellulose and hemicellulose) can be converted to primary product $p \in P$ through technologies $j \in J$. Primary product p and its composition $q' \in Q'$ can be further converted to final products $p' \in P'$ via technologies $j' \in J'$. Besides producing primary and final products p and p', technologies j and j' can also generate energy $e \in E$. In this work, the objective is to develop a systematic approach to synthesize an optimal and robust BTS configuration with maximum economic performance while considering multiple supply and demand scenarios, $s \in S$.

In addition, the proposed approach is able to determine the optimum design capacities of the selected equipment based on the maximum operating capacity of all scenarios. As most of the available equipment sizes in the market are fixed, the proposed approach can also select the equipment based on the available design capacities of technologies $j\left(F_{jn}^{\mathrm{Design}}\right)$ and $j'\left(F_{j'n'}^{\mathrm{Design}}\right)$, respectively. The aforementioned optimization model is then solved via a multi-period optimization whereby each scenario s is assigned a fraction of occurrence, α_s.

The following section further explains parameters and variables involved in the optimization model developed for this work. The equations formulated in the optimization model are clearly presented and described methodically to address the BTS synthesis problem.

5.3 Multi-period Optimization Formulation

The generic superstructure of a BTS is shown in Figure 5.1. Based on Figure 5.1, the following subsections present a detailed formulation of the proposed multi-period optimization model.

5.3.1 Material Balance

As mentioned previously, it is important to consider uncertainties in seasonal raw material supply and energy demand for the synthesis of a BTS. To consider such uncertainties, this work considers multiple raw material supply and energy demand scenarios (as shown in Figure 5.2). Each scenario is a quantitative projection of the relationship between inputs and outputs. With such consideration, decision makers are able to synthesize a feasible BTS configuration which is robust to potential changes (or uncertainties). In this work, each scenario considered is represented by index s. For each scenario s, each biomass i consists of lignocellulosic components q such as lignin, cellulose and hemicellulose. Note that different types of biomass can be represented with the different composition of lignocellulosic components. Therefore, in this work, a robust BTS which handles multiple feedstocks can be synthesized.

Equation 5.1 shows the component balance for biomass i where f_{iq}^{BIO} and x_{iq} are the component flow rates and compositions, respectively. Besides, F_i^{BIO} shows the total flow rate of biomass i:

$$\left(f_{iq}^{\text{BIO}}\right)_s = \left(F_i^{\text{BIO}} x_{iq}\right)_s \quad \forall i \forall q \forall s \tag{5.1}$$

Each biomass i is split into potential technology j with flow rate of F_{ij}^{I} as shown:

$$\left(F_i^{\text{BIO}}\right)_s = \left(\sum_{j=1}^{J} F_{ij}^{\text{I}}\right)_s \quad \forall i \forall s \tag{5.2}$$

In technology j, the components of biomass i (f_{qj}^{I}) from flow rate F_{ij}^{I} are then converted to primary product p with the conversion of $\text{X}_{qjp}^{\text{I}}$, respectively:

$$\left(f_{qj}^{\text{I}}\right)_s = \left(\sum_{i=1}^{I} F_{ij}^{\text{I}} x_{iq}\right)_s \quad \forall q \forall j \forall s \tag{5.3}$$

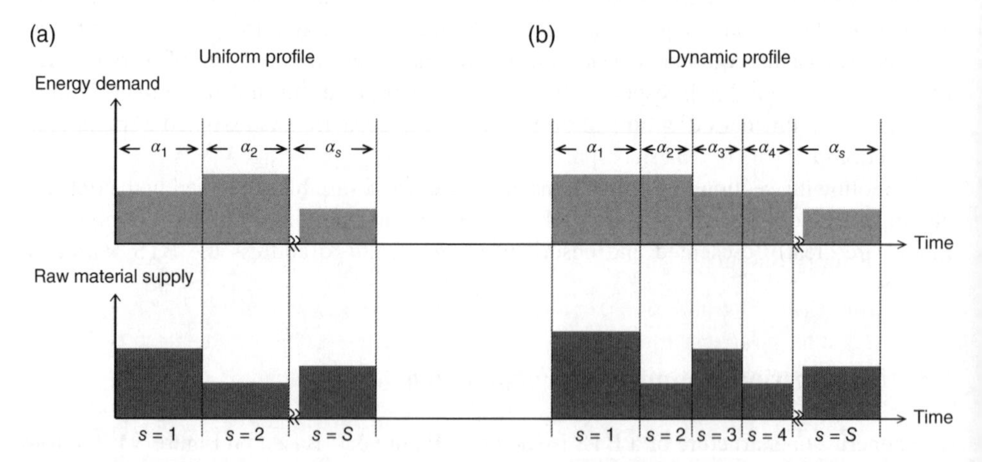

Figure 5.2 Generic representation of scenarios. Reproduced with permission from reference Andiappan et al. (2014), © 2014, American Chemical Society

$$\left(F_{jp}^{\mathrm{I}}\right)_s = \left(\sum_{q=1}^{Q} f_{qj}^{\mathrm{I}} \mathrm{X}_{qjp}^{\mathrm{I}}\right)_s \quad \forall p \forall j \forall s \tag{5.4}$$

The total production rate of primary product *p* for all technologies *j* is given as

$$\left(F_p\right)_s = \left(\sum_{j=1}^{J} F_{jp}^{\mathrm{I}}\right)_s \quad \forall p \forall s \tag{5.5}$$

Next, primary product *p* can be distributed to potential technology *j′* for further processing to produce product *p′*. The splitting of primary product *p* is written as

$$\left(F_p\right)_s = \left(\sum_{j'=1}^{J'} F_{pj'}^{\mathrm{II}}\right)_s \quad \forall p \forall s \tag{5.6}$$

In technology *j′*, the components of primary *p* ($f_{q'j'}^{\mathrm{II}}$) from flow rate $F_{pj'}^{\mathrm{II}}$ (as shown in Equation 5.7) are then converted to final product *p′*. These components are converted with conversion of $\mathrm{X}_{q'j'p'}^{\mathrm{II}}$ given in Equation 5.8:

$$\left(f_{q'j'}^{\mathrm{II}}\right)_s = \left(\sum_{p=1}^{P} F_{pj'}^{\mathrm{II}} x_{pq'}\right)_s \quad \forall j' \forall q' \forall s \tag{5.7}$$

$$\left(F_{j'p'}^{\mathrm{II}}\right)_s = \left(\sum_{q'=1}^{Q'} f_{q'j'}^{\mathrm{II}} \mathrm{X}_{q'j'p'}^{\mathrm{II}}\right)_s \quad \forall j' \forall p' \forall s \tag{5.8}$$

The total production rate of product *p′* for all technologies *j′* is written as

$$\left(F_{p'}\right)_s = \left(\sum_{j'=1}^{J'} F_{j'p'}^{\mathrm{II}}\right)_s \quad \forall p' \forall s \tag{5.9}$$

However, in the case where a single or no technology is required to produce the final product *p′*, biomass *i* and primary product *p* are allowed to bypass technology *j* and *j′* via a 'blank' technology (where no conversion takes place). It is important to note that the representation of final product *p′* is applicable in cases where BTS outputs (e.g. steam, cooling water, etc.) are sold or exported to the end user as raw materials (e.g. in reactions). Alternatively, if BTS outputs are sold or exported as utilities (e.g. for heating, cooling, power, etc.), it would be given by energy *e* and is discussed in the next subsection.

5.3.2 Energy Balance

Apart from material conversions, components *q* in biomass *i* can also be converted into energy *e* via technology *j* with conversion of $\mathrm{V}_{qje}^{\mathrm{I}}$. Moreover, components *q′* in product *p* can also be converted into energy *e* via technology *j′* with conversion of $\mathrm{V}_{q'j'e}^{\mathrm{II}}$. Total energy generated by technologies *j* and *j′* is shown in Equation 5.10:

$$\left(E_e^{\mathrm{Gen}}\right)_s = \left(\sum_{q=1}^{Q}\sum_{j=1}^{J} f_{qj}^{\mathrm{I}} \mathrm{V}_{qje}^{\mathrm{I}} + \sum_{q'=1}^{Q'}\sum_{j'=1}^{J'} f_{q'j'}^{\mathrm{II}} \mathrm{V}_{q'j'e}^{\mathrm{II}}\right)_s \quad \forall e \forall s \tag{5.10}$$

Besides generating energy, some technologies in the BTS may consume energy. The total energy consumption of BTS is determined based on the energy requirement in all technologies j and j' that used to convert biomass i to products p and p' as well as energy e. The total energy consumption is showed in Equation 5.11:

$$\left(E_e^{\text{Con}}\right)_s = \left(\sum_{p=1}^{P}\sum_{j=1}^{J} F_{jp}^{\text{I}} \text{Y}_{ejp}^{\text{I}} + \sum_{p'=1}^{P'}\sum_{j'=1}^{J'} F_{j'p'}^{\text{II}} \text{Y}_{ej'p'}^{\text{II}}\right)_s \quad \forall e \forall s \tag{5.11}$$

where Y_{ejp}^{I} and $Y_{ejp'}^{\text{II}}$ are specific energy consumption for technologies j and j', respectively. In the case where energy generated by the BTS exceeds the total energy consumption and energy demand from process or customer $\left(E_e^{\text{Gen}} > E_e^{\text{Con}} + E_e^{\text{Demand}}\right)$, excess energy E_e^{Exp} can be sold or exported to the power grid. In contrast, import of external energy is required if the total energy consumption of the BTS is more than the energy generated $\left(E_e^{\text{Gen}} < E_e^{\text{Con}} + E_e^{\text{Demand}}\right)$. The overall energy correlation for the BTS can be written as

$$\left(E_e^{\text{Gen}} + E_e^{\text{Imp}}\right)_s = \left(E_e^{\text{Con}} + E_e^{\text{Demand}} + E_e^{\text{Exp}}\right)_s \quad \forall e \forall s \tag{5.12}$$

where E_e^{Imp} is the import of external energy into the BTS.

5.3.3 Economic Analysis

The economic feasibility of the BTS is analysed by determining its economic potential (EP). To determine EP, Equation 5.13 is included in the analysis:

$$\text{EP} = \text{GP} - \text{CRF} \times \text{CAP} \tag{5.13}$$

where CRF and CAP represent the capital recovery factor and total capital costs of the BTS, respectively. The GP can be determined using Equation 5.14:

$$\text{GP} = \text{AOT} \times \left(\sum_{p'=1}^{P'} F_{p'} C_{p'} + \sum_{e=1}^{E} E_e^{\text{Exp}} C_e^{\text{Exp}} - \sum_{e=1}^{E} E_e^{\text{Imp}} C_e^{\text{Imp}} - \sum_{i=1}^{I} F_i^{\text{BIO}} C_i^{\text{BIO}}\right) \tag{5.14}$$

where AOT is the annual operating time, $C_{p'}$ is the selling price of products p', C_e^{Exp} is the selling price of exporting excess energy e, C_e^{Imp} is the cost of importing external energy e, and C_i^{BIO} is the cost of raw material (lignocellulosic biomass i). Note that Equation 5.14 assumes a scenario where uncertainties do not play a role in the economic analysis. However, the solution obtained from Equation 5.14 may not necessarily be the optimal (or may even be an infeasible) in real-world applications where uncertainties are present. In this respect, Equation 5.14 is modified as Equation 5.15 to incorporate the fraction of occurrence for scenario s (α_s):

$$\text{GP} = \sum_{s} \alpha_s \left[\text{AOT} \times \left(\sum_{p'=1}^{P'} F_{p'} C_{p'} + \sum_{e=1}^{E} E_e^{\text{Exp}} C_e^{\text{Exp}} - \sum_{e=1}^{E} E_e^{\text{Imp}} C_e^{\text{Imp}} - \sum_{i=1}^{I} F_i^{\text{BIO}} C_i^{\text{BIO}}\right)\right] \tag{5.15}$$

subject to

$$\sum_{s} \alpha_s = 1 \tag{5.16}$$

With the inclusion of α_s in Equation 5.15, the economic feasibility of the BTS in all s potential scenarios is assessed. Each fraction of occurrence represents the time fraction of which a scenario occurs. This time fraction is calculated by dividing the duration of a scenario s with the total duration (time horizon) considered. Thus, the sum of these fractions must equal to one as shown in Equation 5.16. Note that the duration of scenario s is dependent on the raw material supply and energy demand profiles. For instance, if the raw material supply and energy demand profiles are uniform in terms of duration (e.g. hourly, weekly, monthly basis, etc.), the time fraction of occurrence for scenario s can be obtained as shown in Figure 5.2a. As for dynamic profiles where supply and demand profiles are not uniform, the duration of scenario s can be broken down to smaller intervals and obtained as shown in Figure 5.2b. With this consideration, optimization constraints between scenarios can be established. Such feature is useful when dealing with seasonal variations. In practice, these fractions can be estimated subjectively based on the historical and projected information in variations of raw material supply, energy demand, product prices, etc.

On the other hand, CAP is determined based on the selected technologies j and j' as well as their corresponding design capacities. As shown in Equations 5.17 and 5.18, the design capacities are determined based on the maximum operating capacity among all s scenarios:

$$\sum_{n=1}^{N} z_{jn} F_{jn}^{\text{Design}} \geq \left(\sum_{p=1}^{P} F_{jp}^{\text{I}} \right)_s \quad \forall j \forall s \tag{5.17}$$

$$\sum_{n'=1}^{N'} z_{j'n'} F_{j'n'}^{\text{Design}} \geq \left(\sum_{p'=1}^{P'} F_{j'p'}^{\text{II}} \right)_s \quad \forall j v \forall s \tag{5.18}$$

where F_{jn}^{Design} and $F_{j'n'}^{\text{Design}}$ represent the design capacities available for purchase for technologies j and j', respectively. Besides, z_{jn} and $z_{j'n'}$ are positive integers which represent the number of units of design capacity n and n' selected. Note that design capacities F_{jn}^{Design} and $F_{j'n'}^{\text{Design}}$ can be revised according to current market availability in order to produce an up-to-date economic analysis. With these equations, chances of selecting design capacities that will result in an under-designed system will be eliminated. Once the technologies are selected based on their appropriate design capacities, the total capital cost for technologies j and j' is determined via Equation 5.19:

$$\text{CAP} = \sum_{j=1}^{J} \sum_{n=1}^{N} z_{jn} C_{jn} + \sum_{j'=1}^{J'} \sum_{n'=1}^{N'} z_{j'n'} C_{j'n'} \tag{5.19}$$

where C_{jn} and $C_{j'n'}$ are capital costs of the design capacities n and n' for levels j and j', respectively.

Lastly, CRF is used to annualize capital costs by converting its present value into a stream of equal annual payments over a specified operation lifespan, $t_k^{\max}$, and discount rate, r. The CRF can be determined via Equation 5.20:

$$\text{CRF} = \frac{r(1+r)^{t_k^{\max}}}{(1+r)^{t_k^{\max}} - 1} \quad k \in j, j' \tag{5.20}$$

To illustrate this approach, a case study is presented based on lignocellulosic biomass generated from a palm oil mill (POM). In this case study, the optimal BTS configuration is synthesized and analysed based on variations in biomass supply and energy demand.

5.4 Case Study

Malaysian palm oil industry generates high amounts of palm-based biomasses (e.g. empty fruit bunches (EFB), palm kernel shells (PKS), palm mesocarp fibre (PMF) and palm oil mill effluent (POME)) as by-products or waste (Ng & Ng, 2013). Note that the palm-based biomasses contain useful amount of energy which can be recovered to meet energy demands in the industry and generate additional power for exporting to the national grid. To recover this potential energy source, a palm BTS can be used in POMs. Implementing such systems would provide the industry an opportunity to turn waste products into valuable renewable energy and reduce the importation of energy from external providers (e.g. grid) (Perunding, 2014). However, due to the seasonal operations in the industry, palm-based biomass supply and energy demand tend to vary between each agricultural seasons. Such long-term variations should be taken into account as it would significantly affect the economic characteristics of the BTS. In this respect, this case study presents the synthesis and design of a palm BTS under seasonal variations in biomass supply and energy demand.

In this case study, it is assumed that an investor is interested to implement a new palm BTS which supplies utilities such as power, low-pressure steam (LPS), cooling water and chilled water to an existing POM (as shown in Figure 5.3). The raw material (palm-based biomass) for the BTS is purchased from the POM at the costs shown in Table 5.A.1. Meanwhile, the utilities produced by the BTS would be supplied to the POM at the costs given in Table 5.A.2.

Apart from that, this case study assumes that the existing POM has the similar behaviour of a typical POM presented by Kasivisvanathan et al. (2012). As illustrated in Figure 5.4, the input–output model shows the overall input and output flow rates of the POM normalized to the rate of crude palm oil (CPO) produced. As shown, the amount of biomass generated by the POM is dependent on its CPO production.

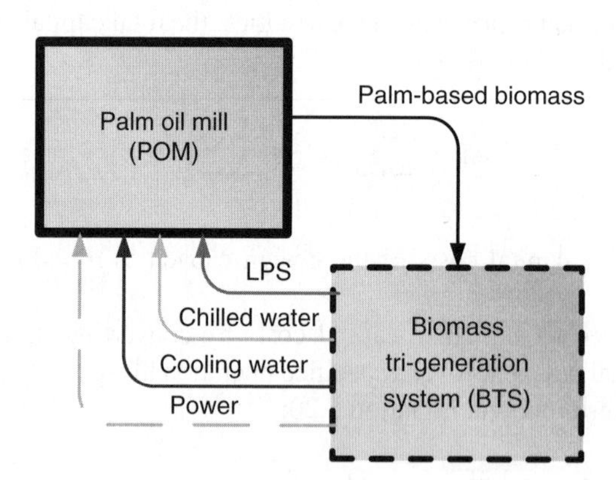

Figure 5.3 Block diagram for case study. Reproduced with permission from reference Andiappan et al. (2014), © 2014, American Chemical Society

Figure 5.4 *Normalized production for POM. Reproduced from Kasivisvanathan et al. (2012)*

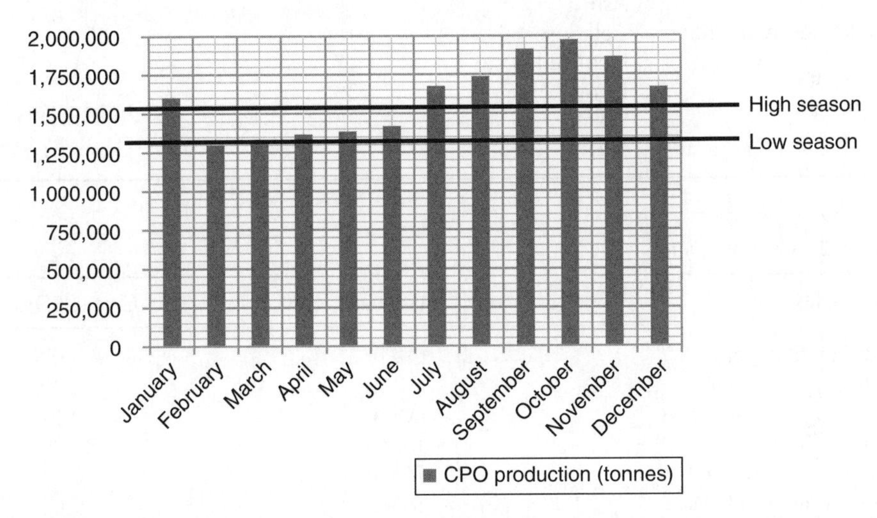

Figure 5.5 *Total CPO production in Malaysia for 2013 (Board, 2015). Reproduced from http://bepi.mpob.gov.my/index.php/statistics/production/125-production-2014/657-annual-forecast-production-of-crude-palm-oil-2013-2014.html*

In the Malaysian palm oil industry, total CPO productions tend to vary between agricultural seasons, as shown in Figure 5.5. Based on national statistics (Malaysian Palm Oil Board, 2015) (Figure 5.5), the production of CPO can be divided into three seasons (low season, midseason and high season). In this case study, total CPO productions less than 1,600,000 tonnes/year are taken as the low season. Meanwhile, total CPO productions falling in between 1,600,000 and 1,800,000 tonnes/year are taken as the midseason; total CPO productions higher than 1,800,000 tonnes/year are then taken as high season. Due to the seasonal operation, it is foreseen that the demand from the POM would vary according to its seasonal period in a calendar year. It is imperative that BTS investor give special attention to the variation as energy demand patterns significantly affect the economic characteristics of the BTS. As such, the fraction of occurrence for each season is taken into consideration

Table 5.1 Lignocellulosic composition of palm-based biomass and price for Cases 1 and 2

Raw materials i	Component q	Composition x_{iq}	Cost, C_i^{BIO} (USD/tonne)
Empty fruit bunches (EFBs)	Cellulose	13	6
	Hemicellulose	12	
	Lignin	8	
	Water	65	
	Ash	2	
Palm mesocarp fibre (PMF)	Cellulose	21	22
	Hemicellulose	19	
	Lignin	15	
	Water	40	
	Ash	5	
Palm kernel shell (PKS)	Cellulose	16	50
	Hemicellulose	17	
	Lignin	39	
	Water	23	
	Ash	4	

Table 5.2 Cost of utilities

Utility sales	Base unit	Cost (USD/unit)
Electricity to the grid, C_e^{Exp}	kW h	0.095
Electricity to POM, C_e^{Exp}	kW h	0.09
Electricity from the grid, C_e^{Imp}	kW h	0.12
LPS to POM, $C_{p'}$	Tonne	25.00
Cooling water to POM, $C_{p'}$	Tonne	0.09
Chilled water to POM, $C_{p'}$	Tonne	0.62

Reproduced with permission from reference Andiappan et al. (2014), © 2014, American Chemical Society.

to synthesize a BTS. The fraction of occurrence can be estimated based on the number of months in a calendar year in which the CPO production falls in the low, mid- and high season (as shown in Table 5.3). It is noted that the seasons and corresponding fraction are calculated based on the actual CPO production of the Malaysian palm oil industry for 2013 (as shown in Figure 5.5). It is worth mentioning that more realistic values may be calculated based on the long-term average. The energy demand for each season is then estimated using the input–output model as shown in Figure 5.4. For the low, mid- and high seasons, the CPO production from the POM studied in this case study is assumed to be 11,000 kg/h, 13,000 kg/h and 15,000 kg/h, respectively. Based on these CPO flow rates, the seasonal utility demand and palm-based biomass generated from the POM are summarized in Tables 5.2 and 5.3. As shown in Tables 5.4 and 5.5, each season has uniform biomass supply and energy demand profiles. In this respect, each season of biomass supply and energy demand is represented by a fraction of occurrence.

Table 5.3 Fraction of occurrence for low, mid- and high seasons

Season	Occurrence	Fraction of occurrence
Low	Less than 1,600,000 tonnes	$\alpha_L = 0.417$
Mid	Between 1,600,000 and 1,800,000 tonnes	$\alpha_M = 0.333$
High	More than 1,800,000 tonnes	$\alpha_H = 0.250$

Reproduced with permission from reference Andiappan et al. (2014), © 2014, American Chemical Society.

Table 5.4 Utility demand of a typical POM

Utility	Unit	Low season	Midseason	High season
Electricity	kW	632.5	747.5	862.5
LPS	kg/h	20,625	24,375	28,125
Cooling water	kg/h	11,000	13,000	15,000
Chilled water[a]	kg/h	1,000	1,000	1,000

Reproduced from reference Kasivisvanathan et al. (2012).
[a] Constant throughout the year as it is for office air conditioning purposes.

Table 5.5 Palm-based biomass availability based on CPO production

Available raw material supply	Low season (kg/h)	Midseason (kg/h)	High season (kg/h)
EFB	12,375	14,625	16,875
PMF	6,875	8,125	9,375
PKS	3,437.5	4,062.5	4,687.5
POME	40,700	48,100	55,500

A superstructure is developed to include all possible technologies and configurations (as shown in Figure 5.6). Note that although only one unit is shown in the superstructure, it is possible to have several pieces in the case where units with smaller design capacities are selected (as described in Equations 5.17 and 5.18). All these alternatives are then mathematically modelled (according to Equations 5.1–5.14 and 5.16–5.20), allowing quantitative analysis and optimization. A list of available design capacities considered in this case study is provided in the Supporting Information Sheet. For each respective design capacity, the capital costs are estimated using correlations presented by Peters et al. (2002). In addition, detailed conversion data, mass and energy balance calculations for this case study are also provided in the appendix section.

To demonstrate the proposed work, three cases are taken into consideration. In the first case, the optimization objective is to synthesize a BTS with the maximum economic performance. As for the second case, the optimization objective is to synthesize a BTS with the maximum economic performance subject to a maximum capital investment of USD 3,703,704 (RM 12,000,000, whereby 1 USD = RM 3.24). Similar to Case 1, the optimization objective of Case 3 is to synthesize a BTS configuration with maximum economic performance. However, this case considers biomass of different quality as

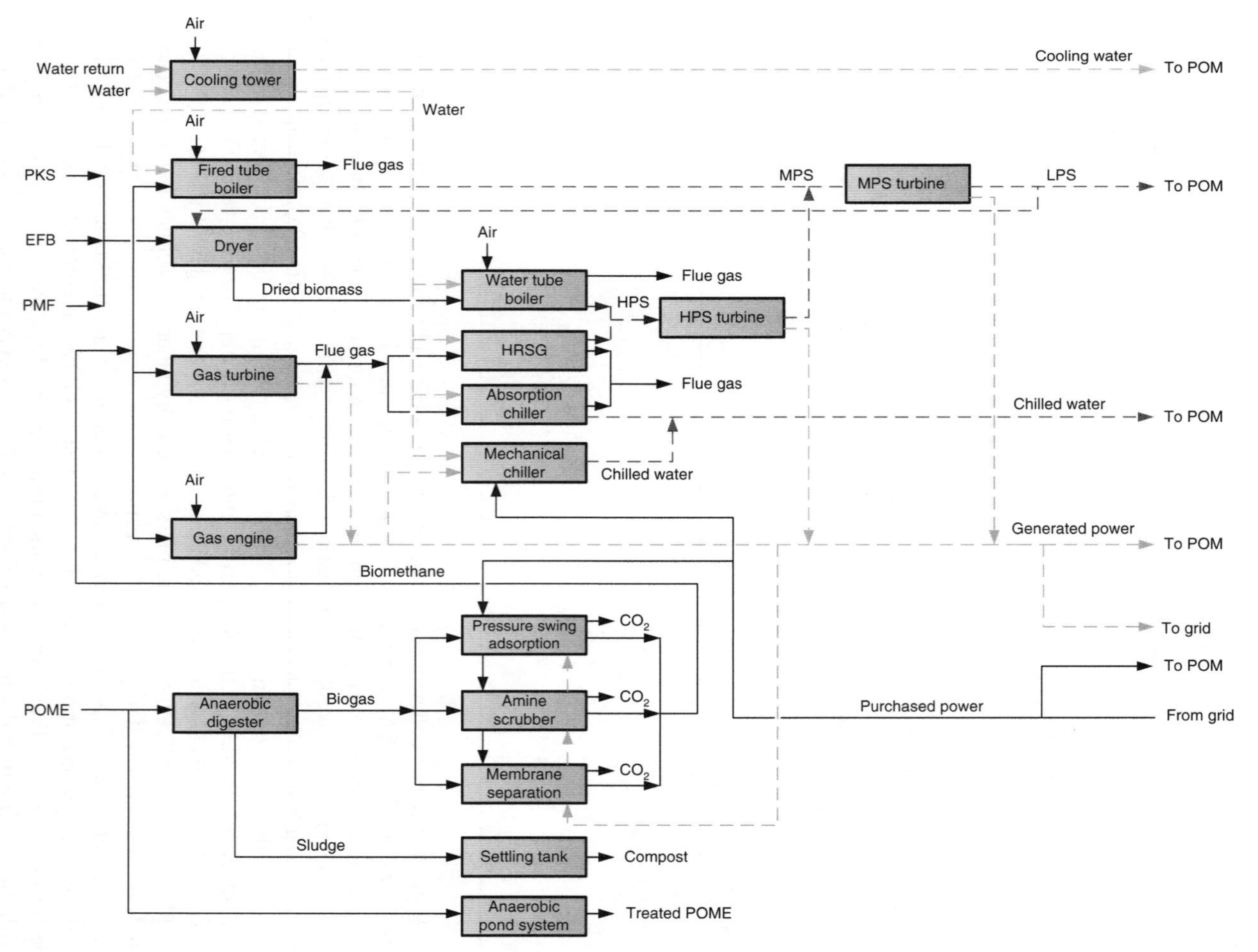

Figure 5.6 *Superstructure for palm BTS*

compared to Case 1 (shown in Table 5.6). Table 5.7 shows the economic parameters that are considered in this case study. The MILP model for all cases are solved via LINGO v13 (branch-and-bound solver) (LINDO Systems Inc., 2011) in an average 0.01s of CPU time (Dell Vostro 3400 with Intel Core i5 (2.40 GHz) and 4 GB DDR3 RAM).

Case 1

To synthesize a BTS with maximum economic performance while meeting the POM demand shown in Table 5.4, the developed model is optimized based on Equation 5.21 subject to Equations 5.1–5.14 and 5.16–5.20. The model formulated for this case consists of 357 continuous variables, 40 integer variables and 348 constraints:

$$\text{Maximise EP} \tag{5.21}$$

The optimized results for this case are summarized in Table 5.8. Note that the optimal configuration of BTS is shown in Figure 5.7. As shown in Table 5.6, the GP is found as USD 2,736,173 (RM 8,865,200) per year over its operational lifespan of 15 years. In addition, the total CAP of the system is found as USD 5,884,475 (RM 19,065,700) with the

Table 5.6 *Lignocellulosic composition of palm-based biomass and price for Case 3*

Raw materials i	Component q	Composition x_{iq}	Cost, C_i^{BIO} (USD/tonne)
Empty fruit bunches (EFBs)	Cellulose	10	6
	Hemicellulose	11	
	Lignin	5	
	Water	72	
	Ash	2	
Palm mesocarp fibre (PMF)	Cellulose	21	22
	Hemicellulose	19	
	Lignin	15	
	Water	40	
	Ash	5	
Palm kernel shell (PKS)	Cellulose	17	50
	Hemicellulose	18	
	Lignin	41	
	Water	21	
	Ash	3	

Table 5.7 *Economic parameters for case study*

Operational hours, AOT	5000/year
Operation lifespan, t_k^{max}	15 years
CRF	0.13/year
Discount rate, r	10%
Currency conversion rate	1 USD (RM 3.24)

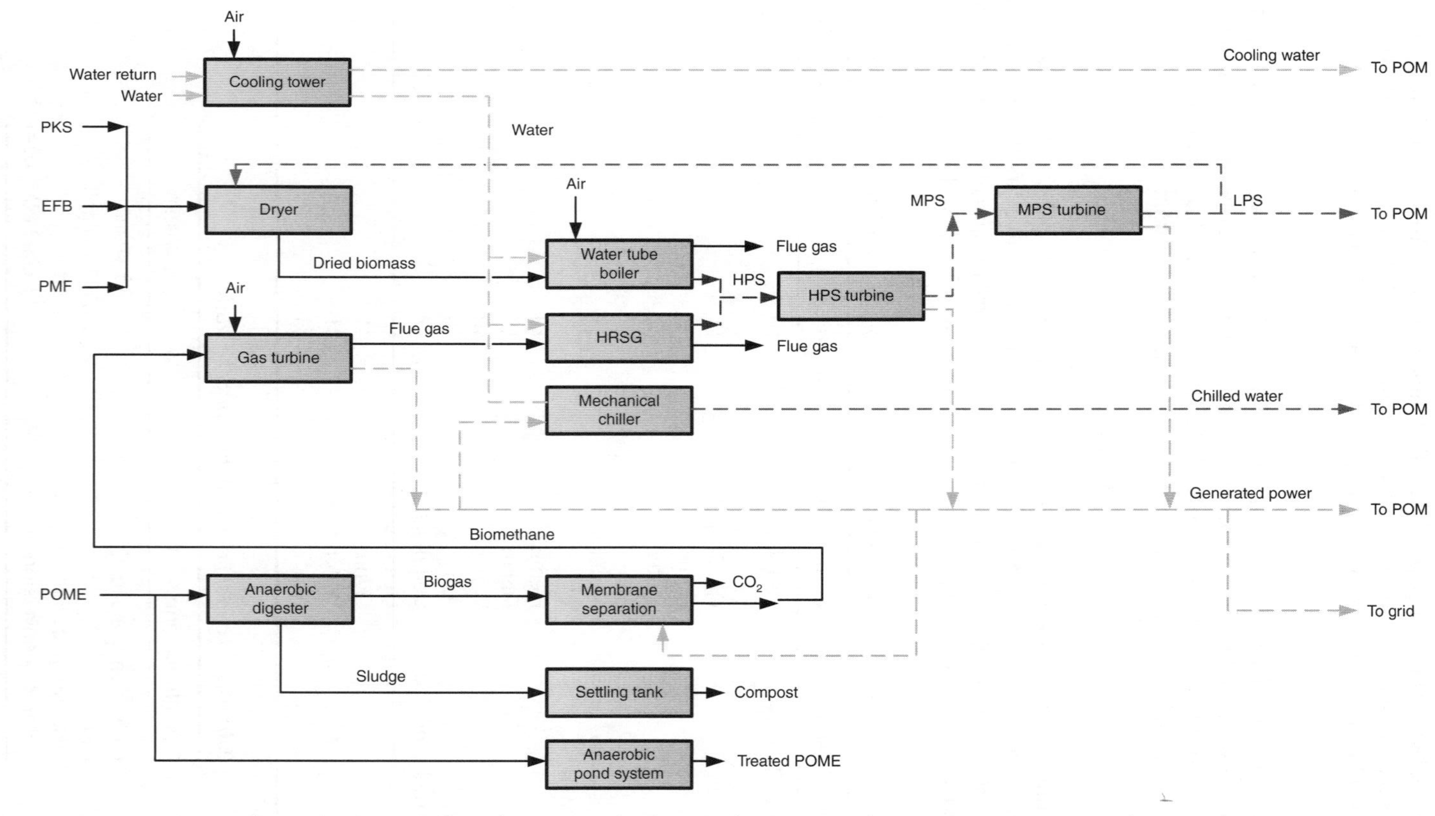

Figure 5.7 Optimal configuration of palm BTS for Case 1 (of maximum economic performance)

Table 5.8 *Optimization output for Cases 1–3*

Model output	Case 1	Case 2	Case 3
Average power generated (kW)	2,503.5	1,452.0	2,438.6
Average power to grid (kW)	1,694.4	723.7	1,629.5
Average power to palm oil mill (kW)	728.3	728.3	728.3
Average power from grid (kW)	0	0	0
Gross profit (USD/year)	2,736,173	2,152,335	2,324,051
Capital cost (USD)	5,884,475	3,698,886	5,672,645
Payback period (year)	2.15	1.72	2.44

Table 5.9 *Chosen technologies for Cases 1–3*

Technology	Design capacity	Case 1	Case 2	Case 3
WT boiler	40 kg/s	1	1	1
HPS turbine	1,000 kW	1	1	1
	500 kW	1	1	0
	250 kW	0	0	1
MPS turbine	500 kW	1	1	1
Gas turbine	1,000 kW	1	0	1
	250 kW	1	0	1
Dryer	10,000 kg/h	2	2	1
Membrane separation	180 kg/h	1	0	1
	70 kg/h	2	0	2
Heat recovery steam generator	12 kg/s	1	0	1
Mechanical chiller	250 kW	1	1	1
Cooling tower	25 kg/s	1	1	1
Anaerobic digester	Based on available flow	1	0	1
Anaerobic pond	Based on available flow	0	1	0
Total units		15	8	14

corresponding design capacities for the technologies shown in Table 5.9. Note also that the simple payback period of the synthesized BTS is determined as 2.15 years.

Case 2

In this case, the investor would unlikely implement a BTS if the required capital cost is more than USD 3,703,704 (RM 12,000,000). Thus, EP is maximized (Eq. 5.21) subject to additional constraints as shown in Equation 5.22 along with other constraints from Case 1. The model formulated for this case consists of 357 continuous variables, 40 integer variables and 349 constraints:

$$\text{CAP} \leq \text{USD } 3{,}703{,}704 \tag{5.22}$$

The optimal configuration synthesized for Case 2 is shown in Figure 5.8. Based on the optimization results in Table 5.8, the GP is found as USD 2,152,335 (RM 6,973,566) per year, while the CAP is found to be USD 3,698,886 (RM 11,984,390). Table 5.9 shows the corresponding design capacities of the selected technologies. Meanwhile, the payback period is determined as 1.72 years.

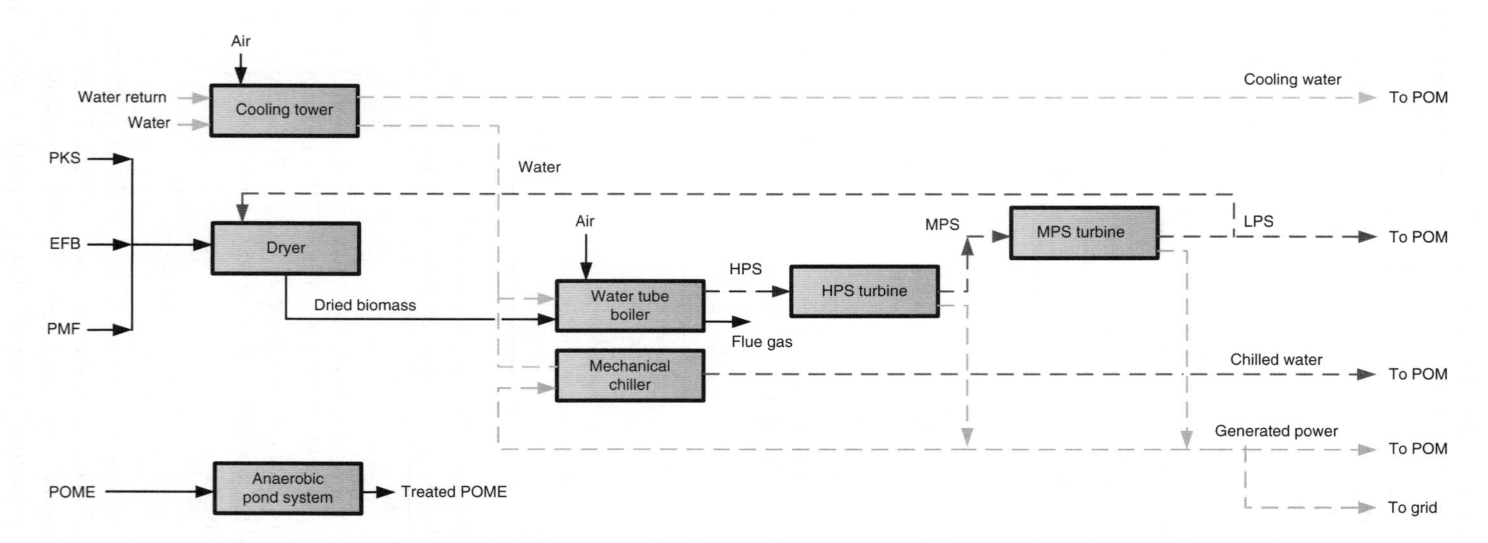

Figure 5.8 *Optimal configuration of palm BTS for Case 2 (with limit on capital investment)*

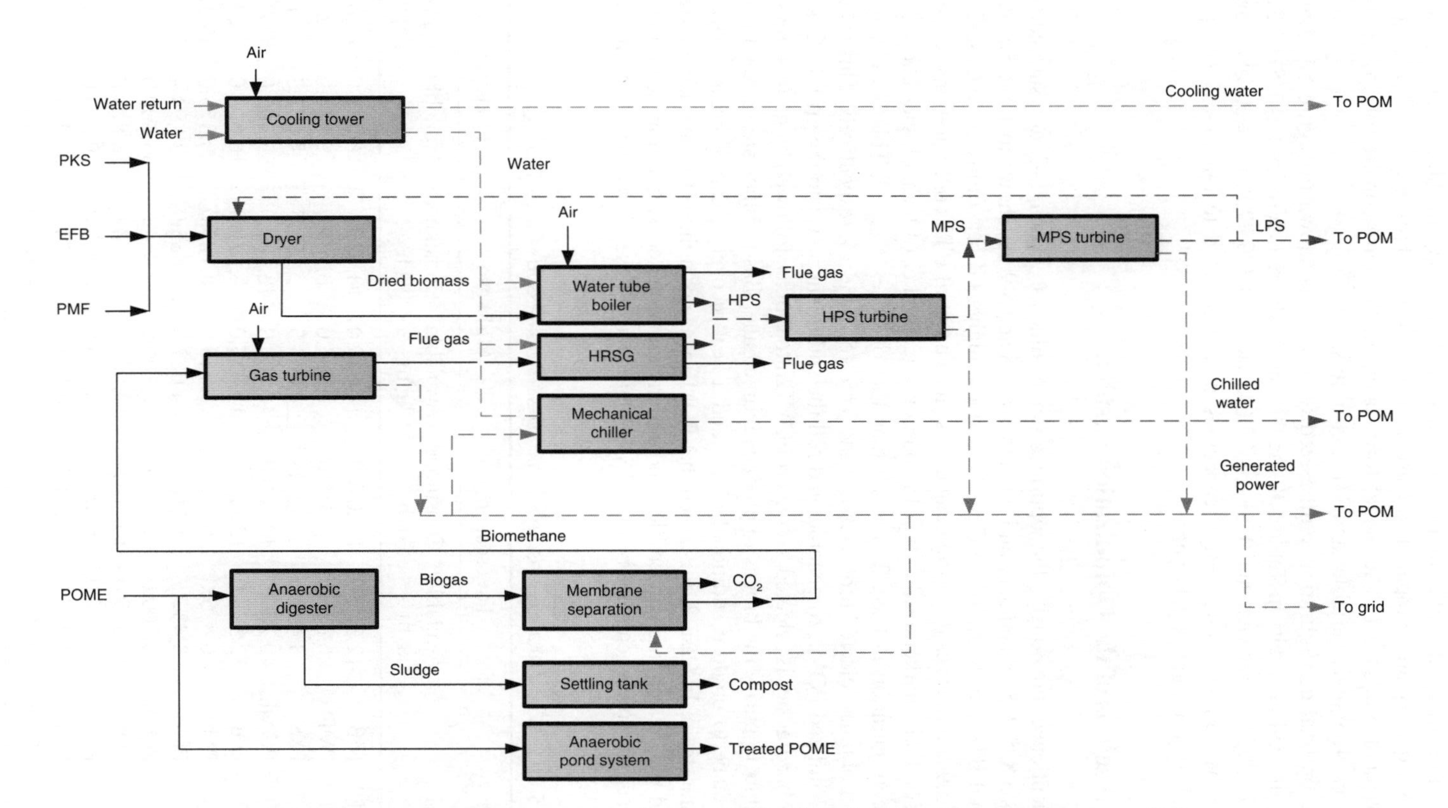

Figure 5.9 *Optimal configuration of palm BTS for Case 3 (with different biomass quality)*

Case 3

In this case, different quality biomass is considered. As shown in Table 5.6, biomasses EFB and PKS quality are compared to Case 1. Thus, EP is maximized (Eq. 5.21) subject to constraints from Case 1. The model formulated for this case consists of 357 continuous variables, 40 integer variables and 348 constraints.

The optimal configuration synthesized for Case 3 is shown in Figure 5.9. Based on the optimization results in Table 5.8, the GP and CAP are determined as USD 2,324,051 (RM 7,529,925) per year and USD 5,672,645 (RM 18,379,370), respectively. Table 5.9 shows the corresponding design capacities of the selected technologies. The payback period is determined as 2.44 year.

5.5 Analysis of the Optimization Results

As mentioned previously, the optimization results for Cases 1–3 are summarized in Tables 5.8, 5.9, 5.10 and 5.11 and Figures 5.7, 5.8 and 5.9. As shown in Table 5.8, the synthesized BTS configuration in Case 1 yielded a higher GP as compared to Case 2 and 3. Table 5.9 summarizes the design capacities chosen for all BTS configurations (Cases 1–3) based on their market availability. In addition, the number of technologies chosen for the BTS configuration in Case 2 is lesser than that of Cases 1 and 3. This is evident as the number of units selected for Cases 1, 2 and 3 are 15, 7 and 14, respectively. This is because Case 1 utilized POME to generate and sell the additional power to the grid. In Case 1, an anaerobic digester is selected to convert POME into biogas, followed by membrane separation units to purify the biogas to produce biomethane. Gas turbines are selected to utilize biomethane to produce additional power and a heat recovery steam generator (HRSG) to generate steam. In Case 2, it is noted that POME was not utilized for producing biogas but instead was treated in an anaerobic pond system due to the constraint imposed on the CAP. Thus, surplus power generated by the BTS configuration in Case 2 is much lower than in Case 1.

Table 5.10 *Available and consumed palm-based biomass for Cases 1–3*

Biomass		Low season		Midseason		High season	
		Available (kg/h)	Consumed (kg/h)	Available (kg/h)	Consumed (kg/h)	Available (kg/h)	Consumed (kg/h)
Case 1	EFB	12,375.0	12,375.0	14,625.0	14,625.0	16,875.0	16,875.0
	PMF	6,875.0	6,875.0	8,125.0	8,125.0	9,375.0	9375.0
	PKS	3,437.5	132.4	4,062.5	156.5	4,687.5	180.6
	POME	40,700.0	40,700.0	48,100.0	48,100.0	55,500.0	55,500.0
Case 2	EFB	12,375.0	12,375.0	14,625.0	14,625.0	16,875.0	16,875.0
	PMF	6,875.0	6,875.0	8,125.0	8,125.0	9,375.0	9,375.0
	PKS	3,437.5	563.4	4,062.5	665.9	32,93.4	768.3
	POME	40,700.0	0.0	48,100.0	0.0	55,500.0	0.0
Case 3	EFB	12,375.0	10,781.3	14,625.0	10,074.6	16,875.0	8,574.1
	PMF	6,875.0	6,875.0	8,125.0	8,125.0	9,375.0	9,375.0
	PKS	3,437.5	1,756.7	40,62.5	2,275.4	3,293.4	2,853.3
	POME	40,700.0	40,700.0	48,100.0	48,100.0	55,500.0	55,500.0

Table 5.11 Power distribution for Cases 1–3

Season	Power distribution	Case 1 (kW)	Case 2 (kW)	Case 3 (kW)
Low	To mill	632.5	632.5	632.5
	Internally	70.2	0.0	70.2
	To grid	1471.5	628.5	1474.4
Mid	To mill	747.5	747.5	747.5
	Internally	83.0	0.0	83.0
	To grid	1739.0	742.8	1667.3
High	To mill	862.5	862.5	862.5
	Internally	95.7	0.0	95.7
	To grid	2006.6	857.0	1837.8

As for Case 3, the amount of utilized is lesser biomass than in Case 1. As such, only one dryer was chosen in Case 3 instead of two dryers as shown in Case 1.

On the other hand, Table 5.10 shows that the amount of EFB, PMF and PKS biomasses utilized in Case 2 are higher than in Cases 1 and 3. Since POME is utilized to generate power in Case 1, less amount of EFB, PMF and PKS would be required to achieve maximum GP. In contrast, Case 2 utilized more of EFB, PMF and PKS biomasses to compensate for not generating power from POME. Besides that, Case 3 utilized much higher amount of PKS biomass than in Cases 1 and 2 due to the low-quality EFB biomass considered. Despite such difference in all three cases, it is noted the available palm-based biomass supply in each season was sufficient for the BTS to be energy self-sustained and to meet the POM demands (shown in Table 5.11).

5.6 Conclusion and Future Work

BTS produce heat, power and cooling simultaneously from a single fuel source. Such systems offer significant benefits to the industry. These benefits include reduction in operating costs, waste generation, dependency on grid power, etc. Apart from that, the use of tri-generation systems improves utilization of the overall resources, reduces environmental emissions and improves reliability of the energy system. Furthermore, such system increases sustainability and overall energy security. However, like most of the renewable energy resources, availability of biomass is seasonal in nature. To address this issue, a systematic multi-period optimization approach was presented to synthesize a BTS configuration with maximum economic performance considering seasonal variations in biomass supply and energy demand. Besides determining the optimal BTS configuration, selection of design capacities based on the available size in the market is also performed simultaneously. To illustrate the proposed approach, a palm-based biomass case study was solved. In the case study, three different cases were considered. In Case 1, the BTS configuration is synthesized based on three seasons with different biomass supply and energy demand from POM. In Case 2, the BTS configuration is synthesized with a constraint imposed on the capital investment. As for Case 3, biomass of lower quality (than in Case 1) was considered. Based on the results obtained for all three cases, it is noted that the available biomass is sufficient for use in the BTS. Future research work is directed towards incorporating reliability of equipment to consider short-term uncertainties such as equipment failure in the synthesis of an optimal BTS.

Appendix A

Table 5.A.1 *Capital costs for each technology based on available design capacity*

Technology j and j'	Design capacity $\left(F_{jn}^{\text{Design}} \text{ and } F_{jn}^{\text{Design}}\right)$	Capital costs[a] (C_{jn} and $C_{j'n'}$ (USD))
Water tube boiler	84 kg/s	3,841,076
	70 kg/s	3,353,655
	40 kg/s	2,211,185
Fired tube boiler	12 kg/s	902,498
	10 kg/s	787,974
	6 kg/s	538,753
High pressure	1000 kW	165,401
Steam turbine	500 kW	122,380
	250 kW	90,549
Medium pressure	500 kW	122,380
Steam turbine	250 kW	90,549
	200 kW	82,180
Gas turbine	1000 kW	274,778
	500 kW	179,647
	250 kW	117,451
Internal combustion engine	315 kW	77,077
	400 kW	93,828
Dryer	40,000 kg/h	720,000
	25,000 kg/h	450,000
	10,000 kg/h	180,000
Amine scrubbing	700 kg/h	3,166,744
	500 kg/h	2,261,960
	200 kg/h	904,784
Pressure swing adsorption	500 kg/h	1,546,605
	200 kg/h	618,642
	100 kg/h	309,321
Membrane separation	180 kg/h	501,111
	100 kg/h	278,395
	70 kg/h	194,877
Heat recovery steam	12 kg/s	902,498
generator	40 kg/s	2,211,185
Absorption chiller	250 kW	42,518
	300 kW	48,926
	350 kW	55,092
Mechanical chiller	250 kW	42,518
	300 kW	48,926
	350 kW	55,092
Cooling tower	50 kg/s	25,867
	30 kg/s	17,762
	25 kg/s	15,532
Anaerobic digester	Based on available flow	11.9/(kg/h) POME

[a] Capital costs for each technology are estimated based on design capacity using correlations presented by Peters et al. (2002).

Table 5.A.2 *Conversions for technologies considered*

Technology	Conversion	Inlet	Product/by-product
Water tube boiler	0.5556 kg H_2O/kg cellulose[a] 0.5556 kg H_2O/kg hemicellulose[a] 1.674 kg H_2O/kg lignin[a] 1.63 kg CO_2/kg cellulose[a] 1.63 kg CO_2/kg hemicellulose[a] 3.9 kg CO_2/kg lignin[a] Efficiency = 0.55 1.19 kg O_2/kg cellulose[a] 1.19 kg O_2/kg hemicellulose[a] 4.13 kg O_2/kg lignin[a] Steam enthalpy (30 bar, 350°C) (Rogers & Mayhew, 1964) = 2858 kJ/kg Water enthalpy (1.01 bar, 30°C) (Rogers & Mayhew, 1964) = 104.92 kJ/kg Heating value (cellulose) = 17,000 kJ/kg cellulose(BISYPLAN, 2014) Heating value (hemicellulose) = 16,000 kJ/kg hemicellulose (Murphy & Masters, 1978) Heating value (lignin) = 25,000 kJ/kg lignin (BISYPLAN, 2014)	Water (30°C) Dried palm biomass Air	HPS (30 bar, 350°C)
Fired tube boiler	2.25 kg H_2O/kg biomethane[a] 2.75 kg CO_2/kg biomethane[a] 4 kg O_2/kg biomethane[a] Efficiency = 0.60 Steam enthalpy (20 bar, 284°C) (Rogers & Mayhew, 1964) = 2736 kJ/kg Water enthalpy (1.01 bar, 30°C) (Rogers & Mayhew, 1964) = 104.92 kJ/kg Heating value (biomethane) (The Engineering ToolBox, 2014) = 22,000 kJ/kg biomethane	Water (30°C) Biomethane Air	MPS (20 bar, 284°C)
HPS turbine	Steam enthalpy (30 bar, 350°C) (Rogers & Mayhew, 1964) = 2858 kJ/kg Steam enthalpy (20 bar, 284°C) (Rogers & Mayhew, 1964)= 2736 kJ/kg Efficiency = 0.98	HPS (30 bar, 350°C)	MPS (20 bar, 284°C) Electricity

(continued overleaf)

Table 5.A.2 (continued)

Technology	Conversion	Inlet	Product/by-product
MPS turbine	Steam enthalpy (20 bar, 284°C) (Rogers & Mayhew, 1964) = 2736 kJ/kg Steam enthalpy (3 bar, 134°C) (Rogers & Mayhew, 1964) = 2707 kJ/kg Efficiency = 0.98	MPS (20 bar)	LPS (3 bar, 134°C) Electricity
Gas turbine	2.25 kg H_2O/kg biomethane[a] 2.75 kg CO_2/kg biomethane[a] 4 kg O_2/kg biomethane[a] Flue gas enthalpy (16 bar)[b] = –1142 kJ/kg Flue gas enthalpy (1.01 bar)[b] = 41.09 kJ/kg Efficiency = 0.60	Biomethane Air	Flue gas Electricity
Internal combustion engine	2.25 kg H_2O/kg biomethane[a] 2.75 kg CO_2/kg biomethane[a] 4 kg O_2/kg biomethane[a] Heating value (biomethane) (The Engineering ToolBox, 2014) = 22,000 kJ/kg biomethane Efficiency = 0.20 Operating conditions: 16 bar, 600°C	Biomethane Air	Flue gas Electricity
Dryer	Outlet moisture content (wt%) = 10	Wet palm biomass	Dried palm biomass
Amine scrubbing	0.9994 kg biomethane/kg raw biomethane (Bauer et al., 2013) 0.998 kg CO_2/kg raw CO_2 (Bauer et al., 2013) Power required = 0.14 kW h/kg Biogas (Bauer et al., 2013)	Biomethane (65 wt%) CO_2 (35 wt%)	Biomethane (97 wt%) CO_2 (98 wt%)
Pressure swing adsorption	0.98 kg CO_2/kg raw biomethane (Bauer et al., 2013) 0.98 kg CO_2/kg raw CO_2 (Bauer et al., 2013) Power required = 0.3 kW h/kg biogas (Bauer et al., 2013)	Biomethane (65 wt%) CO_2 (35 wt%)	Biomethane (97 wt%) CO_2 (98 wt%)
Membrane separation	0.99 kg biomethane/kg raw biomethane (Bauer et al., 2013) 0.98 kg CO_2/kg raw CO_2 (Bauer et al., 2013) Power required = 0.3 kW h/kg biogas (Bauer et al., 2013)	Biomethane (65 wt%) CO_2 (35 wt%)	Biomethane (97 wt%) CO_2 (98 wt%)

Heat recovery steam generator	Flue gas enthalpy = 2076.29 kJ/kg	Water (30°C)	HPS (30 bar, 350°C)
	Efficiency = 0.35	Flue gas	
	Steam enthalpy (30 bar, 350°C) (Rogers & Mayhew, 1964) = 2858 kJ/kg		
	Water enthalpy (1.01 bar, 30°C) (Rogers & Mayhew, 1964) = 104.92 kJ/kg		
Absorption chiller	COP (Chartered Institution of Building Services Engineers, 2012) = 0.7 output kW/input kW	Cool water (30°C)	Chilled water (7°C)
	Flue gas enthalpy[b] = 799.12 kJ/kg	Flue gas (600°C)	Flue gas (30°C)
	Cooling water (Air-Conditioning Heating & Refrigeration Institute, 2000) = 7.247 kg/kW h		
	Chilled water (Air-Conditioning Heating & Refrigeration Institute, 2000) = 0.00722 kg/kW h		
Mechanical chiller	COP (Hondeman, 2000) = 6.1 output kW/input kW	Cool water (30°C)	Chilled water (7°C)
	Cooling water (Air-Conditioning Heating & Refrigeration Institute, 2000) = 7.247 kg/kW h	Electricity	
	Chilled water (Air-Conditioning Heating & Refrigeration Institute, 2000) = 0.00722 kg/kW h		
Cooling tower	Air required = 20.1 kg/kW h	Hot water (40°C)	Cool water (30°C)
	Cooling load = 42 kJ/kg water	Air	
Anaerobic digester	Biomethane (Chin et al., 2013) = 0.184 kg/kg COD	POME	Biomethane (65 wt%)
	CO_2 (Chin et al., 2013) = 0.0992 kg/kg COD		CO_2 (35 wt%)
	COD (Poh et al., 2013) = 50,000 mg/L		
	POME density (Poh et al., 2013) = 1600 kg/m (Stojkov et al., 2011)		
	Sludge (Poh et al., 2013) = 0.06 kg/kg POME		

[a] Based on stoichiometric balance for complete combustion reaction.
[b] Based on Aspen HYSYS Simulation.

Nomenclature

Sets

i	Index for biomass
j	Index for technologies at level j
j'	Index for technologies at level j'
p	Index for primary products
p'	Index for final products
q	Index for component balance of biomass i
q'	Index for component balance of primary product p
n	Index for available design capacities for technology j
n'	Index for available design capacities for technology j'
s	Index for scenarios
e	Index for energy

Variables

F_i^{BIO}	Flow rate of biomass i in kg/h
f_{iq}^{BIO}	Component flow rate of biomass i in kg/h
F_{ij}^{I}	Flow rate of biomass i to technology j in kg/h
f_{qj}^{I}	Component flow rate of biomass i to technology j in kg/h
F_{jp}^{I}	Production rate of primary product p in kg/h at technology j
F_p	Total production rate of primary product p in kg/h at technology j
$F_{pj'}^{\text{II}}$	Flow rate of primary product p to technology j' in kg/h
$f_{q'j'}^{\text{II}}$	Component flow rate of product p to technology j' in kg/h
$F_{j'p'}^{\text{II}}$	Production rate of final product p' in kg/h at technology j'
$F_{p'}$	Total production rate of final product p' in kg/h at technology j'
E_e^{Gen}	Total energy generated by technology j and j' in kW h
E_e^{Con}	Total energy consumption for technology j and j' in kW h
E_e^{Imp}	Total external energy imported in kW h
E_e^{Exp}	Total excess energy exported in kW h
E_e^{Demand}	Total energy demand in kW h
EP	Total economic potential in USD per year (RM per year)
GP	Total gross profit in USD per year (RM per year)
CAP	Total capital cost in USD (RM)
PBP	Payback period (year)
$t^{\max}$	Operation lifespan (year)
r	Discount rate
z_{jn}	Number of units of design capacity n selected
$z_{j'n'}$	Number of units of design capacity n' selected

Parameters

X_{iq} Composition q of biomass i
$X_{pq'}$ Composition q' of product p
X^{I}_{qip} Component mass conversion of biomass i
$X^{\mathrm{II}}_{q'j'p'}$ Component mass conversion of primary product p
V^{I}_{qje} Component energy conversion at technology j
$V^{\mathrm{II}}_{q'j'e}$ Component energy conversion at technology j'
Y^{I}_{ejp} Component energy consumption conversion at technology j
$Y^{\mathrm{II}}_{ej'p'}$ Component energy consumption conversion at technology j'
AOT Annual operating time in h/year
C^{BIO}_{i} Cost of biomass i in USD/kg
$C_{p'}$ Revenue from primary product p' in USD/kg
C^{Imp}_{e} Purchase cost of importing energy in USD/kW h
C^{Exp}_{e} Selling cost of exporting excess energy in USD/kW h
C_{jn} Capital cost for technology j in USD
$C_{j'n'}$ Capital cost for technology j' in USD
α_s Fraction of occurrence for scenario s
CRF Capital recovery factor
F^{Design}_{jn} Available design capacity for technology j in kg/h
$F^{\mathrm{Design}}_{j'n'}$ Available design capacity for technology j' in kg/h

References

Aaltola, J. Simultaneous Synthesis of Flexible Heat Exchanger Network. *Appl. Therm. Eng.* 2002, **22**, 907.

Aguilar, O.; Perry, S. J.; Kim, J.-K.; Smith, R. Design and Optimization of Flexible Utility Systems Subject to Variable Conditions. Part 1: Modelling Framework. *Chem. Eng. Res. Des.* 2007a, **85**, 1136.

Aguilar, O.; Perry, S. J.; Kim, J.-K.; Smith, R. Design and Optimization of Flexible Utility Systems Subject to Variable Conditions. Part 2: Methodology and Applications. *Chem. Eng. Res. Des.* 2007b, **85**, 1149.

Air-Conditioning Heating & Refrigeration Institute. *Standard for Absorption Water Chilling and Water Heating Packages*; Arlington, VA, 2000.

Andiappan, V.; Ng, D. K. S.; Bandyopadhyay, S. Synthesis of Biomass-based Trigeneration Systems with Uncertainties. *Ind. Eng. Chem. Res.* 2014, **53**, 18016–28.

Arcuri, P.; Florio, G.; Fragiacomo, P. A Mixed Integer Programming Model for Optimal Design of Trigeneration in a Hospital Complex. *Energy* 2007, **32**, 1430.

Bauer, F.; Hulteberg, C.; Persson, T.; Tamm, D. *Biogas Upgrading – Review of Commercial Technologies*; 2013.

BISYPLAN. BISYPLAN Web-based handbook. Available at http://bisyplan.bioenarea.eu/html-files-en/04-01.html (accessed on 17 September 2014).

Bridgwater, A. V. Renewable Fuels and Chemicals by Thermal Processing of Biomass. *Chem. Eng. J.* 2003, **91**, 87.

Buoro, D.; Casisi, M.; Pinamonti, P.; Reini, M. Optimal Lay-Out and Operation of District Heating and Cooling Distributed Trigeneration Systems. In *ASME Turbo Expo 2010: Power for Land, Sea, and Air*; ASME, Glasgow; June 14–18, 2010; pp. 157–166.

Buoro, D.; Casisi, M.; Pinamonti, P.; Reini, M. Optimization of Distributed Trigeneration Systems Integrated with Heating and Cooling Micro-Grids. *Distrib. Gener. Altern. Energy J.* 2011, **26**, 7.

Buoro, D.; Casisi, M.; Pinamonti, P.; Reini, M. Optimal Synthesis and Operation of Advanced Energy Supply Systems for Standard and Domotic Home. *Energy Convers. Manag.* 2012, **60**, 96.

Buoro, D.; Casisi, M.; De Nardi, a.; Pinamonti, P.; Reini, M. Multicriteria Optimization of a Distributed Energy Supply System for an Industrial Area. *Energy* 2013, **58**, 128.

Cardona, E.; Piacentino, A.; Cardona, F. Energy Saving in Airports by Trigeneration. Part I: Assessing Economic and Technical Potential. *Appl. Therm. Eng.* 2006, **26**, 1427.

Carpaneto, E.; Chicco, G.; Mancarella, P.; Russo, A. Cogeneration Planning under Uncertainty. Part I: Multiple Time Frame Approach. *Appl. Energy* 2011a, **88**, 1059.

Carpaneto, E.; Chicco, G.; Mancarella, P.; Russo, A. Cogeneration Planning under Uncertainty. Part II: Decision Theory-Based Assessment of Planning Alternatives. *Appl. Energy* 2011b, **88**, 1075.

Carvalho, M.; Serra, L. M.; Lozano, M. A. Optimal Synthesis of Trigeneration Systems Subject to Environmental Constraints. *Energy* 2011, **36**, 3779.

Carvalho, M.; Lozano, M. A.; Serra, L. M. Multicriteria Synthesis of Trigeneration Systems Considering Economic and Environmental Aspects. *Appl. Energy* 2012a, **91**, 245.

Carvalho, M.; Lozano, M. A.; Serra, L. M.; Wohlgemuth, V. Modeling Simple Trigeneration Systems for the Distribution of Environmental Loads. *Environ. Model. Softw.* 2012b, **30**, 71.

Carvalho, M.; Lozano, M. A.; Ramos, J.; Serra, L. M. Synthesis of Trigeneration Systems: Sensitivity Analyses and Resilience. *ScientificWorld J.* 2013, **2013**, 604852.

Chartered Institution of Building Services Engineers. Absorption Cooling; England, 2012.

Chen, Y.; Ii, T. A. A.; Barton, P. I. Optimal Design and Operation of Flexible Energy Polygeneration Systems. *Ind. Eng. Chem. Res.* 2011, **50**, 4553.

Chicco, G.; Mancarella, P. Distributed Multi-Generation: A Comprehensive View. *Renew. Sustain. Energy Rev.* 2009, **13**, 535.

Chin, M. J.; Poh, P. E.; Tey, B. T.; Chan, E. S.; Chin, K. L. Biogas from Palm Oil Mill Effluent (POME): Opportunities and Challenges from Malaysia's Perspective. *Renew. Sustain. Energy Rev.* 2013, **26**, 717.

Cho, H.; Mago, P. J.; Luck, R.; Chamra, L. M. Evaluation of CCHP Systems Performance Based on Operational Cost, Primary Energy Consumption, and Carbon Dioxide Emission by Utilizing an Optimal Operation Scheme. *Appl. Energy* 2009, **86**, 2540.

Dimopoulos, G. G.; Kougioufas, A. V.; Frangopoulos, C. A. Synthesis, Design and Operation Optimization of a Marine Energy System. *Energy* 2008, **33**, 180.

Fumo, N.; Mago, P. J.; Chamra, L. M. Emission Operational Strategy for Combined Cooling, Heating, and Power Systems. *Appl. Energy* 2009, **86**, 2344.

Gamou, S.; Yokoyama, R.; Ito, K. Optimal Unit Sizing of Cogeneration Systems in Consideration of Uncertain Energy Demands as Continuous Random Variables. *Energy Convers. Manag.* 2002, **43**, 1349.

Hastings, D.; McManus, H. A Framework for Understanding Uncertainty and Its Mitigation and Exploitation in Complex Systems. In *Engineering Systems Symposium*; Massachusetts Institute of Technology, Cambridge, MA; March 29–31, 2004; pp. 1–19.

Herder, P. M.; Weijnen, M. P. C. A Concurrent Engineering Approach to Chemical Process Design. *Int. J. Prod. Econ.* 2000, **64**, 311.

Hondeman, H. Electrical Compression Cooling versus Absorption Cooling – A Comparison. *IEA Heat Pump Centre Newsletter* 2000, 23–25.

Iyer, R. R.; Grossmann, I. E. Synthesis and Operational Planning of Utility Systems for Multiperiod Operation. *Comput. Chem. Eng.* 1998, **22**, 979.

Kasivisvanathan, H.; Ng, R. T. L.; Tay, D. H. S.; Ng, D. K. S. Fuzzy Optimisation for Retrofitting a Palm Oil Mill into a Sustainable Palm Oil-Based Integrated Biorefinery. *Chem. Eng. J.* 2012, **200–202**, 694.

Kavvadias, K. C.; Tosios, a. P.; Maroulis, Z. B. Design of a Combined Heating, Cooling and Power System: Sizing, Operation Strategy Selection and Parametric Analysis. *Energy Convers. Manag.* 2010, **51**, 833.

Li, C. Z.; Shi, Y. M.; Huang, X. H. Sensitivity Analysis of Energy Demands on Performance of CCHP System. *Energy Convers. Manag.* 2008, **49**, 3491.

Li, C.-Z.; Shi, Y.-M.; Liu, S.; Zheng, Z.; Liu, Y. Uncertain Programming of Building Cooling Heating and Power (BCHP) System Based on Monte-Carlo Method. *Energy Build.* 2010, **42**, 1369.

LINDO Systems Inc. *LINDO User's Guide*; Chicago, IL, 2011.

Lozano, M. A.; Ramos, J. C.; Carvalho, M.; Serra, L. M. Structure Optimization of Energy Supply Systems in Tertiary Sector Buildings. *Energy Build.* 2009a, **41**, 1063.

Lozano, M. A.; Carvalho, M.; Serra, L. M. Operational Strategy and Marginal Costs in Simple Trigeneration Systems. *Energy* 2009b, **34**, 2001.

Lozano, M. A.; Ramos, J. C.; Serra, L. M. Cost Optimization of the Design of CHCP (combined Heat, Cooling and Power) Systems under Legal Constraints. *Energy* 2010, **35**, 794.

Lozano, M. A.; Carvalho, M.; Serra, L. M. Allocation of Economic Costs in Trigeneration Systems at Variable Load Conditions. *Energy Build.* 2011, **43**, 2869.

Malaysian Palm Oil Board. Annual and Forecast of Crude Palm Oil 2013 & 2014. Available at http://bepi.mpob.gov.my/index.php/statistics/production/125-production-2014/657-annual-forecast-production-of-crude-palm-oil-2013-2014.html (accessed on 4 September 2015).

Maréchal, F.; Kalitventzeff, B. Process Integration: Selection of the Optimal Utility System. *Comput. Chem. Eng.* 1998, **22**, S149.

Mehleri, E. D.; Sarimveis, H.; Markatos, N. C.; Papageorgiou, L. G. Optimal Design and Operation of Distributed Energy Systems: Application to Greek Residential Sector. *Renew. Energy* 2013, **51**, 331.

Mitra, S.; Sun, L.; Grossmann, I. E. Optimal Scheduling of Industrial Combined Heat and Power Plants under Time-Sensitive Electricity Prices. *Energy* 2013, **54**, 194.

Murphy, W. K.; Masters, K. R. Gross Heat of Combustion of Northern Red Oak (*Quercus Rubra*) Chemical Components. *Wood Sci.* 1978, **10**, 139.

Ng, R. T. L.; Ng, D. K. S. Applications of Process System Engineering in Palm-Based Biomass Processing Industry. *Curr. Opin. Chem. Eng.* 2013, **2**, 448.

Papaulias, S. A.; Grossmann, I. E. A Structural Optimisation Approach in Process Synthesis. I: Utility Systems. *Comput. Chem. Eng.* 1983, **7**, 695.

Majutek Perunding. The Trigeneration Advantage. Available at http://majutekperunding.tripod.com/cogenerationanddistrictcoolingsystemspecialist/id7.html (accessed 5 May 2014).

Peters, M.; Timmerhaus, K.; West, R. *Plant Design and Economics for Chemical Engineers*; New York: McGraw-Hill Science/Engineering/Math; 5 ed., 2002; p. 1008.

Rezvan, A. T.; Gharneh, N. S.; Gharehpetian, G. B. Optimization of Distributed Generation Capacities in Buildings under Uncertainty in Load Demand. *Energy Build.* 2013, **57**, 58.

Rogers, G. F. C.; Mayhew, Y. R. *Thermodynamic and Transport Properties of Fluids: SI Units*; 5th ed.; Cambridge, MA: Wiley-Blackwell, 1964.

Shah, N. Single- and Multisite Planning and Scheduling: Current Status and Future Challenges. *In AIChE Symposium Series: Proceedings of the Third International Conference of the Foundations of Computer-Aided Process Operations*; Snowbird, UT; July 5–10, 1998.

Stojkov, M.; Hnatko, E.; Kljajin, M.; Hornung, K. CHP and CCHP Systems Today. In *Development of Power Engineering in Croatia*; Osijek: Faculty of Electrical Engineering, 2011; Vol. **2**, pp. 75–79.

Subrahmanyam, S.; Pekny, J. F.; Reklaitis, G. V. Design of Batch Chemical Plants Under Market Uncertainty. *Ind. Eng. Chem. Res.* 1994, **33**, 2688.

Sugiartha, N.; Tassou, S. A.; Chaer, I.; Marriott, D. Trigeneration in Food Retail: An Energetic, Economic and Environmental Evaluation for a Supermarket Application. *Appl. Therm. Eng.* 2009, **29**, 2624.

The Engineering ToolBox. Fuel Gases – Heating Values. Available at http://www.engineeringtoolbox.com/heating-values-fuel-gases-d_823.html (accessed 17 September 2014).

Voll, P. *Automated Optimization-Based Synthesis of Distributed Energy Supply Systems*, PhD thesis, RWTH Aachen University, 2014.

Wang, J.-J.; Jing, Y.-Y.; Zhang, C.-F. Optimization of Capacity and Operation for CCHP System by Genetic Algorithm. *Appl. Energy* 2010, **87**, 1325.

Wu, D. W.; Wang, R. Z. Combined Cooling, Heating and Power: A Review. *Prog. Energy Combust. Sci.* 2006, **32**, 459.

Yokoyama, R.; Shinano, Y.; Taniguchi, S.; Ohkura, M.; Wakui, T. *Optimization of Energy Supply Systems by MILP Branch and Bound Method in Consideration of Hierarchical Relationship between Design and Operation. Energy Convers. Manag.* 2015, **92**, 92.
Yoshida, S.; Ito, K.; Yokoyama, R. Sensitivity Analysis in Structure Optimization of Energy Supply Systems for a Hospital. *Energy Convers. Manag.* 2007, **48**, 2836.
Yunt, M.; Chachuat, B.; Mitsos, A.; Barton, P. I. Designing Man-Portable Power Generation Systems for Varying Power Demand. *AIChE J.* 2008, **54**, 1254.
Ziher, D.; Poredos, A. Economics of a Trigeneration System in a Hospital. *Appl. Therm. Eng.* 2006, **26**, 680.

Part 2
Regional Biomass Supply Chains and Risk Management

6

Large-Scale Cultivation of Microalgae for Fuel

Christina E. Canter[1], *Luis F. Razon*[2], *and Paul Blowers*[3]

[1] *Department of Mechanical Engineering, University of Alberta, Edmonton, Canada*
[2] *Department of Chemical Engineering, De La Salle University, Manila, Philippines*
[3] *Department of Chemical and Environmental Engineering, The University of Arizona, Tucson, AZ, USA*

6.1 Introduction

Over the years, there has been much interest in the large-scale cultivation of microalgae for use as energy sources and the bulk production of various products. Microalgae have been investigated for energy in the case of biofuels and cofiring with coal for electricity production. For biofuels, microalgae are viewed as a potential source of renewable energy that can produce less greenhouse gas emissions than conventional fuels (Lardon et al., 2009; Clarens et al., 2010). They have an advantage over terrestrial crops in that they use less land area for production and can sometimes use salt water for growth, limiting the amount of freshwater required (Chisti, 2007; Mata et al., 2010). Soil quality is a concern for terrestrial crops but not for microalgae because only water and nutrients are required for growth. Some energy fuels that can be produced from these aquatic species include biodiesel, methane, hydrogen, ethanol, and butanol (Mata et al., 2010).

Microalgae can also produce high-value products like beta-carotene and animal feed or can be consumed directly as a health food product (Spolaore et al., 2006). Some microalgal species, like *Arthrospira* and *Chlorella*, are sold as nutritional supplements (Spolaore et al., 2006). Docosahexaenoic acid (DHA), an omega-3 fatty acid, can be extracted from some species for use in infant formula. When produced for animal feed, they can be used to

Process Design Strategies for Biomass Conversion Systems, First Edition. Edited by
Denny K. S. Ng, Raymond R. Tan, Dominic C. Y. Foo, and Mahmoud M. El-Halwagi.
© 2016 John Wiley & Sons, Ltd. Published 2016 by John Wiley & Sons, Ltd.

replace necessary proteins, as in the case of poultry. In aquaculture, microalgae are used to provide nutrients and the coloring of farm-raised salmon. Cosmetics are another area this aquatic organism is utilized. There are many additional applications in both skin and hair care products.

The desired final product used to determine which production methods are best for microalgae. This chapter discusses the large-scale production of these aquatic organisms for fuel. There are three main areas that address this topic: cultivation, harvesting, and conversion to products. For cultivation, the types of species and their selection, growth systems, chemicals required for growth, and commercial production are discussed. Harvesting addresses the difficulties in separating the microalgae species from the growth medium, which includes microalgal physical traits that ease separation. Also discussed are some common separation technologies with their advantages and disadvantages. When considering conversion to products, various parts of the microalgae cell are utilized. This chapter discusses the products that can be made from the lipids, carbohydrates, and proteins, along with methods for converting those constituents into fuel. Figure 6.1 provides an overall framework illustrating how all of the major inputs, outputs, and operations fit together. Variations from this framework will occur for specific cases.

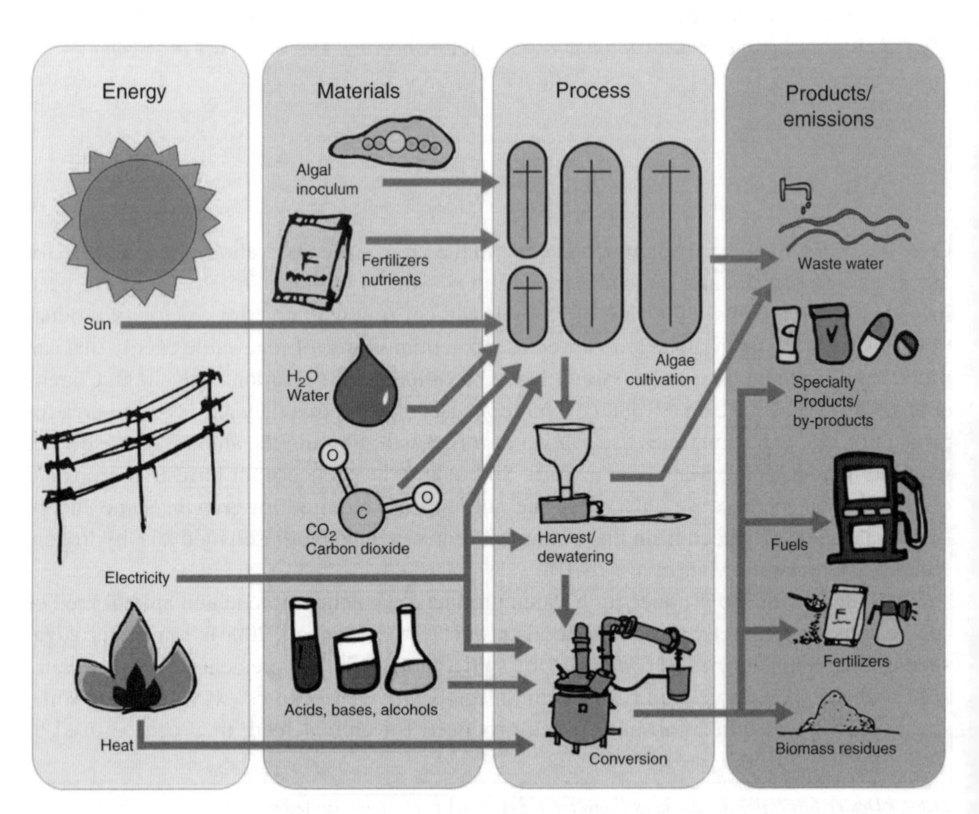

Figure 6.1 *An overall framework showing major operations for obtaining fuels from microalgae. Reproduced with permission from R. Ubando*

6.2 Cultivation

6.2.1 Organisms for Growth

The two categories of microorganisms that can be used for cultivation are prokaryotic cyanobacteria and eukaryotic microalgae (Zhu et al., 2013) (Figure 6.2). Even though prokaryotic cyanobacteria, or blue-green algae, are not actually microalgae, they still undergo photosynthesis (Scott et al., 2010) which turns light into chemical energy. Eukaryotic microalgae are unicellular aquatic organisms that also undergo photosynthesis for energy. It is estimated that there are more than 50,000 microalgae species in the world (Mata et al., 2010), which range in size from 2 to 200 μm (Greenwell et al., 2010). Prokaryotic cyanobacteria are high in proteins and low in lipids, while eukaryotic microalgae contain more lipids. For the rest of the chapter, the term microalgae will include both types of microorganisms.

Microalgae are made up of carbohydrates, proteins, lipids, ash, fiber, RNA, DNA, pigments (like chlorophyll and carotenoids), and water. The amount of each of the constituent within the algal cells depends on both the algal species and the growing conditions like light, temperature, and cell growth stage (Gatenby et al., 2003). There have been numerous representations of the elemental composition in literature, and the Redfield ratio was developed as an average for marine algae which gives the atomic C:N:P ratio at 106:16:1 (Geider and La Roche, 2002). Another representation is for the average elemental composition of microalgae at $CH_{1.7}O_{0.4}N_{0.15}P_{0.0094}$ (Greenwell et al., 2010).

Microalgae can grow in either heterotrophic or autotrophic conditions. Heterotrophic microalgae do not require light but do need a carbon source like glucose for growth (Scott et al., 2010). Autotrophic microalgae require light, carbon dioxide, and nutrients for growth. Also called the "blue-green algae," the cyanobacteria are diazotrophic.

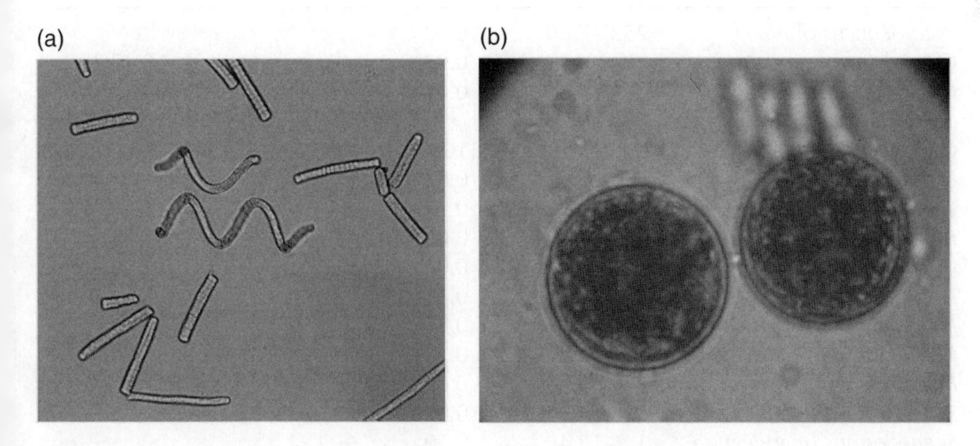

Figure 6.2 Examples of two aquatic microorganisms used for biomass growth: (a) prokaryotic cyanobacteria Spirulina (Simon, 1994) (Photo taken by Joan Simon / CC-BY-SA-2.5). Reproduced from http://commons.wikimedia.org/wiki/File:Spirul.jpg and (b) eukaryotic microalgae Haematococcus (Taka, 2006) (Photo taken by Taka.) Reproduced from http://commons.wikimedia.org/wiki/File:Haematococcus_(2006_02_27).jpg

That is, they are able to convert atmospheric nitrogen to reactive or soluble nitrogen compounds. Because of this capability, supplemental nitrogen fertilizer is unnecessary, and they even have the potential to provide fertilizer for other microalgae. Cyanobacteria have been previously studied for wastewater treatment (Markou et al., 2011) and for biogas generation (Converti et al., 2009). The possibility of harvesting these cyanobacteria for their ammonium compounds has also been investigated (Razon, 2012, 2014).

6.2.2 Selection of a Species for Growth

The end product is important for selecting an algal species for growth. For instance, if looking for a specific product, like the carotenoid beta-carotene, *Dunaliella salina* has been a successful algal species for large-scale cultivation (Lundquist et al. 2010). Some species grow faster than others (Mata et al., 2010), like *Chlorella*, which can range from 3.5 to 13.9 g/m²/day versus *Botryococcus braunii* at 3.0 g/m²/day (Table 6.1). When lipids are the desired product, as is the case with some biofuels, like biodiesel, species are chosen that have both higher lipid content and fast growth rates. Bioethanol and biogas would need to be sourced from microalgae that are high in digestible carbohydrates. Unfortunately, this selection can become a zero-sum game. High carbohydrate content means that the other major components like protein and fat will be reduced. These "other" fractions must likewise be utilized in some desirable product lest they become a waste stream to be treated. We turn now to biofuels, instead of general products attainable from microalgae.

Table 6.1 *Lipid content and productivities of various microalgae species*

Algal species	Lipid content (% dry weight)	Volumetric productivity (g/L/day)	Areal productivity (g/m²/day)
Botryococcus braunii	25.0–75.0	0.02	3
Chlorella sp.	28–32	0.02–2.5	1.61–16.47/25
Chlorella vulgaris	5.0–58.0	0.02–0.20	0.57–0.95
Chlorella	18.0–57.0		3.5–13.9
Crypthecodinium cohnii	20.0–51.1	10	
Dunaliella salina	6.0–25.0	0.22–0.34	1.6–38
Dunaliella tertiolecta	16.7–71.0	0.12	
Haematococcus pluvialis	25	0.05–0.06	10.2–36.4
Isochrysis sp.	7.1–33	0.08–0.17	
Nannochloropsis sp.	20.0–56.0	0.17–0.51	
Nannochloropsis oculata	22.7–29.7	0.37–0.48	
Nannochloropsis sp.	12.0–68	0.17–1.43	1.9–5.3
Phaeodactylum tricornutum	18.0–57.0	0.003–1.9	2.4–21
Scenedesmus obliquus	11.0–55.0	0.004–0.74	
Scenedesmus quadricauda	1.9–18.4	0.19	
Scenedesmus sp.	19.6–21.1	0.03–0.26	2.43–13.52
Spirulina maxima	4.0–9.0	0.21–0.25	25
Tetraselmis suecica	8.5–23.0	0.12–0.32	19

Reproduced from Mata et al. (2010) with permission from Elsevier.

The selection of an appropriate species of microalgae for conversion to biofuels is a more interesting problem than if terrestrial plants were used because there is a larger variety of species that may be chosen from. The now-defunct Aquatic Species Program of the US Department of Energy (Sheehan et al., 1998) accumulated a collection of 3000 species that it tested solely for the purpose of being used as sources of biodiesel. There is therefore a potentially immense number of other species that may be tested for other applications also.

What makes the selection of the appropriate species more challenging is that the composition of microbial species can vary significantly with growing conditions. Light incidence, temperature, nutrient levels, and residence time have all been shown to have an impact, not just on yield and productivity but also the chemical composition of the microalgae. Not only the ratios of carbohydrates, fats, and proteins may change in the biomass, but the composition of each individual fraction is also influenced by growing conditions. For example, the fatty acid profile of the lipids (or the composition of the lipid fraction) in microalgae is altered by the growing conditions (Ben-Amotz et al., 1985). These variations in the chemical composition affect the ultimate products in different ways. These are discussed in more detail in later sections concerning product processing after harvesting. Another choice that has to be made is the growing location. For example, when looking to produce biofuels in the United States, the gulf coast region is a promising area for growth due to warm temperatures that allow year-round growth (ANL et al., 2012). We turn in the next section to discuss different growing conditions.

6.2.3 Types of Growth Systems

6.2.3.1 Open systems

Systems are considered "open" when the surface of the water is in direct contact with the outdoor atmosphere. The microalgae are grown in ponds with large surface areas. There are numerous pond configurations, which include circular designs, in the case of wastewater treatment systems. In a "raceway" pond (Figure 6.3), the water flows around the pond with the aid of a paddle wheel, and the culture can be from 5 to 100 cm deep with flow rates ranging from 5 to 40 cm/s to keep the culture in laminar flow (Ben-Amotz, 2013). Keeping the pond moving prevents settling of the algae and ensures that each cell has the opportunity to obtain sunlight. Baffles can be put in at the ends of the raceways to provide greater mixing (Greenwell et al., 2010). However, even with mixing, raceways are not efficient in maximizing light opportunities for all cells (Chisti, 2007).

Raceway ponds can be constructed in a variety of ways. For wall supports, either earthen berms, which are made by moving around and compacting the earth with construction equipment (Lundquist et al., 2010), or concrete walls and floors can be implemented

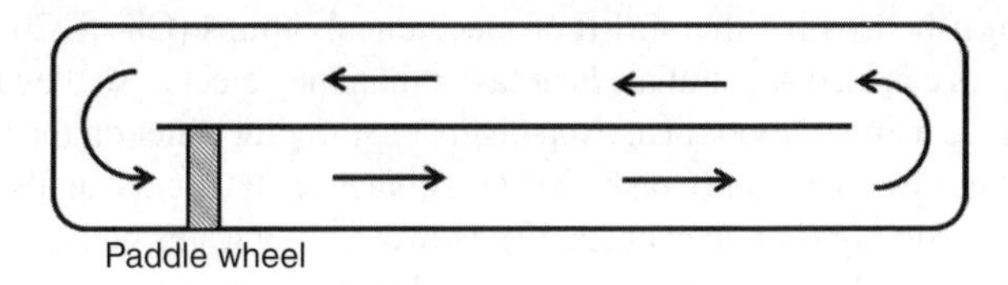

Figure 6.3 *Raceway pond*

(Ben-Amotz, 2013). The ponds are lined to prevent leakage of the culture (Ben-Amotz, 2013). Numerous materials can be used as a liner, one of which is clay where the material clay is compacted along the bottom of the pond and can either come from the pond site or be brought in. Liners can also be made from a variety of plastics, including high-density polyethylene (HDPE), polypropylene (PP), or polyvinyl chloride (PVC). An advantage of plastic liners over clay is that they are less likely to leak and they lower energy costs due to the bottom of the pond being smoother for flow (ANL et al., 2012). However, when looking at algae for biofuel production processes, the cost of the pond liner can significantly raise the price of fuel and perhaps make it too costly (Lundquist et al. 2010; ANL et al., 2012). Paddle wheels can be constructed from plastic or marine plywood (Ben-Amotz, 2013) and require a concrete base to sit upon. Carbon dioxide can be injected into the pond with carbonation sumps (Lundquist et al., 2010), which can be constructed from concrete. All of these choices impact the growing conditions and economic and environmental sustainability of the process for making biofuels.

6.2.3.2 Closed systems

Systems are considered "closed" when the algal culture does not make contact with the atmosphere and are called photobioreactors. They are designed to have large surface areas for distribution of light. There are many different reactor types, some of which are tubular, flat plate, and bubble column (Schenk et al., 2008). Reactors can also have different configurations like horizontal, vertical, and helical (Greenwell et al., 2010). The culture flows through the system in the presence of natural or artificial light, nutrients, water, and carbon dioxide. Carbon dioxide is injected into the reactor and oxygen is removed because it can inhibit photosynthesis (Rawat et al., 2013), or even damage cells (Chisti, 2007). To ensure this does not occur, the oxygen concentration should be less than 400% of the air saturation value (Chisti, 2007). Outdoor tubular reactors are typically run north and south (Schenk et al., 2008; Rawat et al., 2013) because direct sunlight on the tubes can bleach the algae. Sometimes, the ground beneath the tubes is painted white to increase the amount of light reflected from below onto the tubes (Rawat et al., 2013).

Tubular reactors (Figure 6.4) are made from transparent materials like polyethylene or fiberglass (Greenwell et al., 2010) and range in diameter from 2.5 to 5.0 cm (Zhu et al., 2013). Flat plate reactors can be made from glass, Plexiglas, or polycarbonate (Kumar et al., 2011). An advantage of tubular reactors is that their design makes them suitable for continuous operation (Chisti, 2007). Tubular photobioreactors are better for biofuel production because flat plate reactors may use more energy than they produce (Jorquera et al., 2010).

Using artificial light is more expensive than sunlight (Rawat et al., 2013), but if it is done, then the system should supply lights with wavelengths between 400 and 700 nm, which is the active region for photosynthesis (Greenwell et al., 2010). Flow within the system is accomplished with either airlift or mechanical pumps (Chisti, 2007).

Photobioreactors can also accumulate heat faster than they eject it, so they must be cooled. Possibilities include using outdoor heat exchangers or spraying water on the tubes so cooling can occur by surface evaporation (Chisti, 2007). Tubular exchangers can also be submerged in water to assist in cooling (Quinn et al., 2012). However, any water savings due to removing evaporation in open pond systems by using a closed system can then be lost.

Figure 6.4 *Tubular photobioreactor (IGV Biotech, 2013). Photo taken by IGV Biotech. Reproduced from http://commons.wikimedia.org/wiki/File:Photobioreactor_PBR_4000_G_IGV_Biotech.jpg*

6.2.3.3 Comparison of both systems

Both systems have advantages and disadvantages. Determining which system to use will depend on the final product produced. Open ponds are advantageous because they are cheaper for construction materials and operating costs (Scott et al., 2010) when compared to photobioreactors. Commercial applications of photobioreactors include producing high-value products (Greenwell et al., 2010; Rawat et al., 2013) because the process is not constrained by energy use and the money recovered from the products is large. Even though they use more energy, photobioreactors provide savings in dewatering costs because they can produce higher algal concentrations. Open systems produce algal concentrations up to 0.5 g/L, while closed systems can reach up to 8 g/L. Another advantage of closed systems over open ones is that they use less water and nutrients (Schenk et al., 2008).

Closed systems also control contamination much better than open ones (Mata et al., 2010). However, there are some solutions to controlling contamination in open systems, like choosing a resilient algal species. In addition, extremophiles can survive in environments that most others cannot. For instance, culturing *Spirulina* at high pH values from 9 to 11.5 and *D. salina* in hypersaline conditions reduces the amount of grazers and other species that could harm the system (Schenk et al., 2008; Lundquist et al., 2010).

6.2.4 Nutrients, Water, and Carbon Dioxide for Growth

6.2.4.1 Nutrients

The main nutrients required for microalgae growth are nitrogen, phosphorus, and potassium. Micronutrients are also required and include calcium, magnesium, iron, sulfur, and various trace elements (Zhu et al., 2013). Nitrogen can be supplied in numerous forms, which can affect both the absorption by the microalgae and the accumulation of lipids within the cells. The ordering of nitrogen compounds on the absorption ability from greatest to least is ammonia, urea, nitrate, and nitrite. The nitrogen concentration also has an effect on the accumulation of lipids. When there is a nitrogen deficiency in the system, algal cells go into stationary growth and accumulate lipids (Gatenby et al., 2003; Mata et al., 2010) or carbohydrates (Dragone, 2011). When this occurs, the composition of carbon, hydrogen, and oxygen in the cells increases, resulting in a decrease of cell proteins.

Microalgae use phosphorus in metabolic processes like energy conversion and photosynthesis (Zhu et al., 2013). Like nitrogen, some forms are more easily absorbed than others, with the most easily absorbed form as orthophosphate. Phosphorus must be added to the culture in excess of the stoichiometric requirement because not all of it will be bioavailable (Chisti, 2007). However, if the concentration is too high, growth is inhibited (Zhu et al., 2013).

Additional sources of nutrients and methods for recycling them are being evaluated in the case of biodiesel production from microalgae. Any large-scale production of microalgae will use a considerable amount of nutrients, so investigations are being conducted to evaluate the effect of lowering process costs and the impact on the fertilizer market. Recycling of nutrients can be either returning process water after harvesting and dewatering back to the system or in extracting leftover nutrients from algal cells after the desired product is removed and then recycling those recovered compounds.

6.2.4.2 Water

Microalgae require large amounts of water and can grow in waters that are fresh, marine, brackish, or hypersaline, depending on the species (Zhu et al., 2013). During cultivation, input water is required to make up for any water lost to evaporation to ensure that salinity is kept in a range where the algal species can grow (Rawat et al., 2013). Water loss in open systems occurs by evaporation, and in closed systems, it can occur during photobioreactor cooling (Mata et al., 2010). Wastewater can also be used as a water source for some microalgae species (Mata et al., 2010; Sturm and Lamer, 2011), which has the benefit of providing nutrients for growth and reducing fertilizer costs. There is concern, though, with nutrient recycling using wastewater due to the possibility of recycling and concentrating heavy metals (Greenwell et al., 2010).

6.2.4.3 Carbon dioxide

Microalgae use carbon dioxide (CO_2) during photosynthesis (Schenk et al., 2008). To prevent an uptake limitation, CO_2 must be supplied in sufficient amounts, 0.15 kPa in the case of photobioreactors (Schenk et al., 2008). For cultivation, approximately 1.83 g of CO_2 is required per gram of microalgae produced (Chisti, 2007). The gas can also be used to regulate pH in the culture. Sources for CO_2 include an air- and CO_2-enriched mixture or flue gas, which contains high enough concentrations of CO_2 for algal growth (Mata et al., 2010; Zhu et al., 2013). Sometimes, microalgae grow faster with flue gas than with the enriched gas, as in the case of *Chlorella* (Zhu et al., 2013).

6.2.5 Large-Scale Commercial Microalgae Growth

Chlorella is the most abundantly produced algal species at 5×10^9 tons per year, followed by *Dunaliella* at 1×10^9 tons per year and *Haematococcus* at 0.1×10^9 per year (Lundquist et al. 2010). There are numerous large-scale microalgae systems around the world, and Hutt Lagoon in Australia is the largest microalgae production location (Department of Environment and Conservation (DEC), 2009). It contains 250 ha of shallow, unmixed ponds that produce *D. salina*. The water in these ponds (Figure 6.5) is pink due to the production of beta-carotene (Orchard, 2009). Cyanotech is a company located in Kona, Hawaii, that produces nutritional supplements from *Spirulina* and *Haematococcus* (Cysewski, 2013) where production takes place in outdoor raceway ponds at a 36 ha facility.

We turn now to how microalgae are concentrated. After cultivation, the microalgae are suspended in water at concentrations between 0.02 and 0.06 total suspended solids (TSS) (Uduman et al., 2010). Later, processing requires the microalgae to be concentrated to 5–25% TSS, with the final concentration depending on the requirements of later unit operations. Bringing the microalgae to higher TSS values can require a large amount of energy, depending on the unit operations chosen. This makes harvesting and dewatering a major challenge for biofuel production from microalgae (Uduman et al., 2010; Sturm and Lamer, 2011; Rawat et al., 2013). An important factor in determining the effectiveness of a

Figure 6.5 Dunaliella salina *farms located in Hutt Lagoon, Australia (Orchard, 2009). The ponds are pink due to the beta-carotene produced by the microalgae. Photo taken by Samuel Orchard. Reproduced from http://commons.wikimedia.org/wiki/File:Hutt_Lagoon,_Western_Australia.jpg*

particular technology is recovery rate, which is the amount of microalgae cells that can be recovered by a process. Each method is presented individually in this chapter, but they can be used in various combinations to achieve desired results.

6.3 Harvesting and Dewatering

6.3.1 Separation Characteristics of Various Species

Microalgae can be difficult to separate due to having similar densities as water, negative surface charges that keep them suspended in the culture medium, and the size and shape of the species' cells (Greenwell et al., 2010; Pahl et al., 2013; Rawat et al., 2013). Some shapes are easier to harvest, like *D. salina*, because they are spiral shaped (Schenk et al., 2008). In the water, microalgae repel each other due to their negative surface charge (Pahl et al., 2013). The strength of this interaction and what the cells will do in the culture (stay suspended or agglomerate) can be predicted by the zeta potential. The closer the value is to zero, the easier it is for cells to agglomerate. The growth phase affects the numeric value of the zeta potential, with low growth rate phases having a number closer to zero than high growth rate phases (Danquah et al., 2009).

6.3.2 Gravity Sedimentation

In gravity sedimentation, a process common in wastewater treatment, the microalgae is sent to a gravity thickener and allowed to settle to the bottom over time, where it is removed. The settling rate is influenced by the size, shape, and density of cells (Pahl et al., 2013). Settling rates of 0.1–2.6 cm/h are a major drawback of this technology (Greenwell et al., 2010). Time of day can also affect settling, with faster rates occurring at night than during the day (Danquah et al., 2009). During the day, the cells undergo photosynthesis, which increases their metabolism, resulting in an increase in the net electronegative shielding effect. At night, the metabolism of the cells slows down, which lowers this affect and allows the cells to aggregate. The rate is also affected by the growth phase, with faster settling rates during the low growth phase due to the zeta potential being closer to zero.

6.3.3 Flocculation

Flocculation works by adding a chemical called a flocculent to the microalgae suspension that allows the cells to agglomerate together (Uduman et al., 2010). The agglomerated cells will then settle due to an increase in their density (Rawat et al., 2013). This process is applicable to many types of algae and can be done as a pretreatment to concentrate the microalgae before other dewatering processes (Christenson and Sims, 2011). Flocculants can be either inorganic or organic (or polymeric) flocculants. Common inorganic flocculants include alum, ferric chloride, ferric sulfate, ferrous sulfate, and lime (Uduman et al., 2010; Pahl et al., 2013). Each species has an optimal pH range and dosage, with some values listed in Table 6.2. Inorganic compounds will work on both freshwater and marine microalgae, with marine requiring 5–10 times the dosage of freshwater species (Uduman et al., 2010). Organic flocculants require a lower dosage than inorganic and are less sensitive to pH, as shown in Table 6.2, but are more expensive (Greenwell et al., 2010; Harun et al., 2010).

Table 6.2 Optimal dose and optimal pH range for inorganic and organic flocculants (Uduman et al., 2010; Pahl et al., 2013)

Type	Flocculant	Optimal dose (mg/L)	Optimal pH
Inorganic	Alum	80–250	5.3–5.6
		0–225[a]	4.0–7.0
	Ferric chloride	15–120[b]	3.5–6.5 and >8.5
	Ferric sulfate	50–90	3.0–9.0
			3.5–7.0 and >9.0
	Ferrous sulfate		>8.5
	Lime	500–700	11.5–11.5
Organic	Purifloc	35	3.5
	Zetag 51	10	>9
	Dow 21 M	10	4.0–7.0
	Dow C-31	1–5	2.0–4.0
	Chitosan	100	8.4

Reproduced from Uduman et al. (2010) with permission from AIP Publishing LLC.
[a] Dosage for *Isochrysis galbana* at a pH of 5.5 and a range of 0.1–0.7 M solutions.
[b] Dosage for *I. galbana* at a pH of 5.0 and a range of 0.1–0.7 M solutions.

6.3.4 Dissolved Air Flotation

Dissolved air flotation is another common technology used in wastewater treatment. Freshwater microalgae culture is sent to a flotation tank where air is injected at atmospheric pressure to the point of supersaturation into a flotation tank (Greenwell et al., 2010; Uduman et al., 2010). Bubbles ranging from 10 to 100 μm in diameter stick to the cells and carry them to the surface of the water where they are skimmed off (Greenwell et al., 2010; Uduman et al., 2010; Sturm and Lamer, 2011). To enhance the process, approximately 10 ppm of polyelectrolyte salts can be added (as in the case of wastewater treatment), and a portion of the effluent is recycled back to the system (Greenwell et al., 2010; Pahl et al., 2013). The process can treat more than 10,000 m^3 of water per day, but there is concern of flocculants contaminating downstream processes (Greenwell et al., 2010).

6.3.5 Centrifugation

Centrifugation uses centrifugal force to separate microalgae cells from water based on their density differences (Pahl et al., 2013). This method is the most rapid and effective separation technique (Uduman et al., 2010; Rawat et al., 2013) but is energy intensive. In larger centrifuges, the centrifugal force can range from 5,000 to 10,000 g, and 95% cell recoveries are possible (Greenwell et al., 2010). There are many different types of centrifuges that can be used for microalgae, including disk stack, perforated basket, imperforated basket, decanter, and hydrocyclones (Pahl et al., 2013). Again, high costs due to energy use and machine maintenance can be a major drawback of this technology (Greenwell et al., 2010; Uduman et al., 2010). Energy requirements vary by machine type and operating conditions, but values can be as low as 0.3 kW/m^3 for hydrocyclones or as high as 8 $kW/m^3$for decanters (Pahl et al., 2013).

6.3.6 Filtration

During filtration, either freshwater or marine microalgae are separated from the culture based on their size, and the fluid flows either through, or tangentially across, a membrane. The liquid flows through the membrane with the aid of gravity, pressure, vacuum, or centrifugal forces, while the algal cells remain on or in the membrane. Cells are recovered by scraping them off the surface of the membrane. Pressure and vacuum filtration are effective for microalgae greater than 30 μm in diameter (Uduman et al., 2010; Rawat et al., 2013). Both ultrafiltration (pore sizes from 1 to 100 nm) and microfiltration membranes (pore sizes from 10 to 10,000 nm) can be used (Pahl et al., 2013), with microfiltration leading to clogging earlier than the larger pore sizes. Tangential flow filtration is seen as a promising technology for large-scale microalgae production (Uduman et al., 2010). Benefits of this technology include recoveries from 70 to 89%, ease of scale-up, and maintaining cell structure when operating at lower pressures (Uduman et al., 2010). One drawback of filtration is membrane fouling, which reduces the efficiency of the process. Tangential flow filtration recycles retentate back through the system to reduce fouling (Uduman et al., 2010). Energy requirements due to pumping are high at 3–10 kWh/m^3 and are comparable to centrifugation (Pahl et al., 2013). Membranes need to be replaced periodically, but their lifetime can be improved by operating systems at low transmembrane pressures and tangential flow velocities (Uduman et al., 2010).

6.3.7 Electrocoagulation

Electrocoagulation is a process that uses electricity to produce metal ions to coagulate microalgae (Uduman et al., 2010). Electrodes are made from metals like aluminum or iron, and more ions are formed by increasing the amount of electricity supplied (Uduman et al., 2010). The benefit of this technology is that the process can recover 95% of the microalgae (Uduman et al., 2010) and concentrate the cells from 0.5 to 5 kg dry weight/m^3 or 15 to 20% solids (Greenwell et al., 2010). The process is not as energy intensive as centrifugation or filtration at 0.8–1.5 kWh/m^3 but does require continual replacement of electrodes, which may be costly (Uduman et al., 2010).

6.4 Conversion to Products

6.4.1 Utilization of the Lipid Fraction (Biodiesel)

Biodiesel has an interesting characteristic: its fuel properties depend on the feedstock (Knothe, 2005). Unlike ethanol and hydrogen, biodiesel is not a pure substance but is a mixture of alkyl esters that result from the transesterification reactions between the triglycerides and an alkyl alcohol. The transesterification reactions proceed via the following general reaction.

$$\begin{array}{c} H_2C-O-C(=O)-R' \\ | \\ HC-O-C(=O)-R'' \\ | \\ H_2C-O-C(=O)-R''' \end{array} + 3\,ROH \xrightleftharpoons{\text{Catalyst}} \begin{array}{c} R-O-C(=O)-R' \\ R-O-C(=O)-R'' \\ R-O-C(=O)-R''' \end{array} + \begin{array}{c} H_2C-OH \\ | \\ HC-OH \\ | \\ H_2C-OH \end{array} \quad (6.1)$$

Triglycerides (Vegetable oil) | Alcolhol | Alkyl Ester (biodiesel) | Glycerol

The chain lengths and degree of unsaturation of the alkyl groups attached to the glycerin molecule R′, R″, and R‴ determine the volatility and reactivity of the mixture and also, consequently, the fuel properties of the biodiesel. This is commonly called the fatty acid profile of the triglyceride and of the biodiesel. One of the lengthier lists of fatty acid profiles of microalgal lipids is from Knothe (2011), and in the interest of brevity, only some of the more interesting ones are reproduced in Table 6.3. It can be seen that there is a great variety in the fatty acid profiles. Highly unsaturated fatty acids commonly occur in microalgae, and unusual fatty acids can also be found. Table 6.4 illustrates how the microalgal fatty acid profiles change with growing conditions. Choice of the microalgal species and the growing conditions is therefore a critical factor even before any processing of the biomass can be done.

A simplified process flow diagram for the biodiesel process is shown in Figure 6.6. Prior to the transesterification reactions and after oil extraction, the oil may need to be refined to remove gums and free fatty acids. Free fatty acids may develop under the influence of heat, water, and sometimes naturally occurring enzymes. Because a base catalyst is usually used for the transesterification reaction, the free fatty acids may react with the catalyst to form soap via the reaction:

$$R-COOH+NaOH \leftrightarrow H_2O+R-COO-Na \tag{6.2}$$

The formation of soap consumes catalyst and may cause problems downstream during the separation of the biodiesel from the glycerin phase. If the oil contains free fatty acids with concentrations above 1%, then the oil is usually first reacted with methanol and an acid catalyst to esterify the free fatty acids and form fatty acid methyl esters. The possibility that the saponification reaction may occur also necessitates that the oil be dry, thus requiring the dewatering steps described earlier in this chapter. Oil drying is another significant contributor to the energy consumption of the overall process.

If the oil is of acceptable quality, the oil may be taken directly to the transesterification reactor where it is contacted with an excess of the chosen alcohol (usually methanol) and a base catalyst (usually NaOH or KOH). The transesterification reaction is conducted at about 60°C, which is a constraint set by the boiling point of methanol at atmospheric pressure. The oil is immiscible with the shorter-chain alcohols: therefore, maintaining intimate contact between the two phases by providing adequate mixing is an important design and operating criterion for the reactor. It is also an important energy input that must be accounted for.

The reaction product consists of two phases which are separated by gravity in a settling tank. The upper phase consists of the biodiesel product, impurities, and any soap that may have been formed during the transesterification reaction. The impurities and soap are washed out in a wash column. Depending on the level of impurities, the wash operation can be extensive and become a significant consumer of resources. Quality standards usually require a clear product, and this is another area where the feedstock has a significant influence. Different oils have significantly different colors. Figure 6.7 shows the color of oils and biodiesel from different sources. The color by itself has no impact on the fuel combustion qualities; however, a manufacturer would, in general, want to have a uniform color for its product, especially if it is to be blended into a petrodiesel. A dark color would complicate efforts to attain uniformity. While the figure does not include any microalgal oils, it serves to illustrate the wide variance in colors. Some microalgal oils have been observed to have a dark color (Sanford, 2009).

Table 6.3 *Fatty acid profiles of some microalgae*

Algae	Fatty acid profile					
	Saturated	Monounsaturated	Polyunsaturated, two double bonds	Polyunsaturated, three double bonds	Polyunsaturated, four or more double bonds	Other
Chlorella vulgaris	16 : 0/15–18% 18 : 0/3–5% 20 : 0/0.1–0.2%	14 : 1/0.5% 16 : 1/0.5–1% 18 : 1/20–25% 20 : 1/0.2%	14 : 2/0.1% 16 : 2/3–5% 18 : 2/13–18% 20 : 2/0.2%	14 : 3/0.1% 16 : 3/10–12% 18 : 3/13–18%		ai-15 0.1% ai-17 1–1.5% ai-19 0.2%
Dunaliella salina	14 : 0/0.5% 15 : 0/0.2% 16 : 0/17.8% 18 : 0/1.5%	14 : 1ω7/0.1% 16 : 1ω7/0.8% 16 : 1ω13/1.7% 18 : 1ω9/2.8% 18 : 1ω7/0.6%	16 : 2ω6/1.5% 18 : 2ω6/6.1%	16 : 3ω6/1.7% 16 : 3ω4/0.3% 16 : 3ω3/2.1% 18 : 3ω6/2.5% 18 : 3ω3/36.9%	16 : 4ω3/18.2% 18 : 4ω3/0.7%	anteiso-15 : 0/0.9%
Dunaliella tertiolecta	14 : 0/0.5–0.8% 15 : 0/1.0–2.0% 16 : 0/19.3–24.6% 18 : 0/1.3–1.9%	14 : 1/0.3–0.4% 15 : 1/0–1.7% 16 : 1/1.3–1.9% 17 : 1/0.4–0.9% 18 : 1/5.8%	16 : 2/0.3% 18 : 2ω-6/10.1–10.7%	18 : 3ω6/2.0–4.2% 18 : 3ω-3/38.3–39.3%	18 : 4/0.5–0.8%	
Nannochloropsis oculata	14 : 0/3.9% 15 : 0/0.5% 16 : 0/20.5% 17 : 0/0.4% iso-17 : 0/0.7% 18 : 0/1.8%	16 : 1ω7/25.2% 18 : 1ω9/3.6% 18 : 1ω7/0.5%	16 : 2ω6/0.6% 16 : 2ω4/0.2% 18 : 2ω6/3.5% 20 : 2ω6/0.1%	18 : 3ω6/0.7% 18 : 3ω3/0.2% 20 : 3ω6/0.3%	18 : 4ω3/0.1% 20 : 4ω6/5.3% 20 : 5ω3/29.7%	
Scenedesmus obliquus	14 : 0/1.5% 16 : 0/21.8% 18 : 0/0.4%	16 : 1/6.0% 18 : 1/17.9%	18 : 2/21.7%	18 : 3ω6/3.7%	16 : 4/0.4% 18 : 4/0.2%	
Tetraselmis suecica	12 : 0/0.7% 14 : 0/6.3% 16 : 0/22.8%	16 : 1ω7/10.2% 18 : 1ω9/6.9%	18 : 2ω6/6.9%	18 : 3ω3/14.9% 18 : 3ω6/1.6%	18 : 4ω3/21.6% 20 : 4ω6/2.0% 20 : 5ω3/6.2%	

Reproduced from Knothe (2011) with permission from the Royal Society of Chemistry.

Table 6.4 *Change in fatty acid profile of Isochrysis sp. as salinity and supply of N are varied*

Algae	Fatty acid profile				
	Saturated	Monounsaturated	Polyunsaturated, two double bonds	Polyunsaturated, three double bonds	Polyunsaturated, four or more double bonds
Isochrysis sp., 0.5 M NaCl; N sufficient	14 : 0 + 14 : 1/13.3% 16 : 0/11.7%	16 : 1/6.3% 18 : 1/15.0%	18 : 2/3.7%	18 : 3/5.6%	18 : 4/16.6% 20 : 5/1.6% 22 : 4/1.3% 22 : 6/12.8%
Isochrysis sp., 0.5 M NaCl; N deficient	14 : 0 + 14 : 1/18.6% 16 : 0/1.4%	16 : 1/2.9% 18 : 1/17.2% 20 : 1/1.1%	18 : 2/5.0%	18 : 3/6.3%	18 : 4/14.0% 22 : 4/2.7% 22 : 6/14.9%
Isochrysis sp., 1 M NaCl; N sufficient	14 : 0 + 14:1/11.5% 16 : 0/8.5% 19 : 0/0.9%	16 : 1/3.8% 18 : 1/21.6%	18 : 2/4.3%	18 : 3/4.4%	18 : 4/17.7% 22 : 4/6.0% 22 : 6/14.6%

Reproduced from Knothe (2011) with permission from the Royal Society of Chemistry.

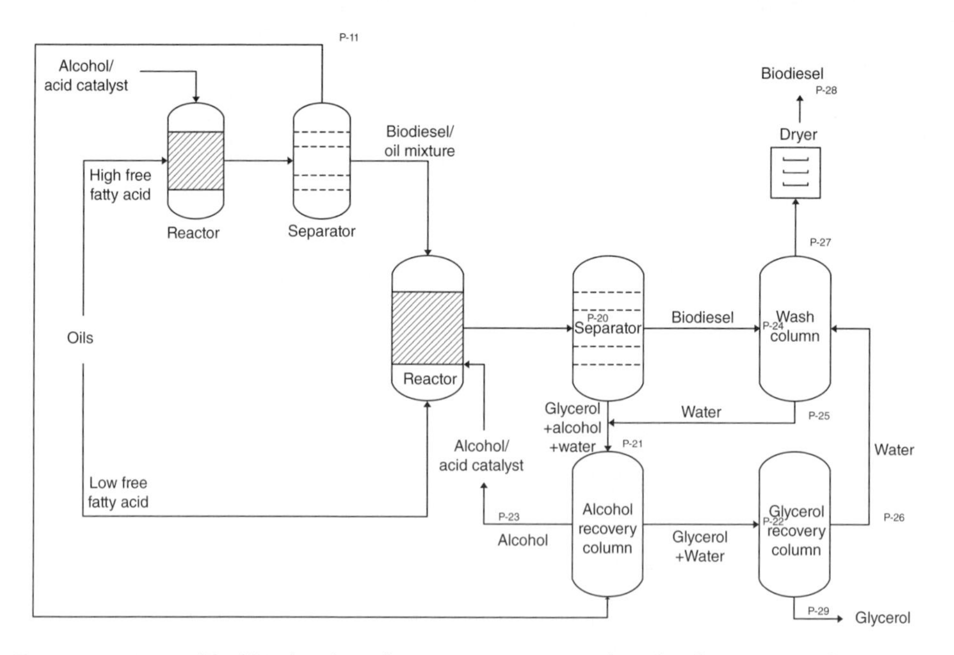

Figure 6.6 *A simplified biodiesel production process. Reproduced with permission from D. Co.*

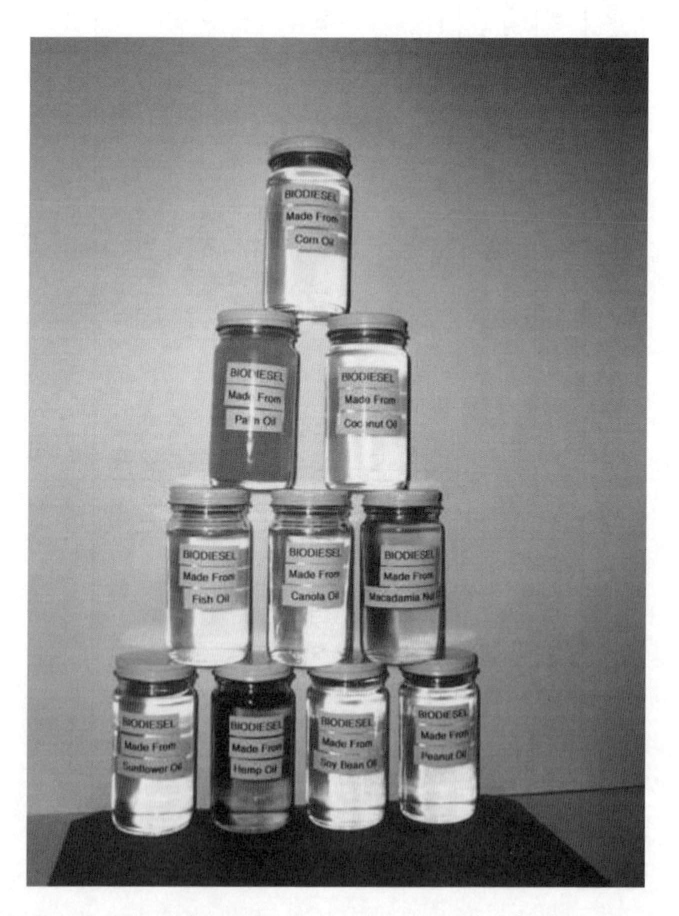

Figure 6.7 *Biodiesels from different sources vary greatly in clarity and color. Photograph courtesy of Utah Biodiesel Supply, with permission*

Most of the excess methanol used in the prior described process is distributed in the heavier glycerin phase, and this is recovered in a distillation process. The glycerin is further separated from any water that may be formed and may be either sold as crude glycerin or refined further. Because of the large volume of glycerin being produced by biodiesel manufacturers, the price of glycerin has dropped considerably and it is currently an interesting challenge to find other uses for glycerin (Singhabhandhu and Tezuka, 2010).

6.4.2 Utilization of the Carbohydrate Fraction (Bioethanol and Biogas)

There are two major options for the utilization of the carbohydrate fraction of the microbial biomass: bioethanol and biogas (sometimes called biomethane). An excellent review on the possible utilization of the carbohydrate fraction has been recently published by Markou et al. (2013). When compared to the work being done on biodiesel, there are fewer studies that have been done on microalgal bioethanol and biogas.

Biogas has the advantage of being a relatively low technology process which has been used for a long time for the utilization of agricultural wastes (Deublein and Steinhauser, 2008). Energy inputs are low except for occasional agitation and heating, if the location is in one of the colder climates. Because of this, it has been used as a catch-all technology for making use of residues and wastes of all kinds. After extraction of the more valuable components, like the lipids, then the residue can be fermented into biogas since the conversion to biogas is less sensitive to the feedstock. This improves the overall process efficiency and furthermore mitigates a waste stream. It has been suggested that utilization of the carbohydrate fraction is essential for improving the energy and resource efficiency of the overall process (Sialve et al., 2009).

The yield of biogas may be estimated according to the equation

$$C_aH_bO_dN_eS_f + yH_2O \rightarrow xCH_4 + eNH_3 + fH_2S + (a-x)CO_2 \qquad (6.3)$$

where $x=(4a+b-2d-3e-2f)/8$ and $y=(4a-b-2d+3e+2f)/4$ (Deublein and Steinhauser, 2008). It must be noted that there is necessarily some CO_2 generation during the generation of the methane. In many proposed biorefinery concepts, the biogas is fed back to the microalgal culture process for utilization by the equipment supporting microalgae production.

In theory, practically any living matter can be converted to biogas, although, in reality, yields vary according to the feedstock. Carbohydrates consist of chains of the simple sugars, and the relative digestibility depends on the length of the chain and the stability of the bonds between the sugars. Lignocellulosic feedstocks are more difficult to digest, while the sugars and the starchy portions are easier. Microalgal feedstocks are known to have comparatively low amounts of lignin and cellulose (Markou and Georgakakis, 2011). Table 6.5 shows the composition of the carbohydrate fraction of some microalgal species. It has also been well established that the starchy fraction can be increased by changing the growing conditions such as by changing the light, temperature, and nutrient levels (Dragone et al., 2011; Fernandes et al., 2013), as discussed earlier.

A critical parameter in the generation of biogas is the C/N ratio. Too much nitrogen may be inhibitory to the methanogenic bacteria, whereas too little nitrogen could lead to an

Table 6.5 *Carbohydrate composition of selected microalgae*

Sugar	Microalgae					
	Chloroccum sp.	*Spirulina platensis*	*Chlamydomonas reinhardtii*	*Nitzschia closterium*	*Phaeodactylum tricornutum*	*Dunaliella tertiolecta*
Xylose (%)	27	7.0		7.0	7.5	1.0
Mannose (%)	15	9.3	2.3	16.8	45.9	4.5
Glucose (%)	47	54.4	74.9	32.6	21.0	85.3
Galactose (%)	9		4.5	18.4	8.9	1.1
Rhamnose (%)		22.3	1.5	7.7	8.6	5.5
Reference	Harun and Danquah (2011a)	Shekharam et al. (1987)	Choi et al. (2010)	Brown (1991)	Brown (1991)	Brown (1991)

Adapted from Markou et al. (2012).

insufficient amount of biogas generation (Deublein and Steinhauser, 2008). To improve the C/N ratio, codigestion of the microalgae with other feedstocks can be done to improve yield (Samson and LeDuy, 1983; Markou et al., 2013). Thermal pretreatment has been shown to be advantageous for methane generation from *Nannochloropsis salina* biomass (Schwede et al., 2013), although a study of the effects of thermal, alkali, and ultrasonic pretreatment on methane generation from a mixed *Chlorella–Scenedesmus* sp. biomass was not found to offer much advantage (Cho et al., 2013).

The steps after biogas generation differ according to the intended use. Direct combustion may be done in engines designed specifically for that purpose. If it is desired that the biogas be fed into the natural gas distribution network, then further purification is necessary. In addition to a few other requirements, the main criterion is that the purity of the methane must be at least 96% (Deublein and Steinhauser, 2008).

Bioethanol is a more complex product to produce than biogas. Unlike biogas, which can be produced by indigenous microorganisms, production of ethanol requires a specific microorganism. It has been commonly accepted that *Saccharomyces cerevisiae* is the most efficient organism for the conversion of carbohydrates to ethanol (John et al., 2011), although other organisms like *Zymomonas mobilis* have also been tested (Ho et al., 2013). As with other biomass feedstocks, some pretreatment may be necessary prior to the anaerobic fermentation by *S. cerevisiae*. Unlike most terrestrial plant biomass, microalgal biomass contains no lignin. However, it does contain cellulosic carbohydrates, and these need to be broken up to obtain the highest yield. Harun et al. (Harun and Danquah, 2011a, 2011b; Harun et al., 2011) have investigated the use of enzymes, alkali, and acids for pretreatment of *Chlorococcum infusionium* and found significant yield improvements for each one. No comparisons were made, however, so it is difficult to know which pretreatment scheme is more desirable.

6.4.3 Utilization of the Protein Fraction (Nitrogenous Compounds)

The third major component of biomass is protein, which consists of chains of amino acids. Generally speaking, this third component would contain almost all of the elemental nitrogen present in the biomass. The nitrogen in the proteins has been "fixed." In other words, while the nitrogen in the atmosphere can be considered essentially inert, the proteins in living material are now reactive. This provides an opportunity to "harvest" some of this fixed nitrogen. Conceptually, then, the process can be similar to that shown in Figure 6.8.

After oil extraction and conversion to biodiesel, the remaining residue would contain the carbohydrate and protein fractions. There are then two options for the utilization of the carbohydrate fraction. The biodiesel residue can be converted to bioethanol, and then the resulting residue can be converted to biogas, as suggested by Park et al. (2012). Alternatively, the entire carbohydrate fraction can be converted to biogas. In either case, the residue from the biogas digester, commonly called the digestate, would contain most of the nitrogenous compounds.

It has been suggested that the nitrogen-rich digestate may be directly used as a fertilizer (e.g., see Skenhall et al., 2013). A possibly better alternative is to recover the ammonia and utilize it as a substitute for ammonia produced by traditional means. The ammonia from the digestion of the biomass is distributed between the biogas and the digestate. Strik et al. (Strik et al., 2006) found that the distribution of nitrogen between the gas and liquid phases in an anaerobic digester roughly follows Henry's law and that the Henry's law coefficient (K_H) is 310 M atm^{-1}. This is much larger than the Henry's law coefficient for pure water which is 51 M atm^{-1} and is expected to be sensitive to pH (Deublein and Steinhauser, 2008).

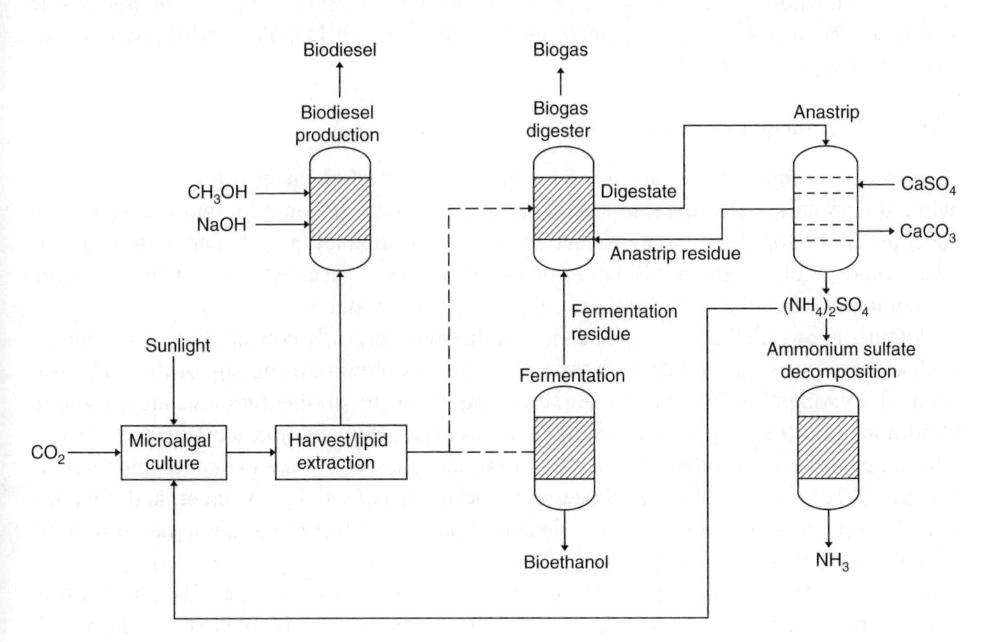

Figure 6.8 A possible process flow diagram for utilization of microalgal biomass. Reproduced with permission from D. Co.

The Henry's law constant estimate will probably have to be determined for each specific situation because its value can be masked by mass transfer limitations.

The ammonia compounds in the digestate can be recovered using a process called ANASTRIP® (Deublein and Steinhauser, 2008). In ANASTRIP, the ammonia-bearing digestate is stripped with a gas, which is subsequently scrubbed with a gypsum ($CaSO_4 \cdot H_2O$) solution which produces ammonium sulfate [$(NH_4)_2SO_4$] (Deublein and Steinhauser, 2008) via the reaction

$$CaSO_4 + 2NH_3 + CO_2 + H_2O \rightarrow CaCO_3 + (NH_4)_2 SO_4 \tag{6.4}$$

This process for converting the ammonia in the digestate to a solid form has advantages which include the relative stability of ammonium sulfate as compared to anhydrous ammonia or ammonium nitrate and a low operating temperature (80°C) which enables the use of waste heat from other processes. The end products consist of a 40% solution of ammonium sulfate and a slurry consisting of a 70% calcium carbonate solution. The ammonium sulfate solution can be separated from the calcium carbonate by gravity settling and further processed into crystals, if desired. The 70% slurry of calcium carbonate can be used for agricultural "liming" or for addition to algal ponds. In addition, there is a residue consisting of the digestate stripped of ammonium compounds. This is a very dilute solution consisting of undigested biomass and exopolysaccharides. This stream may be recycled back to the digester.

The resulting ammonium sulfate may be decomposed to NH_3 and H_2SO_4. The ammonia can be recovered and liquefied, while the sulfuric acid can be absorbed in a sulfuric acid solution to produce a more concentrated sulfuric acid solution. A process to recover biogas and ammonia products from a nitrogen-fixing cyanobacteria has been analyzed from a life-cycle viewpoint, and it was shown that there are possibly great savings in nonrenewable energy use and global warming potential (Razon, 2012, 2014). The entire process flow diagram is shown in Figure 6.9.

6.4.4 Thermochemical Conversion

The processes shown in Figure 6.8 are generally undertaken under mild conditions, and while the process looks ideal, it is unlikely that an organism can be found that can be an ideal feedstock for all possible products: biodiesel, bioethanol, biogas, and ammonia. It is much more likely that a combination of two products, or at most three, would be more economically effective since costs are shared between products.

Another approach that has been taken is to thermochemically convert the entire biomass under higher pressures and temperatures. One of the more promising approaches is hydrothermal treatment where the temperature and pressure of the biomass are raised to conditions near the critical point and the biomass is reacted with the accompanying water. The concept has been previously applied to other types of biomass (Kroger and Muller-Langer, 2012). In hydrothermal processing, the water is used as a reactant, and thus, the expensive process of drying the microalgae is bypassed. However, there is a trade-off in the higher operating temperatures and pressures.

In general, lower temperatures (160–220°C) and pressures (less than 20 atm) result in hydrothermal carbonization which gives a solid, carbonaceous product (biochar), which can be further gasified and used for Fischer–Tropsch fuel synthesis. The concept has been demonstrated on a lab scale using *Chlamydomonas reinhardtii* (Heilmann et al., 2010).

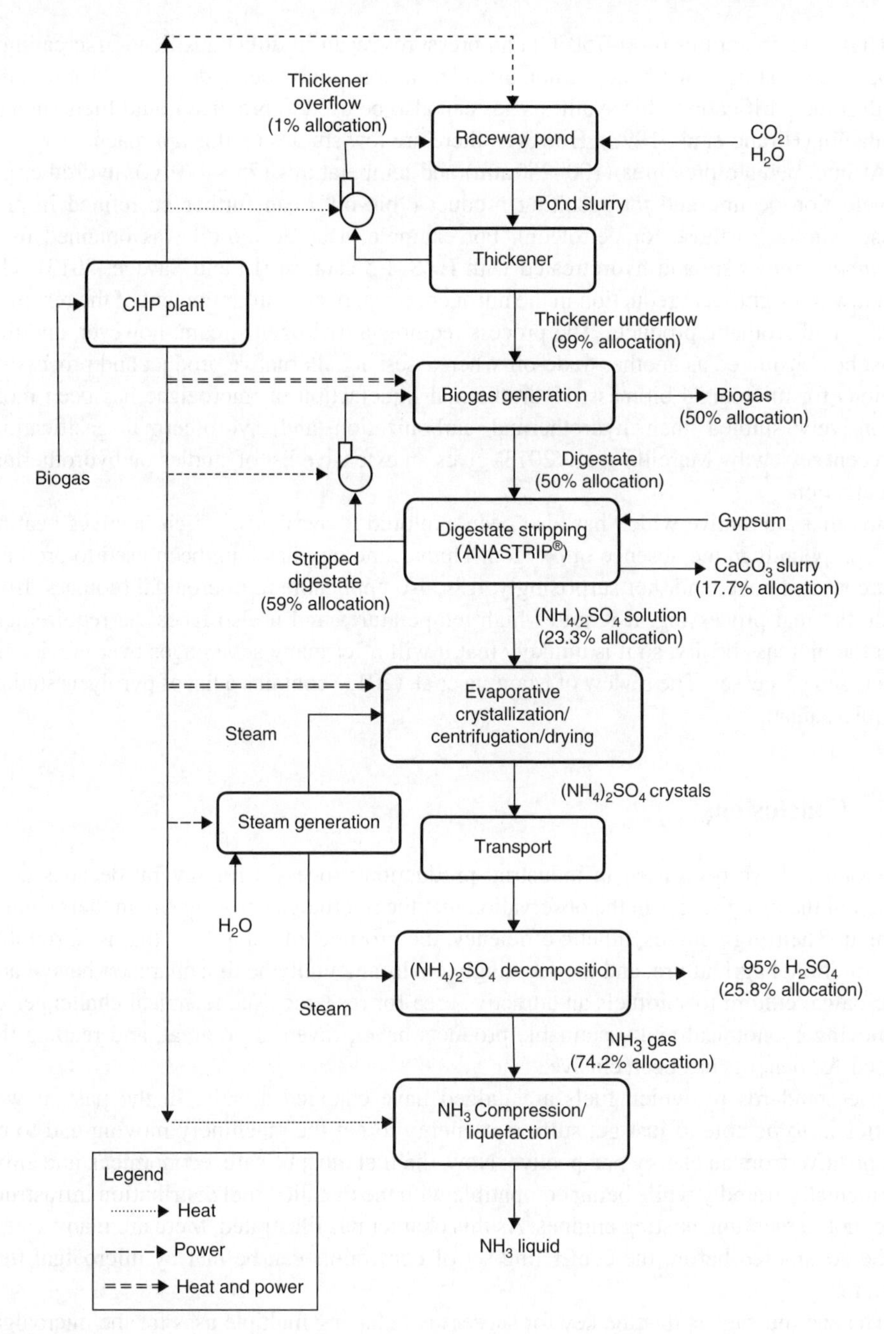

Figure 6.9 *A possible process flow diagram for the production of ammonia from cyanobacteria. Reproduced from Razon (2014) with permission from John Wiley*

Higher temperatures (400–750°C) and pressures result in direct gasification, creating a product consisting of methane, carbon dioxide, carbon monoxide, hydrogen, and ammonia. With some purification, this synthesis gas can also be used to produce liquid fuels such as methanol (Hirano et al., 1998). However, there are few studies on this approach.

At intermediate pressures (100–250 atm) and temperatures (280–370°C), hydrothermal liquefaction occurs, and the resulting product ("bio-oil") can further be refined in processes similar to those for petroleum. For example, a crude bio-oil was obtained from *Nannochloropsis* sp. and hydrotreated with HZSM-5 catalyst (Li and Savage, 2013). The result was a significant reduction in the nitrogen, oxygen, and sulfur content of the resulting alkane and aromatic products. The process requires a hydrogen stream, however, and this must be considered as another trade-off when assessing alternative product and processing options for microalgal biomass. Hydrothermal liquefaction of microalgae has been more extensively studied than hydrothermal carbonization and hydrothermal gasification. A recent review by Marcilla et al. (2013) gives an extensive list of studies on hydrothermal liquefaction.

Another alternative which has also been explored is pyrolysis, which involves heating the compounds in the absence of oxygen. This technology has long been used to produce charcoal and coke, and, not surprisingly, it is also applicable to microalgal biomass. Like hydrothermal processing, it requires high temperatures and it also faces the requirement that the biomass be dry, so it is unlikely that it will offer many advantages over previously discussed processes. The review of Marcilla et al. (2013) contains a list of pyrolysis studies on microalgae.

6.5 Conclusions

Microalgae have been used in industrial production processes for several decades now. Much of the promise lies in the observation that the microalgae grow easily in their natural habitat. Their high photosynthetic efficiency, the prospect of using land that is unsuitable for traditional agriculture, and the seemingly environmentally benign character have made microalgal culture for biofuels an attractive area for research. The technical challenges of achieving economical and sustainable products have proven to be great, and reaping the hoped-for benefits has been elusive.

The standards by which fuels are judged have changed greatly. In the past, it was sufficient to be able to just get sufficient energy to get the machinery moving and to be net positive from an energy perspective. Now, the fuel must be safe, economical, and environmentally friendly while being compatible with the installed fuel distribution infrastructure, not to mention existing engines. As this chapter has illustrated, there are many things to be considered before the conflicting set of constraints can be met by microalgal fuel systems.

The end message is that the key for success is in having multiple uses for the microalgal biomass, such that there are many products which may share the financial and environmental costs. The other advantage then is that there would not be as much waste generated. This model is not new. Most successful industrial crops, for example, soybean, corn, coconut, have many and varied downstream products. It is the same with petroleum itself and the same must be done for microalgal systems if they are to succeed.

Acknowledgments

L. F. Razon would like to thank the Sustainability Studies Program of the Philippines' Commission on Higher Education (CHED) Philippine Higher Education Research Network (PHERNet) for funding support. The authors would also like to thank Ms. Dana Mae Co and Ms. Rose Ubando for their help in drawing some of the figures.

References

ANL; NREL; PNNL. (2012). Renewable Diesel from Algal Lipids: An Integrated Baseline for Cost, Emissions, and Resource Potential from a Harmonized Model. 2012, ANL/ESD/12-4; NREL/TP-5100-55431; PNNL-21437. Argonne, IL: Argonne National Laboratory; Golden, CO: National Renewable Energy Laboratory; Richland, WA: Pacific Northwest National Laboratory.

Ben-Amotz, A.; Tornabene, T. G.; Thomas, W. H. Chemical profile of selected species of microalgae with emphasis on lipids. *J. Phycol.* 1985, **21**, 72–81.

Ben-Amotz, A. Large Scale Open Algae Ponds, 2013. http://www.nrel.gov/biomass/pdfs/benamotz.pdf (accessed July 24, 2015).

Brown, M. R. The amino-acid and sugar composition of 16 species of microalgae used in mariculture. *J. Exp. Mar. Biol. Ecol.* 1991, **145**, 79–99.

Chisti, Y. Biodiesel from microalgae. *Biotechnol. Adv.* 2007, **25**, 294–306.

Cho, S.; Park, S.; Seon, J.; Yu, J.; Lee, T. Evaluation of thermal, ultrasonic and alkali pretreatments on mixed-microalgal biomass to enhance anaerobic methane production. *Bioresour. Technol.* 2013, **143**, 330–336.

Choi, S. P.; Nguyen, M. T.; Sim, S. J. Enzymatic pretreatment of *Chlamydomonas reinhardtii* biomass for ethanol production. *Bioresour. Technol.* 2010, **101**, 5330–5336.

Christenson, L.; Sims, R. Production and harvesting of microalgae for wastewater treatment, biofuels, and bioproducts. *Biotechnol. Adv.* 2011, **29**, 686–702.

Clarens, A. F.; Resurreccion, E. P.; White, M. A.; Colosi, L. M. Environmental life cycle comparison of algae to other bioenergy feedstocks. *Environ. Sci. Technol.* 2010, **44**, 1813–1819.

Converti, A.; Oliveira, R. P. S.; Torres, B. R.; Lodi, A.; Zilli, M. Biogas production and valorization by means of a two-step biological process. *Bioresour. Technol.* 2009, **100**, 5771–5776.

Cysewski, G. R. Microalgae Production and Their Use in Animal Feeds, 2013. http://www.oceanicinstitute.org/newsevents/pdf/05CysewskiCyanotech.pdf (accessed July 24, 2015).

Danquah, M. K.; Gladman, B.; Moheimani, N.; Forde, G. M. Microalgal growth characteristics and subsequent influence on dewatering efficiency. *Chem. Eng. J.* 2009, **151**, 73–78.

Department of Environment and Conservation (DEC) (2009). Resource Condition Report for a Significant Western Australia Wetland: Hutt Lagoon. Department of Environment and Conservation, Perth, Western Australia.

Deublein, D.; Steinhauser, A. *Biogas from Waste and Renewable Resources: An Introduction.* Weinheim: Wiley-VCH Verlag GmbH & Co. KgaA, 2008.

Dragone, G.; Fernandes, B. D.; Abreu, A. P.; Vicente, A. A.; Teixeira, J. A. Nutrient limitation as a strategy for increasing starch accumulation in microalgae. *Appl. Energy* 2011, **88**, 3331–3335.

Fernandes, B.; Teixeira, J.; Dragone, G.; Vicente, A. A.; Kawano, S.; Bisova, K.; Pribyl, P.; Zachleder, V.; Vitova, M. Relationship between starch and lipid accumulation induced by nutrient depletion and replenishment in the microalga Parachlorella kessleri. *Bioresour. Technol.* 2013, **144**, 268–274.

Gatenby, C.; Orcutt, D.; Kreeger, D.; Parker, B.; Jones, V.; Neves, R. Biochemical composition of three algal species proposed as food for captive freshwater mussels. *J. Appl. Phycol.* 2003, **15**, 1–11.

Geider, R. J.; La Roche, J. Redfield revisited: variability of C:N:P in marine microalgae and its biochemical basis. *Eur. J. Phycol.* 2002, **37**, 1.

Greenwell, H. C.; Laurens, L. M. L.; Shields, R. J.; Lovitt, R. W.; Flynn, K. J. Placing microalgae on the biofuels priority list: a review of the technological challenges. *J. R. Soc. Interface* 2010, **7**, 703–726.

Harun, R.; Danquah, M. K. Enzymatic hydrolysis of microalgal biomass for bioethanol production. *Chem. Eng. J.* 2011a, **168**, 1079–1084.

Harun, R.; Danquah, M. K. Influence of acid pre-treatment on microalgal biomass for bioethanol production. *Process Biochem.* 2011b, **46**, 304–309.

Harun, R.; Jason, W. S. Y.; Cherrington, T.; Danquah, M. K. Exploring alkaline pre-treatment of microalgal biomass for bioethanol production. *Appl. Energy* 2011, **88**, 3464–3467.

Harun, R.; Singh, M.; Forde, G. M.; Danquah, M. K. Bioprocess engineering of microalgae to produce a variety of consumer products. *Renew. Sustain. Energy Rev.* 2010, **14**, 1037–1047.

Heilmann, S. M.; Davis, H. T.; Jader, L. R.; Lefebvre, P. A.; Sadowsky, M. J.; Schendel, F. J.; von Keitz, M. G.; Valentas, K. J. Hydrothermal carbonization of microalgae. *Biomass Bioenergy* 2010, **34**, 875–882.

Hirano, A.; Hon-Nami, K.; Kunito, S.; Hada, M.; Ogushi, Y. Temperature effect on continuous gasification of microalgal biomass: theoretical yield of methanol production and its energy balance. *Catal. Today* 1998, **45**, 399–404.

Ho, S.; Huang, S.; Chen, C.; Hasunuma, T.; Kondo, A.; Chang, J. Bioethanol production using carbohydrate-rich microalgae biomass as feedstock. *Bioresour. Technol.* 2013, **135**, 191–198.

IGV Biotech. 2013. Photobioreactor PVR 4000G. http://commons.wikimedia.org/wiki/File:Photobioreactor_PBR_4000_G_IGV_Biotech.jpg (accessed February 1, 2014).

John, R. P.; Anisha, G. S.; Nampoothiri, K. M.; Pandey, A. Micro and macroalgal biomass: A renewable source for bioethanol. *Bioresour. Technol.* 2011, **102**, 186–193.

Jorquera, O.; Kiperstok, A.; Sales, E. A.; Embirucu, M.; Ghirardi, M. L. Comparative energy life-cycle analyses of microalgal biomass production in open ponds and photobioreactors. *Bioresour. Technol.* 2010, **101**, 1406–1413.

Jungbluth, N.; Chudacoff, M.; Dauriat, A.; Dinkel, F.; Doka, G.; Faist Emmenegger, M.; Gnansanou, E.; Kljun, N.; Schleiss, K.; Spielmann, M.; Stettler, C.; Sutter, J.Life cycle inventories of bioenergy. Dübendorf, Swiss Centre for Life Cycle Inventories 2007, Ecoinvent Report No. 17.

Knothe, G. A technical evaluation of biodiesel from vegetable oils vs. algae. Will algae-derived biodiesel perform? *Green Chem.* 2011, **13**, 3048–3065.

Knothe, G. Dependence of biodiesel fuel properties on the structure of fatty acid alkyl esters. *Fuel Process. Technol.* 2005, **86**, 1059–1070.

Kroger, M.; Muller-Langer, F. Review on possible algal-biofuel production processes. *Biofuels* 2012, **3**, 333–349.

Kumar, K.; Dasgupta, C. N.; Nayak, B.; Lindblad, P.; Das, D. Development of suitable photobioreactors for CO2 sequestration addressing global warming using green algae and cyanobacteria. *Bioresour. Technol.* 2011, **102**, 4945–4953.

Lardon, L.; Helias, A.; Sialve, B.; Stayer, J.; Bernard, O. Life-cycle assessment of biodiesel production from microalgae. *Environ. Sci. Technol.* 2009, **43**, 6475–6481.

Li, Z.; Savage, P. E. Feedstocks for fuels and chemicals from algae: Treatment of crude bio-oil over HZSM-5. *Algal Res.* 2013, **2**, 154–163.

Lundquist, T. J.; Woertz, I. C.; Quinn, N. W. T.; Benemann, J. R. *A Realistic Technology and Engineering Assessment of Algae Biofuel Production*. Berkeley, CA: Energy Biosciences Institute, 2010 Available at: http://works.bepress.com/tlundqui/5 (accessed July 24, 2015).

Marcilla, A.; Catala, L.; Garcia-Quesada, J. C.; Valdes, F. J.; Hernandez, M. R. A review of thermochemical conversion of microalgae. *Renew. Sust. Energ. Rev.* 2013, **27**, 11–19.

Markou, G.; Angelidaki, I.; Georgakakis, D. Microalgal carbohydrates: An overview of the factors influencing carbohydrates production, and of main bioconversion technologies for production of biofuels. *Appl. Microbiol. Biotechnol.* 2012, **96**, 631–645.

Markou, G.; Angelidaki, I.; Georgakakis, D. Carbohydrate-enriched cyanobacterial biomass as feedstock for bio-methane production through anaerobic digestion. *Fuel* 2013, **111**, 872–879.

Markou, G.; Georgakakis, D. Cultivation of filamentous cyanobacteria (blue-green algae) in agro-industrial wastes and wastewaters: A review. *Appl. Energy* 2011, **88**, 3389–3401.

Mata, T. M.; Martins, A. A.; Caetano, N. S. Microalgae for biodiesel production and other applications: A review. *Renew. Sustain. Energy Rev.* 2010, **14**, 217–232.

Orchard, S. (2009) Hutt Lagoon. http://commons.wikimedia.org/wiki/File:Hutt_Lagoon,_Western_Australia.jpg (accessed February 1, 2014).

Pahl, S. L.; Lee, A. K.; Kalaitzidis, T.; Ashman, P. J.; Sathe, S.; Lewis, D. M. Harvesting, Thickening and Dewatering Microalgae Biomass; Borowittzka, M. A., Moheimani, N. R., Eds., *Algae for Biofuels and Energy*. Springer Science, 2013; pp 165–185.

Park, J.; Yoon, J.; Park, H.; Lim, D. J.; Kim, S. Anaerobic digestibility of algal bioethanol residue. *Bioresour. Technol.* 2012, **113**, 78–82.

Quinn, J.; Catton, K.; Wagner, N.; Bradley, T. Current large-scale US biofuel potential from microalgae cultivated in photobioreactors. *Bioenergy Res.* 2012, **5**, 49–60.

Rawat, I.; Ranjith Kumar, R.; Mutanda, T.; Bux, F. Biodiesel from microalgae: A critical evaluation from laboratory to large scale production. *Appl. Energy* 2013, **103**, 444–467.

Razon, L. F. Life cycle energy and greenhouse gas profile of a process for the production of ammonium sulfate from nitrogen-fixing photosynthetic cyanobacteria. *Bioresour. Technol.* 2012, **107**, 339–346.

Razon, L. F. Life cycle analysis of an alternative to the Haber–Bosch process: Non-renewable energy usage and global warming potential of liquid ammonia from cyanobacteria. *Environ. Prog. Sustainable Energy* 2014, **33**, 618–624.

Samson, R.; LeDuy, A. Improved performance of anaerobic digestion of Spirulina maxima algal biomass by addition of carbon-rich wastes. *Biotechnol. Lett.* 1983, **5**, 677–682.

Sanford, S. D.; White, J. M.; Shah, P. S.; Wee, C.; Valverde, M. A.; Meier, G. R. Renewable Energy Group. 2009. Feedstocks and Biodiesel Characteristic Report. http://www.biodiesel.org/reports/20091117_gen-398.pdf (accessed February 21, 2014).

Schenk, P. M.; Thomas-Hall, S. R.; Stephens, E.; Marx, U. C.; Mussgnug, J. H.; Posten, C.; Kruse, O.; Hankamer, B. Second generation biofuels: High-efficiency microalgae for biodiesel production. *Bioenergy Res.* 2008, **1**, 20–43.

Schwede, S.; Rehman, Z.; Gerber, M.; Theiss, C.; Span, R. Effects of thermal pretreatment on anaerobic digestion of Nannochloropsis salina biomass. *Bioresour. Technol.* 2013, **143**, 505–511.

Scott, S. A.; Davey, M. P.; Dennis, J. S.; Horst, I.; Howe, C. J.; Lea-Smith, D. J.; Smith, A. G. Biodiesel from algae: challenges and prospects. *Curr. Opin. Biotechnol.* 2010, **21**, 277–286.

Sheehan, J.; Dunahay, T.; Bennemann, J.; Roessler, P. (1998) A look back at the U.S. Department of Energy's aquatic species program—biodiesel from algae. Close-out report. Golden, CO. Department of Energy, National Renewable Energy Laboratory. NREL/TP-580-24190.

Shekharam, K. M.; Venkataraman, L. V.; Salimath, P. V. Carbohydrate composition and characterization of two unusual sugars from the blue green alga *Spirulina platensis*. *Phytochemistry* 1987, **26**:2267–2269.

Sialve, B.; Bernet, N.; Bernard, O. Anaerobic digestion of microalgae as a necessary step to make microalgal biodiesel sustainable. *Biotechnol. Adv.* 2009, **27**, 409–416.

Simon, J. 1994. Spirulina. http://commons.wikimedia.org/wiki/File:Spirul.jpg (accessed February 1, 2014).

Singhabhandhu, A.; Tezuka, T. A perspective on incorporation of glycerin purification process in biodiesel plants using waste cooking oil as feedstock. *Energy* 2010, **35**, 2493–2504.

Skenhall, S. A.; Berndes, G.; Woods, J. Integration of bioenergy systems into UK agriculture – New options for management of nitrogen flows. *Biomass Bioenergy* 2013, **54**, 219–226.

Spolaore, P.; Joannis-Cassan, C.; Duran, E.; Isambert, A. Commercial applications of microalgae. *J. Biosci. Bioeng.* 2006, **101**, 87–96.

Strik, D. P. B. T. B.; Domnanovich, A. M.; Holubar, P. A pH-based control of ammonia in biogas during anaerobic digestion of artificial pig manure and maize silage. *Process Biochem.* 2006, **41**, 1235–1238.

Sturm, B. S. M.; Lamer, S. L. An energy evaluation of coupling nutrient removal from wastewater with algal biomass production. *Appl. Energy* 2011, **88**, 3499–3506.

Taka. 2006. Haematococcus. http://commons.wikimedia.org/wiki/File:Haematococcus_(2006_02_27).jpg (accessed February 1, 2014).

Uduman, N.; Qi, Y.; Danquah, M. K.; Forde, G. M.; Hoadley, A. Dewatering of microalgal cultures: A major bottleneck to algae-based fuels. *J. Renew. Sustain. Energy* 2010, **2**, 012701–012701.

Zhu, J.; Rong, J.; Zong, B. Factors in mass cultivation of microalgae for biodiesel. *Chinese J. Catal.* 2013, **34**, 80–100.

7

Optimal Planning of Sustainable Supply Chains for the Production of *Ambrox* based on *Ageratina jocotepecana* in Mexico

Sergio I. Martínez-Guido[1], J. Betzabe González-Campos[2], Rosa E. Del Río[2], José M. Ponce-Ortega[1], Fabricio Nápoles-Rivera[1], and Medardo Serna-González[1]

[1] Chemical Engineering Department, Universidad Michoacana de San Nicolás de Hidalgo, Morelia, Mexico

[2] Institute for Chemical and Biological Researches, Universidad Michoacana de San Nicolás de Hidalgo, Morelia, Mexico

7.1 Introduction

The perfume industry has shown great progress during the past years mainly due to the economic profit that this industry produces every year. The two main components of a perfume are fragrant oils and fixatives, the latter being the most expensive ingredient in a perfume (Fráter et al., 1998). Historically, the ambergris, which comes from some species of sperm whale, has been successfully used as natural fixative, being one of the most valuable raw materials bringing uniqueness to a perfume; however, the environmental implications (such as killing of endangered species such as the *Physeter macrocephalus*) of its use and high cost have promoted the development of synthetic alternatives to ambergris, which may be much more cheaply and reliably produced. Among all the synthetic substitutes, *Ambrox*® [(–)-8α-12-epoxy-13,14,15,16-tetranorlabdane] is the most widely used

Process Design Strategies for Biomass Conversion Systems, First Edition. Edited by Denny K. S. Ng, Raymond R. Tan, Dominic C. Y. Foo, and Mahmoud M. El-Halwagi.
© 2016 John Wiley & Sons, Ltd. Published 2016 by John Wiley & Sons, Ltd.

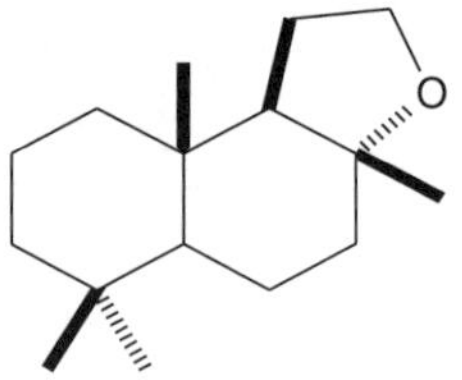

Figure 7.1 *(−)-8α-12-Epoxy-13,14,15,16-tetranorlabdane* "Ambrox." *Reproduced with permission from Martínez-Guido et al. (2014), © 2014, American Chemical Society*

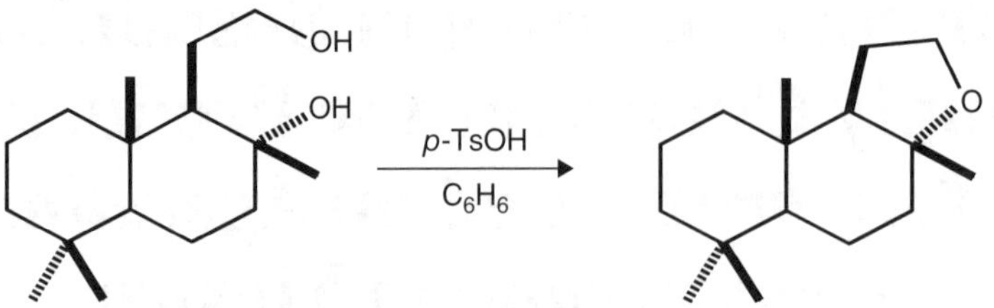

Figure 7.2 *Chemical cyclization to obtain* Ambrox *from tetranorlabdanodiol. Reproduced with permission from Martínez-Guido et al. (2014), © 2014, American Chemical Society*

ambergris-replacement odorant in perfume manufacture and has largely taken its place. It was first synthesized by Stoll and Hinder in 1950 (see Figure 7.1) (Barco et al., 1995; Suwancharoen et al., 2012).

All the chemical routes reported until today involve several chemical steps, having high processing costs, long reaction times, and severe processing conditions such as high pressure and temperature, which leads to unsafe processes with low profits. Recently, the *Ageratina jocotepecana*, an endemic plant of Michoacán state in Mexico, was characterized (Gutiérrez-Pérez et al., 2012), finding that it contains tetranorlabdanodiol, which is a direct precursor for the synthesis of *Ambrox*; tetranorlabdanodiol requires only one chemical cyclization to obtain *Ambrox* (see Figure 7.2).

The synthesis of *Ambrox* by the chemical cyclization of tetranorlabdanodiol obtained from *A. jocotepecana* offers several advantages over current chemical syntheses (Castro et al., 2002; Costa et al., 2005), reducing the synthesis to only one step under mild conditions and higher conversion rates. Additionally, the implementation of this process in industrial scale might contribute to the social and economic development of rural areas in the state of Michoacán, México.

7.2 *Ambrox* Supply Chain

Several factors are involved in the success of an industry; one of the first steps is to develop an efficient supply chain. A supply chain is a network of facilities and distribution mechanisms that perform the functions of material procurement, material transformation to intermediates and final products, and distribution of these products to customers (Shah et al., 2007; Papageorgiou, 2009). Taking these concepts as guidelines for the optimization of the supply chain for the production of *Ambrox* from *A. jocotepecana* is of key importance

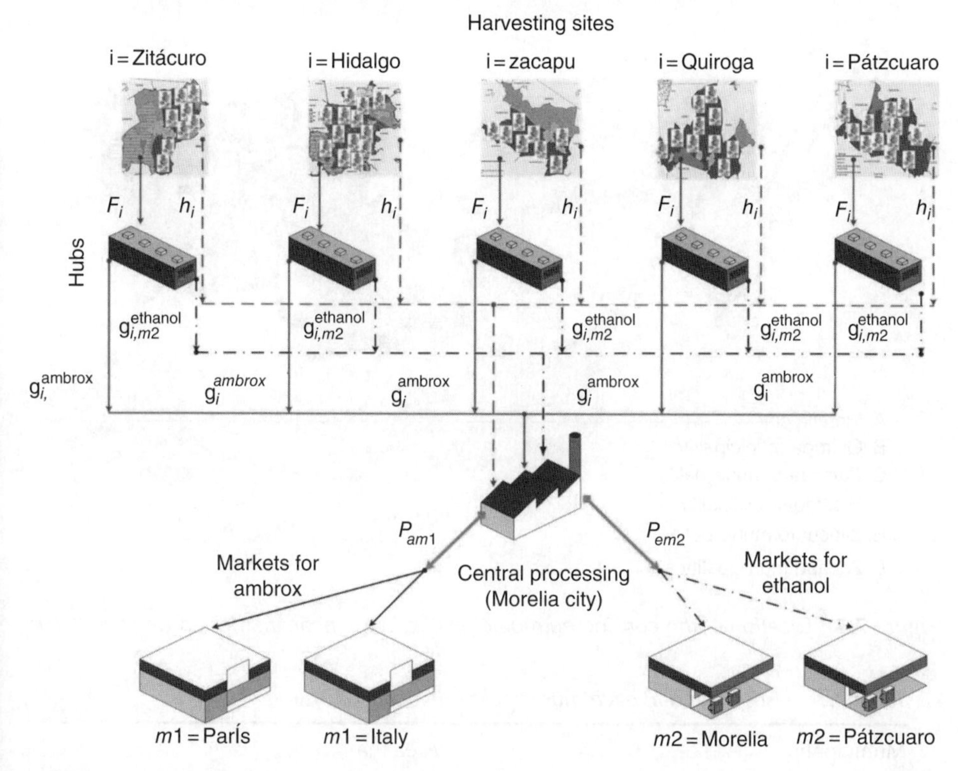

Figure 7.3 *Proposed superstructure for the supply chain of* Ambrox *from* A. jocotepecana. *Reproduced with permission from Martínez-Guido et al. (2014), © 2014, American Chemical Society*

to achieve efficiency as global system and determine if it is economically, environmentally, and socially feasible. The supply chain associated with this process consists of three main parts: the harvesting sites for the raw material production; the processing plants for the synthesis, in this case the existence or not of preprocessing and central processing plants; and the markets for the sale of the final products and subproducts being considered. Figure 7.3 shows a superstructure which represents all the required steps to carry out the *Ambrox* production from *A. jocotepecana* in Michoacán, Mexico.

7.3 Biomass Cultivation

For the raw material cultivation, it is necessary to consider the type of soil and climate conditions that allow the development of the *A. jocotepecana* crops; the plant was endemically found along the road from Morelia to Zacapu, which mainly has mild weather with Andosol soil type. For these reasons, municipalities such as Quiroga, Pátzcuaro, Zacapu, Zitácuaro, and Hidalgo (which have similar weather and soil conditions) are considered potential harvesting sites for *A. jocotepecana*; all these municipalities are located in Michoacán, Mexico (see Figure 7.4). The data for the land used for each municipality are shown in Table 7.1.

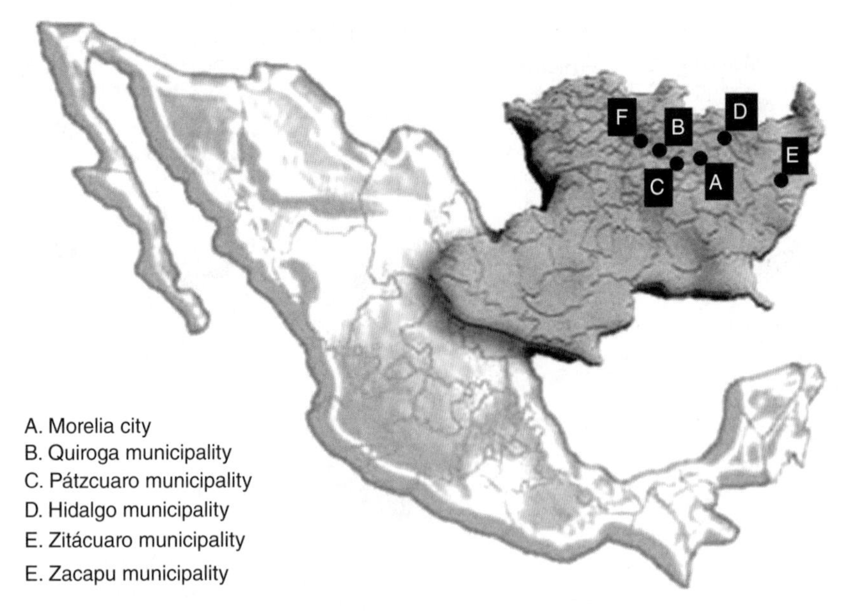

Figure 7.4 *Location of the considered municipalities in Michoacán, Mexico (INEGI, 2010)*

Table 7.1 *Uses of land in each municipality (INEGI, 2010)*

Municipality	Area (ha)			
	Farming	Cultivated with *A. jocotepecana*	Urban zone	Total
Hidalgo	153,921.9	5.59	24,158.15	1,118,522.35
Pátzcuaro	182,621.6	2.09	19,519.34	419,973.15
Quiroga	84,450.9	1.02	6,860.92	204,095.47
Zacapu	180,221.0	2.15	21,156.97	430,055.32
Zitácuaro	209,709.2	2.410	27,733.56	482,077.73

Table 7.2 *Job generation and cost associated with* A. jocotepecana *growing*[a]

Activity	Social	Economic
Crop	0.137 jobs/ha	USD994/ha
Harvest	0.129 jobs/ha	USD556/ha
Transport	0.000,433,2 jobs/ton	USD1.28/ton[b]

[a] (FIRA, 2007) and (Agriculture Commission from Veracruz, 2010).
[b] The distance for this cost is 42 km.

A. jocotepecana cultivation requires several activities such as soil preparation, planting, spraying, crop fertilizing, harvesting, and transportation; the details of the jobs generated for each of these activities along with the associated cost are shown in Table 7.2. It can be seen that growing *A. jocotepecana* might have a positive social impact due to the generation of jobs.

7.4 Transportation System

Raw materials have to be transported from the cropping sites to the processing plants, either central or preprocessing plants. The main difference between these facilities is the location; the central plant is located in Morelia, whereas the preprocessing facilities can be located in the farming sites (i.e., in each municipality). The existence or nonexistence of these plants, as well as their capacity, is a variable of the optimization process and will depend primarily on the amount of raw material that can be produced in each municipality and the transportation cost of raw material and final products and by-products to the customers. Figure 7.5 shows the selected municipalities and their location with respect to Morelia City. The transportation costs are shown in Table 7.3.

7.5 *Ambrox* Production

With the available raw material, the next step is the extraction of the tetranorlabdanodiol and the *Ambrox* production; this is done through the methodology proposed by Gutiérrez-Pérez et al. (2012).

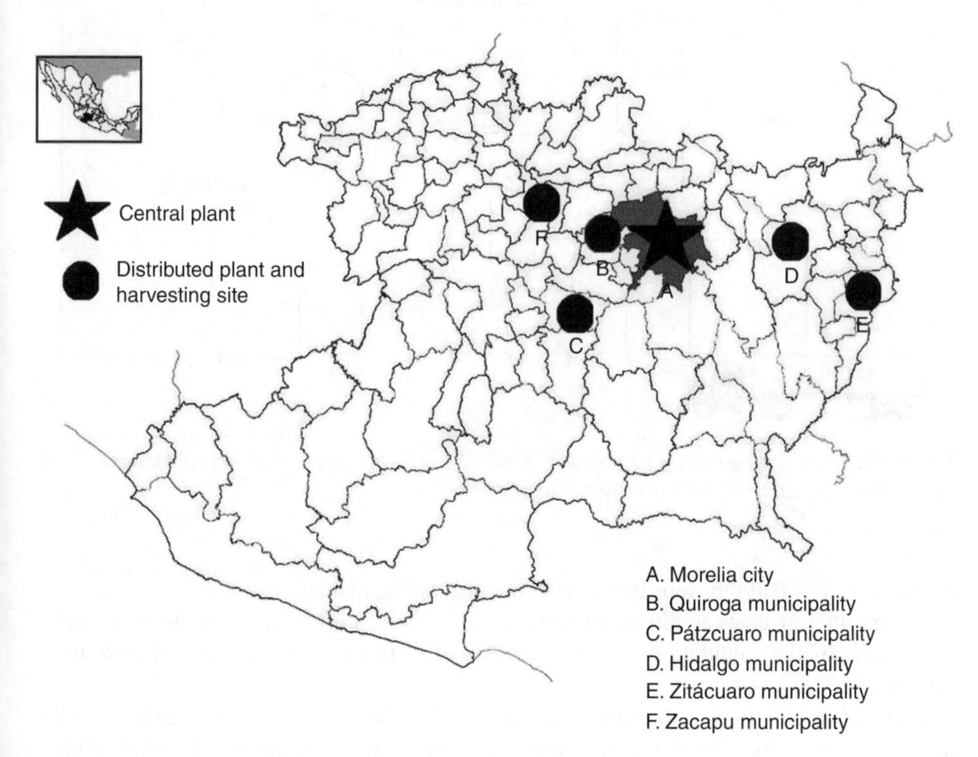

Figure 7.5 *Location of municipalities (preprocessing plants) and Morelia City (central processing plant) in the Michoacán state map (INEGI, 2010)*

Table 7.3 *Transportation costs of raw material, products, and by-products (FIRA, 2007; INEGI, 2010)*

Municipality	Distance (km) to the central plant	Transport of bioethanol (USD/gal)	Transport of *Ambrox* (USD/ton)	Transport of raw material (USD/ton)
Hidalgo	103	0.009,43	3.16	3.16
Pátzcuaro	56.4	0.005,16	1.73	1.73
Quiroga	42	0.003,84	1.28	1.28
Zacapu	82	0.007,50	2.51	2.51
Zitácuaro	152	0.013,90	4.66	4.66

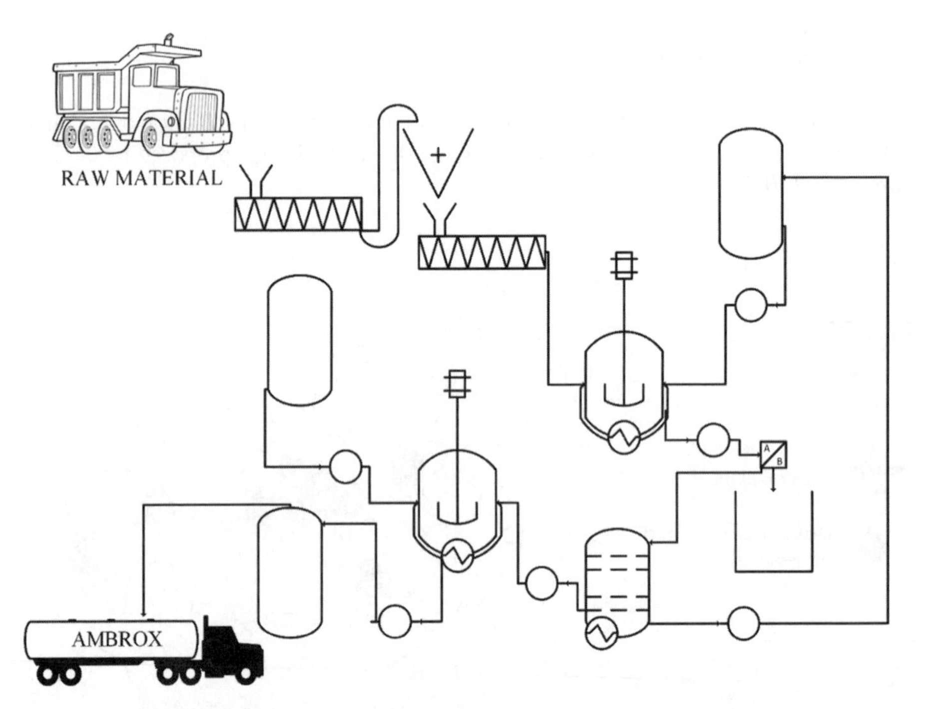

Figure 7.6 *Steps for* Ambrox *processing. Reproduced with permission from Martínez-Guido et al. (2014), © 2014, American Chemical Society*

Figure 7.6 shows all the steps involved in the processing plants.

The production costs are divided into fixed and variable costs. Table 7.4 shows a detailed description of the calculation of variable costs for a plant with a maximum capacity of 40.8 ton/day.

Table 7.5 shows the variable costs for plants with different capacities; it can be seen that the variable cost for central plants is lower than the cost for preprocessing plants; this is

Table 7.4 *Variable costs of processing plants*[a]

Concept	Energy consumed (kWh/ton)	Cost (USD/ton)
Transportation line	5.705	1.990
Drying	18.764	3.252
Strain	5.353	2.020
Filtered	3.757	2.150
Manipulation	3.194	1.606
Heating	1.080	2.173
Evaporating	19.690	3.337
Shake	3	2.247
Pumping	0.3	1.810
Total	60.843	20.585

[a] http://www.tecnologiaslimpias.org/html/central/331104/331104_eca.htm

Table 7.5 *Variable costs associated with the processing plants (SENER, 2006)*

Minimum capacity (ton/day)	Maximum capacity (ton/day)	Central plant	Preprocessing plant
0	0	0	0
0	40.8	15	20
40.8	122.4	13	17
122.4	149.6	10	15

Table 7.6 *Fixed costs associated with the processing plants (SENER, 2006)*

Minimum capacity (ton/day)	Maximum capacity (ton/day)	Cost USD/year
0	0	0
0	40.8	15,000
40.8	122.4	50,000
122.4	149.6	158,000

associated with the fact that the central plant is located in an industrialized zone, and all the utilities and required infrastructure are readily available. The fixed costs are also expressed in terms of the capacity of each plant (ton of *A. jocotepecana* stems processed per day). An economic life of 10 years is considered with a discount rate of 12%. These fixed costs are listed in Table 7.6.

The analysis of the jobs required for the production and transportation of *Ambrox* is shown in Table 7.7.

The cost analysis is the measure of the economic impact generated by the synthesis of *Ambrox*, but it is not the only impact to the process generated. In Mexico, from 1970 to 2003, the industrial sector jobs increased by 104% at national level, while the population

Table 7.7 *Jobs generated for Ambrox production*[a]

Activity	Jobs
Processing in the preprocessing plants	0.2484/ton
Transport of *Ambrox* between plants	1/ton
Processing in the central plant	0.2482/ton
Transport of *Ambrox* to the markets	2/ton

[a]Agriculture Commission from Veracruz (2010) and Gody and Loredo (2010).

Table 7.8 *Production yield*[a]

Municipality	Bioethanol (%/ton) $\left(\beta_i^{\text{ethanol}}\right)$	*Ambrox* (%/ton) $\left(\beta_i^{\text{ambrox}}\right)$	*Ageratina jocotepecana* (ton/ha) (α_i)
Hidalgo	0.65	5.31E−04	13
Pátzcuaro	0.6	6.31E−04	17
Quiroga	0.59	5.21E−04	20
Zacapu	0.55	5.65E−04	15
Zitácuaro	0.62	5.95E−04	10

[a]SAGARPA (2011), SAGARPA (2012a), SAGARPA (2012b), and Gutiérrez-Pérez et al. (2012).

increased by 94%; however, the current situation in the country shows a large lag in job creation in all sectors including the industrial sector. The complete process for *Ambrox* production at industrial level generates some jobs that are needed to carry out the process. These jobs represent one contribution to the employment generation in the state of Michoacán. The number of generated jobs is listed in Table 7.7.

7.6 Bioethanol Production

Due to the large amount of biomass produced in the *Ambrox* process form *A. jocotepecana*, the production of bioethanol is considered as by-product. The yields of *A. jocotepecana* per hectare and from *A. jocotepecana* to *Ambrox* and bioethanol are shown in Table 7.8.

7.7 Supply Chain Optimization Model

Once the *Ambrox* demand has been defined, the entire *Ambrox* supply chain system has to be optimized by means of a modeling framework. The proposed model is a mixed-integer linear programming (MILP) model, which includes all the related aspects of the supply chain, from raw materials to products in the final markets. In order to determine the total extension of land that has to be cultivated with *A. jocotepecana* (A_i), a balance considering the area currently occupied by *A. jocotepecana* (A_i^{exiting}) and the new area required (A_i^{new}) is used:

$$A_i = A_i^{\text{exiting}} + A_i^{\text{new}}, \quad \forall i \in I \tag{7.1}$$

The area cultivated has to be less than the total available area. This way, a maximum limit of the area in each municipality can be defined to avoid excessive change of land or even change of crops; this can be done with the following constraint:

$$A_i \le A_i^{\max}, \quad \forall i \in I \tag{7.2}$$

The amount of stems per hectare is proportional to the cultivated area and the yield factor (α_i):

$$F_i = \alpha_i A_i, \quad \forall i \in I \tag{7.3}$$

The balance for the stems of *A. jocotepecana* indicates that the total flow of stems is equal to the stems sent to the preprocessing plants plus the stems sent to the central plant:

$$F_i \ge f_i + h_i, \quad \forall i \in I \tag{7.4}$$

It can be seen that the preprocessing plants can receive stems only from the harvesting site associated with their location, while on the other hand the central plant can receive stems from any harvesting site, and thus the total flow in the central plant is equal to the summation of all the flows sent from the different municipalities:

$$F_{\text{central}} = \sum_i h_i \tag{7.5}$$

The products and byproducts can be determined with the following relations. In the preprocessing plants, the *Ambrox* and bioethanol produced are a function of the stems processed and the yield factors to *Ambrox* and bioethanol production:

$$g_i^{\text{ambrox}} = f_i \beta_i^{\text{ambrox}}, \quad \forall i \in I \tag{7.6}$$

$$g_i^{\text{ethanol}} = f_i \beta_i^{\text{ethanol}}, \quad \forall i \in I \tag{7.7}$$

In the central plants, a similar relation is used:

$$P^{\text{ambrox}} = F_{\text{central}} \gamma^{\text{ambrox}} \tag{7.8}$$

$$P^{\text{ethanol}} = F_{\text{central}} \gamma^{\text{ethanol}} \tag{7.9}$$

Then, the total *Ambrox* and bioethanol produced are

$$M^{\text{ambrox}} = P^{\text{ambrox}} + \sum_i g_i^{\text{ambrox}} \tag{7.10}$$

$$M^{\text{ethanol}} = P^{\text{ethanol}} + \sum_i g_i^{\text{ethanol}} \tag{7.11}$$

Now, the products must be distributed to the markets:

$$M^{\text{ambrox}} = \sum_{m1} S_{m1}^{\text{ambrox}} \tag{7.12}$$

$$M^{\text{ethanol}} = \sum_{m2} S_{m2}^{\text{ethanol}} \tag{7.13}$$

And a constraint for the demands in the markets must be observed; this implies that the total *Ambrox* and bioethanol produced must be less or equal to the demand in each market:

$$P_{m1}^{\text{ambrox}} \leq P_{m1}^{\text{max ambrox}}, \quad \forall m1 \in M1 \tag{7.14}$$

$$P_{m2}^{\text{ethanol}} \leq P_{m2}^{\text{max ethanol}}, \quad \forall m2 \in M2 \tag{7.15}$$

The total cost of the process is represented by Equation 7.16, which considers the cost of the raw material production, raw material transportation, the fixed and variable costs of processing in the preprocessing and central plants, and the transportation costs of the final products. On the other hand, the social aspect associated with the process is measured through the job generation; this is shown in Equation 7.17. This equation considers the created jobs in the harvesting sites, the processing plants (preprocessing and central), and the required labor for transportation of raw material and final products:

$$\begin{aligned} \text{Cost} = & \sum_i C_i^{\text{harvest}} F_i + \sum_i C_{\text{prepropla},i,p}^{\text{cap}} + \sum C_{\text{cen},p}^{\text{cap}} + \sum_i C_i^{\text{transp-ambrox}} g_i^{\text{ambrox}} + \sum_{m2} \sum_i C_{i,m2}^{\text{trans-ethanol}} g_{i,m2}^{\text{ethanol}} \\ & + \sum_i C_i^{\text{trans-plant}} h_i + \sum_{m1} C_{m1}^{\text{transp-ambrox}} P_{m1}^{\text{ambrox}} + \sum_{m2} P_{m2}^{\text{ethanol}} C_{m2}^{\text{trans-ethanol}} \end{aligned} \tag{7.16}$$

$$\begin{aligned} N_{\text{jobs}} = & \sum_i n_i^{\text{harves}} F + \sum_i n_i^{\text{process}} f_i + \sum_i n_i^{\text{process}} h_i + \sum_i n_i^{\text{transambrox}} + g_i^{\text{ambrox}} \sum_{m2} \sum_i n_{ji,m2}^{\text{transethanol}} g_{i,m2}^{\text{ethanol}} \\ & + \sum_i n_i^{\text{transporplant}} h_i + \sum_{m1} n_{m1}^{\text{transportambrox}} P_{m1}^{\text{ambrox}} + \sum_{m2} P_{m2}^{\text{ethanol}} n_{m2}^{\text{transportethanol}} \end{aligned} \tag{7.17}$$

The capital costs shown in the Equation 7.16 for the preprocessing and central plants depend on the processing capacity; however, due to the fact that the capacity of the plant is an optimization variable, this is handled with two disjunctions. The first disjunction is used to determine the location, capacity, and number of required preprocessing plants:

$$\bigvee_p \begin{bmatrix} Y_{p,i} \\ F_{AJ,i,p}^{\text{lower}} \leq F_{AJ,i} \leq F_{AJ,i,p}^{\text{upper}} \\ C_{\text{prepropla},i}^{\text{cap}} = C_{i,p}^{F} + C_{i,p}^{V} F_{AJ,i} \end{bmatrix}, \quad \forall i \in I$$

If the Boolean variable is true, then a given capacity for the plant will be selected and the appropriate costs will be added. The disjunction is reformulated as a set of algebraic equations as follows. First, only one section must be selected (the first section corresponds to a capacity of zero and the unit costs are zero):

$$\sum_p y_{p,i} = 1, \quad \forall i \tag{7.18}$$

Then, the continuous variables are disaggregated:

$$F_{AJ,i} = \sum_p \text{DF}_{AJ,i,p}, \quad \forall i \tag{7.19}$$

$$C^{\text{cap}}_{\text{prepropla},i} = \sum_{p} \text{DC}^{\text{cap}}_{i,p}, \quad \forall i \tag{7.20}$$

Then, the relationships are stated in terms of the disaggregated variables:

$$y_{p,i} F^{\text{lower}}_{AJ,i,p} \leq \text{DF}_{AJ,i,p} \leq y_{p,i} F^{\text{upper}}_{AJ,i,p}, \quad \forall p, \forall i \tag{7.21}$$

$$\text{DC}^{\text{cap}}_{i,p} = C^{\text{F}}_{i,p} y_{p,i} + C^{\text{V}}_{i,p} \text{DF}_{AJ,i,p}, \quad \forall p, \forall i \tag{7.22}$$

In a similar way, a second disjunction is used to determine the existence or nonexistence of the central plant; in this case, the location is not a decision variable, and only the capacity of the plant is considered:

$$\bigvee_{p} \begin{bmatrix} Y_p \\ F^{\text{lower}}_{AJ,\text{central},p} \leq F_{\text{central}} \leq F^{\text{upper}}_{AJ,\text{central},p} \\ C^{\text{cap}}_{\text{cen}} = C^{\text{F}}_{\text{cen},p} + C^{\text{V}}_{\text{cen},p} F_{\text{central}} \end{bmatrix}$$

The disjunction is again reformulated as a set of algebraic constraints:

$$\sum_{p} y_{\text{cen},p} = 1 \tag{7.23}$$

$$F_{\text{central}} = \sum_{p} \text{DF}_{AJ,\text{cen},p} \tag{7.24}$$

$$C^{\text{cap}}_{\text{cen}} = \sum_{p} \text{DC}^{\text{cap}}_{\text{cen},p} \tag{7.25}$$

$$y_{\text{cen},p} F^{\text{lower}}_{AJ,\text{central},p} \leq \text{DF}_{AJ,\text{cen},p} \leq y_{\text{cen},p} F^{\text{upper}}_{AJ,\text{central},p}, \quad \forall p \tag{7.26}$$

$$\text{DC}^{\text{cap}}_{\text{cen},p} = C^{\text{F}}_{\text{cen},p} y_{\text{cen},p} + C^{\text{V}}_{\text{cen},p} \text{DF}_{AJ,\text{cen},p}, \quad \forall p \tag{7.27}$$

The total sales of the process are calculated as follows:

$$\text{Sales} = \sum_{m1} \text{Price}^{\text{ambrox}}_{m1} P^{\text{ambrox}}_{m1} + \sum_{m2} \text{Price}^{\text{ethanol}}_{m2} P^{\text{ethanol}}_{m2} \tag{7.28}$$

With the production cost and total sales, it is possible to evaluate the net profit:

$$\text{NP} = \text{sale} - \text{cost} \tag{7.29}$$

Additionally, it is important to evaluate the environmental impact associated with the process in order to evaluate not only if it is economically and socially attractive but also if it is environmentally sustainable. The environmental impact is evaluated through the eco-indicator 99 (EI-99) (Geodkoop & Spriensma, 2000) based on the life cycle analysis methodology; this methodology includes the environmental impact caused by a specific substance, process, or activity that is necessary in a process. The EI-99 considers 11 impact categories, which are classified into three main damage categories as shown in Figure 7.7.

The global environmental impact (EI) is the value that is generated for all the supply chain to carry out the process at industrial scale; for the specific case of *Ambrox* production

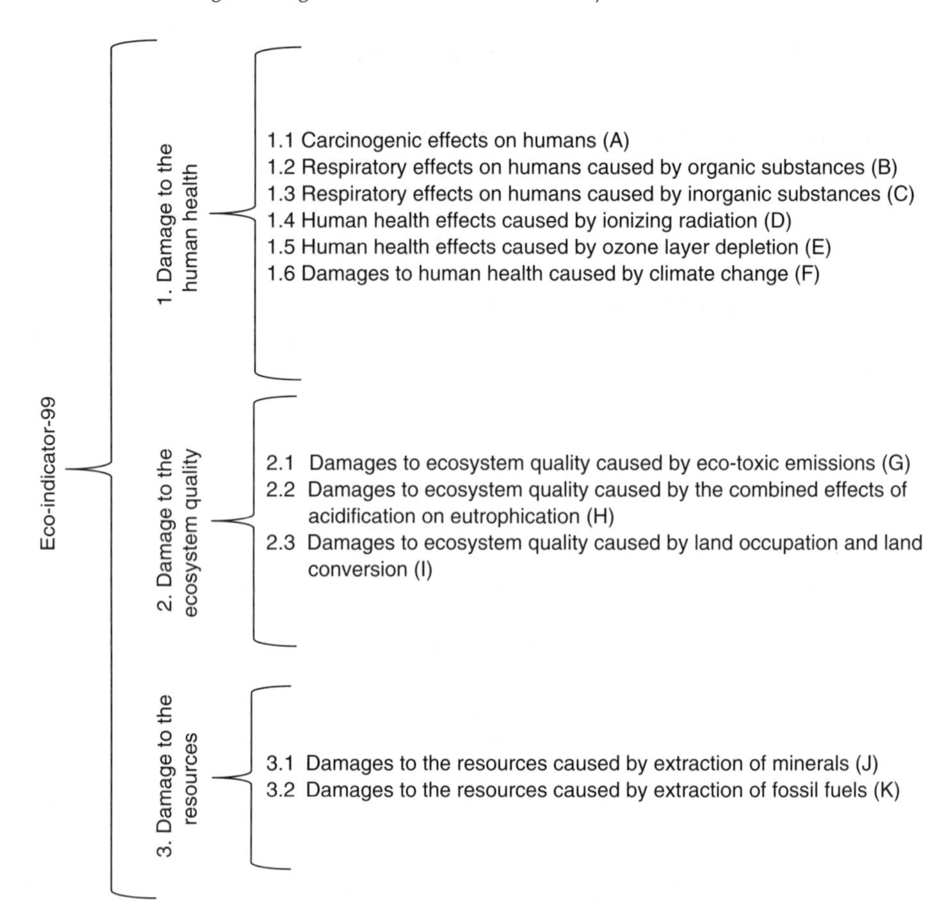

Figure 7.7 *Eco-indicator 99 methodology*

from *A. jocotepecana*, Equation (7.30) applies, and it considers the environmental impact occasioned by the raw material for *Ambrox* and bioethanol production:

$$\mathrm{EI}_{\mathrm{Global}} = \mathrm{EI}_{\mathrm{raw\ material}} + \mathrm{EI}_{\mathrm{BIOETHANOLPRO}} + \mathrm{EI}_{\mathrm{AMBROXPRO}} \tag{7.30}$$

Each of the terms in the previous equation considers the three damage factors considered in the EI-99 methodology—the damage for the human health, the damage for the resources, and the damage for the ecosystem—and finally, the value for each damage is evaluated with the following equations:

$$\mathrm{EI}_{\mathrm{raw\ material}} = D^{\mathrm{RMC}}_{\mathrm{HUMANHEALT}} + D^{\mathrm{RMC}}_{\mathrm{RESOURCES}} + D^{\mathrm{RMC}}_{\mathrm{ECOSYSTEM}} \tag{7.31}$$

$$\mathrm{EI}_{\mathrm{AMBROXPRO}} = D^{\mathrm{AP}}_{\mathrm{HUMANHEALT}} + D^{\mathrm{AP}}_{\mathrm{RESOURCES}} + D^{\mathrm{AP}}_{\mathrm{ECOSYSTEM}} \tag{7.32}$$

$$\mathrm{EI}_{\mathrm{BIOETHANOLPRO}} = D^{\mathrm{BP}}_{\mathrm{HUMANHEALT}} + D^{\mathrm{BP}}_{\mathrm{RESOURCES}} + D^{\mathrm{BP}}_{\mathrm{ECOSYSTEM}} \tag{7.33}$$

And finally, the value for each damage is evaluated with the following equations:

$$D_{\text{HUMANHEALT}}^{\text{RMC}} = \sum_i D_{\text{HE}}^{\text{carcinogenic}} F_i + \sum_i D_{\text{HE}}^{R\text{-organic substances}} F_i + \sum_i D_{\text{HE}}^{R\text{-inorganic substances}} F_i + \sum_i D_{\text{HE}}^{\text{ionzing radiation}} F_i + \sum_i D_{\text{HE}}^{\text{ozone depletion}} F_i + \sum_i D_{\text{HE}}^{\text{climate change}} F_i, \quad \forall i \in I \tag{7.34}$$

$$D_{\text{ECOSYSTEM}}^{\text{RMC}} = \sum_i D_{\text{EC}}^{\text{ecotoxic emissions}} F_i + \sum_i D_{\text{EC}}^{\text{acidification}} F_i + \sum_i D_{\text{EC}}^{\text{land ocupation}} A_i^{\text{existing}} + \sum_i D_{\text{EC}}^{\text{land ocupation}} A_i^{\text{new}}, \quad \forall i \in I \tag{7.35}$$

$$D_{\text{RESOURCES}}^{\text{RMC}} = \sum_i D_{\text{RE}}^{\text{fossil fuels}} F_i + \sum_i D_{\text{RE}}^{\text{mineral extraction}} F_i, \quad \forall i \in I \tag{7.36}$$

$$D_{\text{HUMANHEALT}}^{\text{AP}} = \text{AD}_{\text{HE}}^{\text{carcinogenic}} P^{\text{amrbrox}} + \text{AD}_{\text{HE}}^{R\text{-organic substances}} P^{\text{amrbrox}} + \text{AD}_{\text{HE}}^{R\text{-inorganic substances}} P^{\text{amrbrox}} + \text{AD}_{\text{HE}}^{\text{ionzing radiation}} P^{\text{amrbrox}} + \text{AD}_{\text{HE}}^{\text{ozone depletion}} P^{\text{amrbrox}} + \text{AD}_{\text{HE}}^{\text{climate change}} P^{\text{amrbrox}} \tag{7.37}$$

$$D_{\text{ECOSYSTEM}}^{\text{AP}} = \text{AD}_{\text{EC}}^{\text{ecotoxic emissions}} P^{\text{ambrox}} + \text{AD}_{\text{EC}}^{\text{acidification}} P^{\text{ambrox}} + \text{AD}_{\text{EC}}^{\text{land ocupation}} \text{NAVE} \tag{7.38}$$

$$D_{\text{RESOURCES}}^{\text{AP}} = \text{AD}_{\text{RE}}^{\text{fossil fuels}} P^{\text{amrbrox}} + \text{AD}_{\text{RE}}^{\text{mineral extraction}} P^{\text{amrbrox}} \tag{7.39}$$

$$D_{\text{HUMANHEALT}}^{\text{BP}} = \sum_{m2} \text{BD}_{\text{HE}}^{\text{carcinogenic}} M_{m2}^{\text{ethanol}} + \sum_{m2} \text{BD}_{\text{HE}}^{R\text{-organic substances}} M_{m2}^{\text{ethanol}}{}_i + \sum_{m2} \text{BD}_{\text{HE}}^{R\text{-inorganic substances}} M_{m2}^{\text{ethanol}}{}_i + \sum_{m2} \text{BD}_{\text{HE}}^{\text{ionzing radiation}} M_{m2}^{\text{ethanol}} + \sum_{m2} \text{BD}_{\text{HE}}^{\text{ozone depletion}} M_{m2}^{\text{ethanol}} + \sum_{m2} \text{BD}_{\text{HE}}^{\text{climate change}} M_{m2}^{\text{ethanol}}, \quad \forall m2 \in M2 \tag{7.40}$$

$$D_{\text{ECOSYSTEM}}^{\text{BP}} = \sum_{m2} \text{BD}_{\text{EC}}^{\text{ecotoxic emissions}} M_{m2}^{\text{ethanol}} + \sum_{m2} \text{BD}_{\text{EC}}^{\text{acidification}} M_{m2}^{\text{ethanol}} + \text{BD}_{\text{EC}}^{\text{land ocupation}} \text{NAVE}, \quad \forall m2 \in M2 \tag{7.41}$$

$$D_{\text{RESOURCES}}^{\text{BP}} = \sum_{m2} \text{BD}_{\text{RE}}^{\text{fossil fuels}} M_{m2}^{\text{ethanol}} + \sum_{m2} \text{BD}_{\text{RE}}^{\text{mineral extraction}} M_{m2}^{\text{ethanol}}, \quad \forall m2 \in M2 \tag{7.42}$$

The values for the considered factors are shown in Tables 7.9, 7.10 and 7.11.

Thus, this is a multiobjective optimization problem which can be stated as follows:

$$\text{Objective function} = (\text{max profit; min EI}) \tag{7.43}$$

which was solved using the constraint method (Diwekar, 2008). The model is an MILP problem, which was coded in the software GAMS and solved with CPLEX in a computer with an Intel® Core™ i7 processor at 2.67 GHz with 8 GB of RAM in 00.078 s of CPU time. The size for the model formulation includes 166 continuous variables, 155 single equations, and 24 discrete variables. This GAMS code is provided as a supplementary electronic material.

Table 7.9 *Values for damages associated with the agriculture of A. jocotepecana*

Agriculture				
Emissions to the water				
	Compound	Emissions per ton of *A. jocotepecana*	Associated factor EI-99 (points)	Type of damage
1	PO_4^{3-}	382.1	—	—
2	NO_3^-	194	—	—
3	Pesticides	45 per ha	0.000,395 factor*ha*year	0.0177G
Emissions to the air				
1	CO_2	192	0.00545 factor*ton emissions	1.04F
2	NO_x	1.024	2.30, 0.445	2.3552C, 0.455H
3	SO_x	0.062	1.42, 0.0812	0.088C, 0.00503H
4	N_2O	0.2	1.79	0.358C
5	NH_3	0.0776	2.21, 1.21	0.0171C, 0.093H
Occupation	Factor*m^2*year		0.000,007,49	I
Conversion	Factor*m^2		0.000,268	I

Reproduced with permission from Martínez-Guido et al. (2014), © 2014, American Chemical Society.

Table 7.10 *Values for damages associated with the bioethanol production*

Production of bioethanol				
Emissions to the air	Compound	Emissions per ton of *A. jocotepecana*	Associated factor EI-99 (points)	Type of damage
1	CO_2	3.37	0.005,45 per ton of emissions	0.0183F

Reproduced with permission from Martínez-Guido et al. (2014), © 2014, American Chemical Society.

Table 7.11 *Values for damages associated with the Ambrox production*

Production of *Ambrox*				
Emissions to the air	Compound	Emissions per ton of *A. jocotepecana*	Associated factor EI-99 (points)	Type of damage
1	CO_2	192.551	0.005,45	1.049 F
2	NO_x	1.024	2.30, 0.445	C, H
3	SO_x	0.062	1.42, 0.0812	C, H
Occupation	Factor*m^2*year		0.0655	I
Conversion	Factor*m^2		1.96	I
Energy from coal	Factor*MJ used		0.000,204	K

Reproduced with permission from Martínez-Guido et al. (2014), © 2014, American Chemical Society.

7.8 Case Study

Figure 7.8 shows the optimal Pareto points for the problem; it can be seen that for the maximum profit, also it obtained the maximum environmental impact. This way, if the environmental impact has to be reduced, the profit will also be reduced. Solution A represents the case when a demand of 25 tons of *Ambrox* is produced, and this scenario will be discussed in detail to show the benefits of optimizing the supply chain of the process.

Currently, the land occupied by *A. jocotepecana* in each municipality is 5.59 ha in Hidalgo, 2.090 ha in Pátzcuaro, 1.020 ha in Quiroga, 2.150 ha in Zacapu, and 2.410 ha in Zitácuaro to yield an extension of 13.26 ha of land. In order to satisfy the total demand of 25 tons per year which is represented by the point A, it is necessary to cultivate 2381.224 ha of new land in Quiroga. Quiroga is selected mainly due to the fact that *A. jocotepecana* is endemic in this region (which allows higher yields in the crops) and because it is the closest city to Morelia, which reduces the transportation cost. The use of new land represents the 2.80 % of the total agricultural land in Quiroga. Figure 7.9 shows the comparison between the land destined for agricultural purposes (shown by the dark rectangles), the current land with wild *A. jocotepecana* (shown by smaller boxes) and the new land (shown by the white box in Quiroga), and the urban land represented by the grey rectangles.

Using the aforementioned land (2381.224 ha in total), it is possible to obtain 47,537.05 tons of stems of *A. jocotepecana*. With this amount of raw materials, the production of *Ambrox* can reach 25 tons, whereas 17,462 gallons of bioethanol can be produced. The jobs generated to carry out this process are divided into two sectors, the farm and industrial; considering that it is necessary to cultivate 2381.224 ha with *A. jocotepecana*, activities like site preparation, fertilization, sowing, and harvesting require 633 jobs per year. This value represents 24% of the jobs generated in the farm sector from 1970 to 2003 in Mexico, while the jobs necessary in the plants are 38 direct jobs with three schedules per day; this value

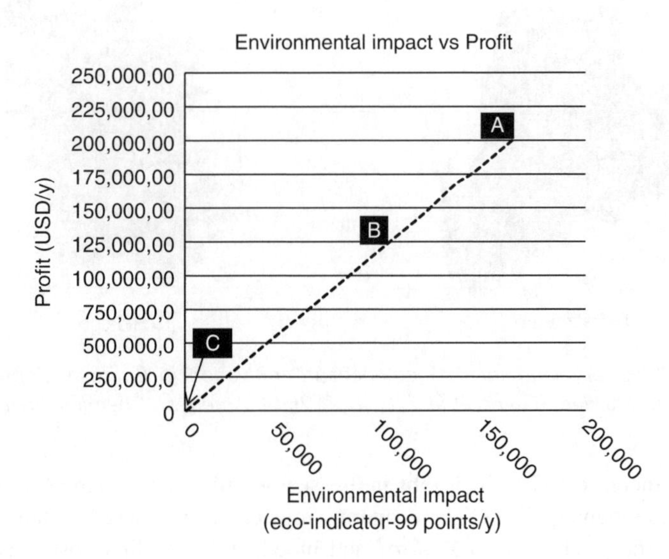

Figure 7.8 *Pareto curve between the profit and environmental impact. Reproduced with permission from Martínez-Guido et al. (2014), © 2014, American Chemical Society*

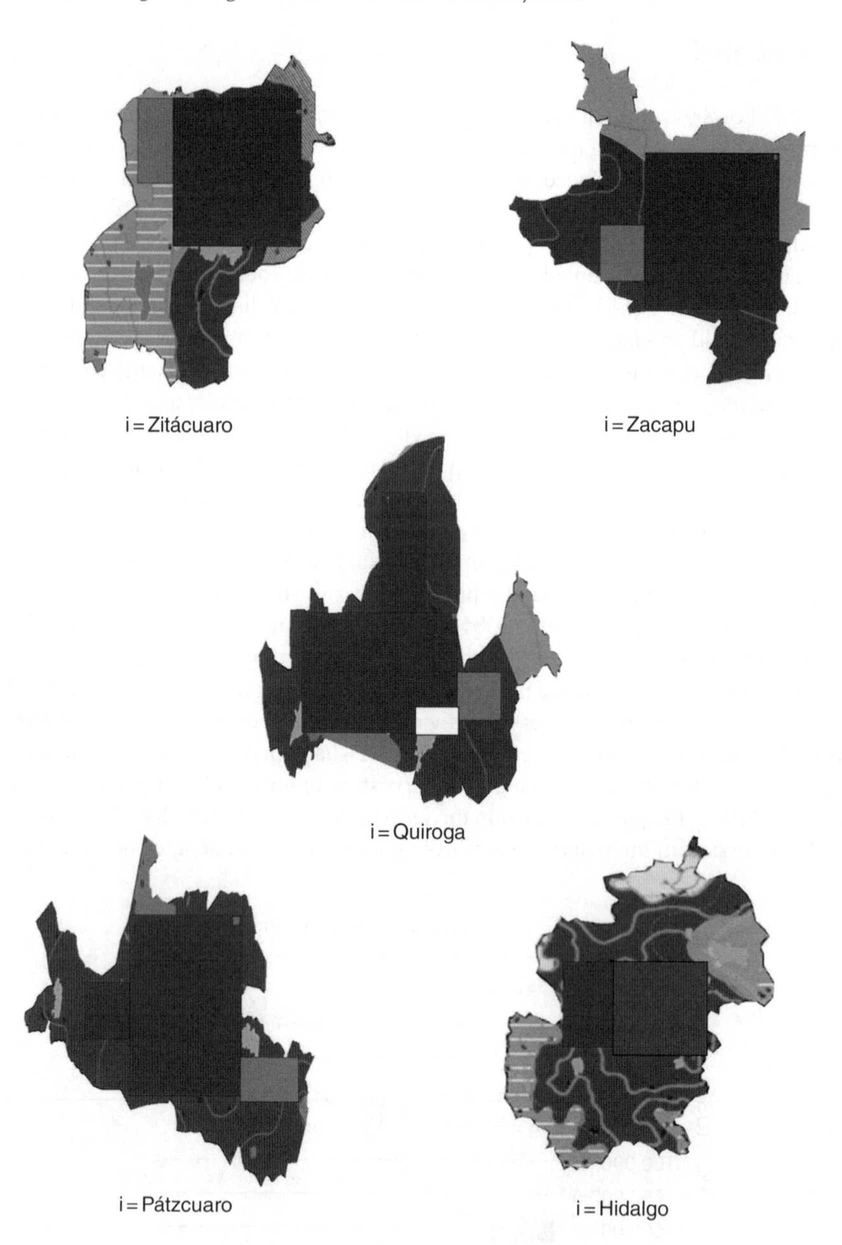

Figure 7.9 *Necessary land for* A. jocotepecana *cultivation in Quiroga. Reproduced with permission from Martínez-Guido et al. (2014), © 2014, American Chemical Society*

represents an increase of 7.61% of jobs in this sector with respect to the period from 1970 to 2003 in these municipalities. The economic impact was measured by the costs that are generated for the production of *Ambrox* and bioethanol. The first cost is generated for activities related to the cultivation of *A. jocotepecana* for a total of USD3,690,122 per year. The next cost evaluated was the transportation of raw material to the processing plant; this cost has a value of USD30,064 per year. The cost of processing *A. jocotepecana* is divided into fixed and variable costs. In solution A, two plants are required: a preprocessing plant located in Quiroga (because it is the largest cultivation site) and the central plant in Morelia. The annualized cost for both plants are USD462,320 and USD362,469, respectively. The cost of transportation of the final product to the markets is USD68,337 per year. The gross profits are USD35,852 per year for *Ambrox* and the sales of bioethanol USD20,132,000 per year.

The total environmental impact was measured based on the EI-99 methodology; the result for scenario A is 164,850 points of EI-99, which encloses all the activities from the cultivation of *A. jocotepecana* to the final product. The aforementioned value is composed by 210,580 points of eco-indicator per year for the cultivation of raw material, 3358 points for the production of bioethanol, and 347 points for the production of *Ambrox*, minus 45,730 points for the positive environmental impact that has the CO_2 fixed by the crops.

Solutions B and C in the Pareto front represent different scenarios in which the environmental impact is reduced; however, the earnings are directly related to the share of the demand that can be satisfied. This means that these solutions cannot cover the 25 tons demanded; on the other hand, solution C is the solution in which no new land is used for the production of *A. jocotepecana*. An alternative to reduce the environmental impact without sacrificing the production is the use of alternative energy sources. Tables 7.12, 7.13, and 7.14 show a comparison of the three evaluated aspects—the social (measured with the job generation), the economic (measured with the processing costs), and the environmental impact (damages for the production in human health, quality of ecosystem and resources)—for the three points in the Pareto front. Figure 7.10 shows the location of the selected plants and the way that raw materials are processed and sent to the final markets.

Table 7.12 *Cost comparison (USD)*

Activity	Farming	Transport raw material	Processing	Transport products	Final earnings
A	3,690,122	30,064	824,789	69,352	20,140,618
B	1,848,047	15,163	452,059	34,668	10,023,401
C	19,778	311	93,126	386	620

Table 7.13 *Comparison of jobs generated*

Sector	Farm	Industrial	Total
A	633	11,799	12,432
B	317	5,898	6,215
C	4	43	47

Table 7.14 *Comparison for environmental impact (EI-99/year)*

Production	Raw material	Bioethanol	*Ambrox*	Subtotal	Fixed CO_2	Total
A	210,580	3358	347	214,285	49,439	164,846
B	105,290	1679	174	107,143	24,719	82,424
C	790	12	4	806	66	740

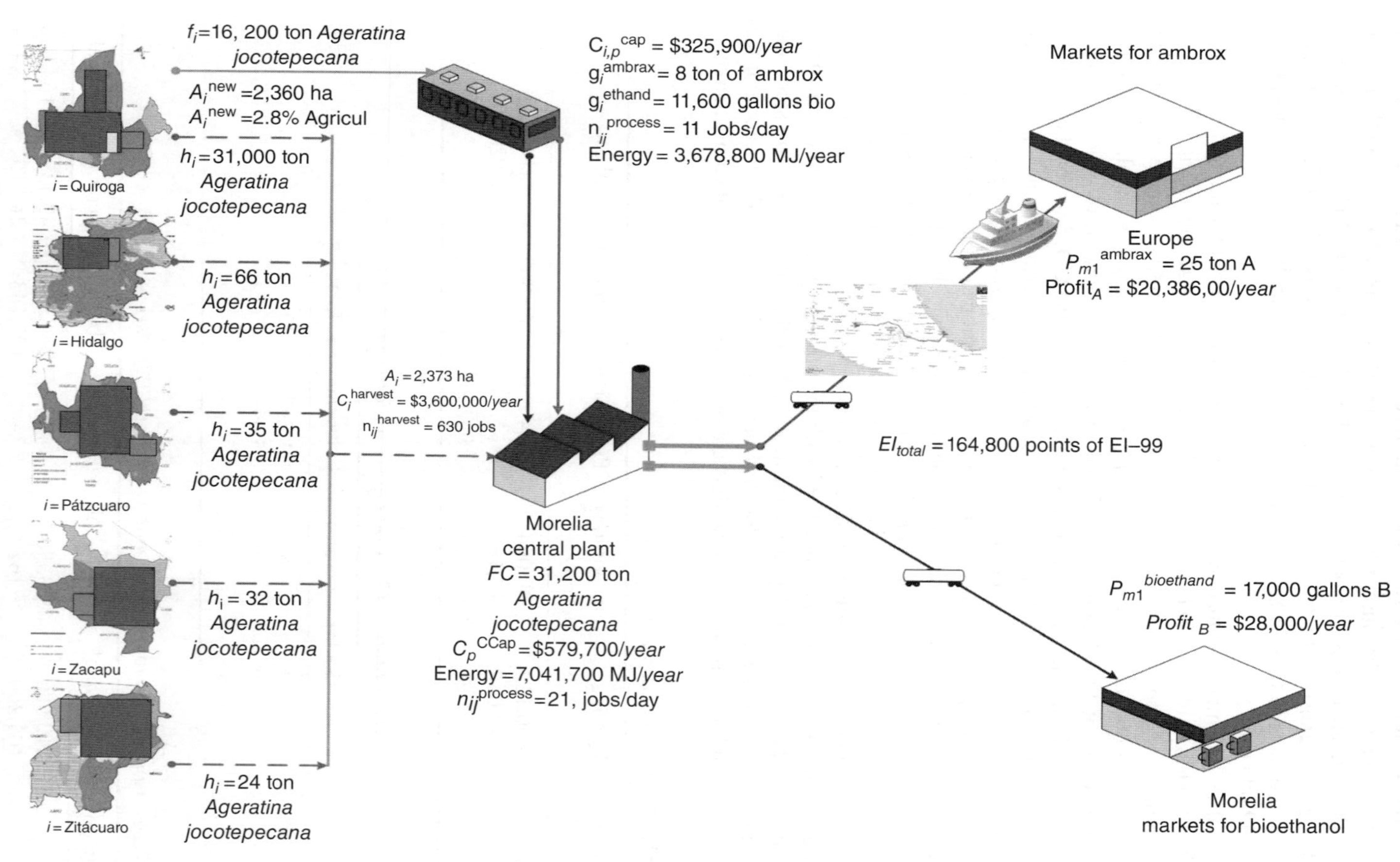

Figure 7.10 *Results for maximizing the net present value. Reproduced with permission from Martínez-Guido et al. (2014), © 2014, American Chemical Society*

7.9 Conclusions

A mathematical programming model for the optimal synthesis of the supply chain for the production of *Ambrox* from *A. jocotepecana* was presented. Results show the potential of the industrial process due to the use of a simplified chemical route coupled with an appropriate supply chain. The implementation of such process in the state of Michoacán, México, due to the availability of the raw materials, impacts positively in the economic and social aspects due to the industrial development and generation of permanent jobs. On the other hand, the major identified concern resides in the environmental impact of the process. The major impact is associated with the production of the raw materials due to the change of land; an alternative to this problem is using existing land designated for agricultural activities as well as sharing crops. Also, the energy associated with the process could be provided by renewable sources, which will also reduce the environmental impact.

Acknowledgments

Authors acknowledge the financial support obtained from CONACyT and CIC-UMSNH.

Nomenclature

Parameters

A_i^{exiting}	Existing area with wild *A. jocotepecana* in site *i*, ha
$A_i^{\max}$	Total area in site *i*
$C_{i,p}^{\text{F}}$	Fixed cost associated with the capacity of each preprocessing facility
$C_{\text{cen},p}^{\text{F}}$	Fixed cost associated with the capacity of central facility
C_i^{harvest}	Unit cost for cultivation of *A. jocotepecana*
C_i^{process}	Unit preprocessing cost
$C_i^{\text{trans-ambrox}}$	Unit processing cost for *Ambrox* in the central processing facility
$C_i^{\text{transpo-ambrox}}$	Unit transportation cost for *Ambrox* from distributed to central processing facilities
$C_{im2}^{\text{transpo-ethanol}}$	Unit bioethanol transportation cost
$C_i^{\text{transpo-plant}}$	Unit transportation cost for raw materials
$C_{m1}^{\text{trans-ethanol}}$	Unit transportation cost for bioethanol from processing facilities to markets
$C_{\text{cen},p}^{\text{V}}$	Variable cost associated with the central facility
$C_{i,p}^{\text{V}}$	Variable cost associated with each preprocessing facility
$F_{AJ,\text{central},p}^{\text{lower}}$	Minimum processed stem of *A. jocotepecana* at central facility
$F_{AJ,i,p}^{\text{lower}}$	Minimum processed stem of *A. jocotepecana* in each preprocessing facility
$F_{AJ,\text{central},p}^{\text{upper}}$	Maximum processed *A. jocotepecana* in the central facility
$F_{AJ,i,p}^{\text{upper}}$	Maximum processed stem flux of *A. jocotepecana* in each preprocessing facility
n_i^{harvest}	Unit generated jobs for cultivation of *A. jocotepecana*
n_i^{process}	Unit processing and transportation jobs for the central processing facility
$n_i^{\text{transambrox}}$	Unit generated jobs for transportation of *Ambrox* from the distributed to the central processing facility

$n_{i,m2}^{\text{transethanol}}$ Unit generated jobs for transportation from the distributed to the central processing facility
$n_i^{\text{transporplant}}$ Unit generated jobs for transportation for *A. jocotepecana* to the central processing facility
$n_{m1}^{\text{transporambrox}}$ Unit generated jobs for the transportation of *Ambrox* to the markets
$n_{m2}^{\text{transporethanol}}$ Unit generated jobs for transportation of bioethanol to markets
$\text{Price}_{m1}^{\text{ambrox}}$ Sale price for *Ambrox* in the market
$\text{Price}_{m2}^{\text{ethanol}}$ Sale price for bioethanol in market
P_{m1}^{ambrox} Demand of *Ambrox*
P_{m2}^{ethanol} Demand of bioethanol
α_i Percent of stalk produced by ha in site i
β_i^{ambrox} Conversion factor from stalks to *Ambrox*
β_i^{ethanol} Conversion factor from stalks to bioethanol
γ^{ambrox} Conversion factor from stalks to *Ambrox* in central plant

Variables

A_i Cultivation area
A_i^{new} New area required
$C_{\text{prepropla},i,p}^{\text{cap}}$ Capital cost associated with each preprocessing facility
$C_{\text{central},p}^{\text{cap}}$ Capital cost associated with central facility
Cost Total capital cost
$\text{DF}_{AJ,i,p}$ Disaggregated variable for flux stem processed in the preprocessing facility
$\text{DC}_{i,p}^{\text{cap}}$ Disaggregated variable for capital cost associated with preprocessing facility
$\text{EI}_{\text{Global}}$ Environmental impact generated by the supply chain
$\text{EI}_{\text{ambrox}}$ Environmental impact generated by *Ambrox* production
$\text{EI}_{\text{bioethanol}}$ Environmental impact generated by the bioethanol production
$\text{EI}_{\text{rawmaterial}}$ Environmental impact generated by the raw material cultivation
F_i Produced *A. jocotepecana* in site i
f_i Stalks sent to the distributed processing facilities
$F_{AJ,i}$ Stem processed in each preprocessing facility
F_{central} Stem processed in the central facility
g_i^{ambrox} *Ambrox* produced in preprocessing facilities
g_i^{ethanol} Bioethanol produced in preprocessing facilities
h_i Flow rate for the stalks sent to the central processing facility
M^{ethanol} Total bioethanol produced
M^{ambrox} Total *Ambrox* produced
N_{jobs} Total generated jobs
NP Gross profit
$P_{m1}^{\text{max ambrox}}$ Demand of *Ambrox* that is satisfied by the sales
$P_{m2}^{\text{max ethanol}}$ Demand of bioethanol that is satisfied by the sales
Profit Total profit
S_{m1}^{ambrox} Amount of *Ambrox* that is sold
S_{m2}^{ethanol} Amount of bioethanol that is sold
Sales Sales
Y_p^c Binary variable for the existence or nonexistence of central facility
$Y_{p,i}$ Binary variable for the existence or nonexistence of preprocessing facility

References

Agriculture Commission from Veracruz. *Monograph sugarcane*. Agriculture Commission: Veracruz, 2010.

Barco, A., Benetti, S., Bianchi, A., Casolari, A., Guarneri, M., Pollini, G.P. Formal synthesis of (±)-*Ambrox*®. *Tetrahedron*. **51** (30),8333–8338, 1995.

Castro, J.M., Salido, S., Altarejos, J., Nogueras, M., Sánchez, A. Synthesis of *Ambrox*® from labdanolic acid. *Tetrahedron*. **52** (29), 5941–5949, 2002.

Costa, M.D.C., Teixeira, S.G., Rodrigues, C.B., Figueiredo, P.R., Curto, M.J.M. A new synthesis of (−)-thio*ambrox* and its 8-epimer. *Tetrahedron*. **61** (18), 4403–4407, 2005.

Diwekar, U.M. *Introduction to applied optimization*, 2nd ed.; Springer-Verlag: Cambridge, MA, 2008.

FIRA. *Impact for the commercialization for the sugar cane*. FIRA: Morelia. 2007.

Fráter, G., Bajgrowicz, J.A., Kraft, P., Fragance chemistry, *Tetrahedron*. **54** (27), 7633–7703, 1998.

Geodkoop, M., Spriensma, R., *The eco-indicator 99 a damage oriented method for life cycle impact assessment: Methodology report and manual for designers*. Technical report. PRé Consultants: Amersfoort, 2000.

Gody, R. and Loredo, J. First Central European mixed with biomass. *The Bioenergy International*. **1** (9), 14–15, 2010.

Gutiérrez-Pérez, A.I., González-Campos, J.B., Del-Río-Torres, R.E.N., Ageratina jocotepecana *fuente de tetranorlbdanodiol*. QFB Thesis. IIQB-UMSNH, Morelia-Michoacán, 2012.

INEGI. *Geographic information for the municipalities of Mexico*. Mexican Institute of Statistics, Geography and Information. Mexico City. 2010.

Martínez-Guido, S. I., González-Campos, J. B., Del-Río, R. E., Ponce-Ortega, J. M., Nápoles-Rivera, F., Serna-González, M., El-Halwagi, M. M. A multiobjective optimization approach for the development of a sustainable supply chain of a new fixative in the perfumery industry. *ACS Sustainable Chemistry & Engineering*. **2** (10), 2380–2390, 2014.

Papageorgiou, L.G. Supply chain optimization for the process industries: Advances and opportunities. *Computers & Chemical Engineering*. **33** (12), 1931–1938, 2009.

SAGARPA. Bioproducts in Mexico. http://www.bioenergeticos.gob.mx/index.php/bioetanol/produccion-a-partir-de-sorgo-dulce.html, 2011. Accessed August 28, 2015.

SAGARPA. Bioethanol from beet. http://www.bioenergeticos.gob.mx/index.php/bioetanol/prouccion-a-partir-de-cana-de-azucar.html, 2012a. Accessed August 28, 2015.

SAGARPA. Bioethanol from sorghum. http://www.bioenergeticos.gob.mx/index.php/bioetanol/produccion-a-partir-de-la-remolacha.html, 2012b. Accessed August 28, 2015.

SENER. *Bioethanol and biodiesel for transportation in Mexico*. Mexican Council of Energy. Mexico City. 2006.

Shah, N., Dunnet, A., Adjiman, C. Biomass to heat supply chains applications of process optimization. *Process Safety and Environmental Protection*. **85** (5), 419–429, 2007.

Suwancharoen, S., Pornpakakul, S., Muangsin, N. Synthesis of en-*Ambrox*® from nidorellol. *Tetrahedron Letters*. **53** (40), 5418–5421, 2012.

8

Inoperability Input–Output Modeling Approach to Risk Analysis in Biomass Supply Chains

Krista Danielle S. Yu[1], *Kathleen B. Aviso*[2], *Mustafa Kamal Abdul Aziz*[3], *Noor Azian Morad*[4], *Michael Angelo B. Promentilla*[2], *Joost R. Santos*[5], *and Raymond R. Tan*[2]

[1] *School of Economics, De La Salle University, Manila, Philippines*

[2] *Chemical Engineering Department, De La Salle University, Manila, Philippines*

[3] *Department of Chemical and Environmental Engineering/Centre of Excellence for Green Technologies, The University of Nottingham, Malaysia Campus, Selangor, Malaysia*

[4] *Malaysia-Japan International Institute of Technology, Universiti Teknologi Malaysia, Johor, Malaysia*

[5] *Engineering Management and Systems Engineering Department, The George Washington University, Washington, DC, USA*

8.1 Introduction

In the recent years, the increasing price of oil and other fossil fuel resources has piqued the interest in developing alternative sources of energy. However, price is not the only factor considered in the search for energy, but these sources have to be sustainable and environmentally friendly as well. Climate change is widely regarded as one of the world's major environmental issues, with the estimated 350 ppm safe limit of atmospheric CO_2 levels having already been exceeded as a result of intensive use of fossil fuels throughout the world (Rockström et al. 2009). Furthermore, the use of alternative energy sources contributes to

Process Design Strategies for Biomass Conversion Systems, First Edition. Edited by Denny K. S. Ng, Raymond R. Tan, Dominic C. Y. Foo, and Mahmoud M. El-Halwagi.
© 2016 John Wiley & Sons, Ltd. Published 2016 by John Wiley & Sons, Ltd.

pollution prevention and ultimately improves societal health. An empirical study by Gauderman et al. (2004) established remarkable correlations between air quality levels and healthy lung development. Based on data from the National Pollution Prevention Roundtable (2009), it has been established that implementing pollution prevention policies could also give rise to significant economic savings. With various governments and international agencies offering lucrative incentives for the use of such resources, the shift toward renewable energy is inevitable. Different countries across the globe have implemented legislations promoting or mandating the blend of biodiesel or bioethanol into commercial fuels to reduce greenhouse gas emissions, as well as reduce dependence on petroleum imports (Balat and Balat 2009). While there are other possible sources of renewable energy, biomass is one of the most widely utilized, accounting for nearly half of renewable energy produced in the United States (US Department of Energy 2013). It is mainly derived from forest and agricultural resources, primarily logging and crop residues, making the production of this type of energy largely dependent on agricultural crop output (Perlack et al. 2005).

Increased demand for agricultural products induces the expansion of agricultural activity into untapped natural ecosystems, which may result in an initial surge in carbon emissions (i.e., "carbon debt"), thus offsetting the advantage of shifting toward bioenergy (Fargione et al. 2008). Stable environment conditions promote sustainable supply of agricultural feedstock. For example, changes in the amount of rainfall, average temperature, and soil conditions can affect the level of crop yield. Furthermore, climate change can also result in shifting plant disease and infestation patterns; it may also cause more frequent extreme weather events that damage crops through wind or flooding. The increasing incidence of climate change-induced disruptions can lead to significant levels of crop failure, resulting to a reduction in biomass available for production of renewable energy (Schaeffer et al. 2012). The effects of climate change also pose risks to various stakeholders and consumers. Aside from the environmental strain that may stem from biofuel production, it can also lead to increased risks pertaining to food security and land-use competition. As an agricultural derivative, farmers will tend to move toward crops that will yield higher returns, reducing the output allocation for food production. This raises the issue of a cointegrated movement between oil prices and food prices in the long run (Baffes 2013).

In evaluating the complex interactions among environmental and economic factors surrounding the use of alternative energy sources, addressing the fundamental questions of risk assessment and risk management is warranted. Kaplan and Garrick (1981) posed the classic triplet of questions for quantitative risk assessment, and their applications to the biomass supply chain are shown in Table 8.1.

Furthermore, to address the risk management aspect, another triplet of questions needs to be addressed (Haimes 1991). An example is given in Table 8.2.

Identification of proper risks is done prior to designing risk management strategies that will efficiently allocate scarce resources. Risk management options are developed in order to satisfy multiple objectives, which are not always commensurate and are often in competition. Hence, a thorough study requires an extensive understanding of the trade-offs faced when deciding among risk management options that require resources to be allocated in an environment when one is seeking to optimize multiple objectives. Multiobjective trade-off analysis provides graphical portrayal of the possible compromises that can be made when evaluating various environmental policy options with respect to multiple objectives (e.g., cost of government subsidy versus environmental benefit) (Zhang et al. 2012, 2014, Luo et al.

Table 8.1 *Triplet of questions in risk assessment for biomass supply chains*

Generic risk assessment question	Bioenergy-specific risk assessment question	Possible risk assessment outcomes
What can go wrong?	What specific events can disrupt supply in a biomass-based supply chain?	Identification of disruptive events, such as: Wind and rain damage from extreme weather events Pest infestation and plant disease Drought and forest fire Seasonal haze
What is the likelihood?	How probable are specific disruptive events based on available information?	Identification of scenario forecasts for specific events, based on: Historical data Computer simulations Expert estimates
What are the consequences?	What are the economic repercussions of the disruption of biomass supply?	Quantification of losses occurring in various economic sectors as a result of upstream and downstream supply chain linkages, measured in terms of: Economic value Physical quantity Normalized risk metrics (e.g., inoperability)

Table 8.2 *Triplet of questions in risk management for biomass supply chains*

Generic risk management question	Bioenergy-specific risk management question	Possible risk management solutions
What can be done and what options are available?	What measures can be taken in order to avoid the disruption of biomass supply?	Identification of programs to encourage biomass production, such as: Government subsidies Tax incentives Carbon trading benefits
What are the trade-offs in terms of costs, benefits, and risks?	What are the economic and environmental costs and benefits of shifting toward bioenergy?	Identification of costs and benefits related with biomass production: Cost of government-funded programs Changes in soil quality and ecological landscape that can affect land prices and crop yield Increased profitability for farmers
What are the impacts of current decisions on future options?	What are objectives that policymakers intend to achieve through implementing such policies?	Determination of the objectives of shifting toward bioenergy: Reduction of adverse environmental impacts of existing technologies Reduction of possible disruptions to the economy (e.g., food security) Welfare improvement for farmers

2014). Some options seek to reduce the adverse environmental impacts of existing products and technologies, while some policy options seek to reduce possible disruptions to the economy (e.g., food supply and land-use issues). For example, there is a tendency to allocate higher acreage of farmland to produce corn ethanol with rising dependence on biofuels.

Several studies have underscored the positive environmental benefits of alternative fuels. With the exception of NOx, studies indicate that using biodiesel (in contrast to conventional diesel) reduces particulate matter, CO, CO_2, SOx, volatile organic compounds, and unburned hydrocarbons. Nevertheless, production of biodiesel and ethanol from corn, soybean, switch grass, and cellulosic ethanol is not cost competitive and would require government subsidy. Hence, risk management options, from the standpoint of government policymakers, may take the form of subsidies. Subsidy of biofuels is motivated by the premise that they are domestic alternatives to fossil fuel and that they generate lower GHG emissions. Government subsidy increases the attractiveness and profitability of biofuel production. In addition to subsidies, another government option to improve mass-scale adoption of alternative fuels is tax incentives. The government can give tax incentives to producers and consumers of biofuels.

In assessing the risks associated with biomass-based systems, the interdependence between economic sectors suggests that "ripple effects" may cascade through interconnected economic sectors. Thus, risks are not unique or isolated to the agriculture and energy sectors only; the economic impact of a disruptive event propagates through other sectors of the economy as well. This chapter introduces a rigorous methodology known as inoperability input–output modeling (IIM) for making quantitative predictions of how sectoral interdependencies lead to collateral risks economy wide that may occur as a result of loss of agricultural crop output in biomass-based systems. The rest of the chapter is organized as follows. Section 8.2 first introduces the general approach of input–output (*I–O*) analysis. Next, Section 8.3 introduces IIM methodology as a variant of *I–O* modeling. A simple tutorial example is then shown in Section 8.4 to illustrate IIM computations to the reader. Then, case studies based on oil palm production in Malaysia and coconut production in the Philippines are shown in Sections 8.5 and 8.6, respectively. Finally, conclusions and outlook for future work are given in Section 8.7.

8.2 Input–Output Model

The production of alternative fuels and adaptation of associated technologies, as well as levels of government support and market response, can significantly impact the economic structure of a region. Strategies to increase the supply and demand for alternative fuels will likely reduce the economic activities of sectors involved in the processing, storage, and transportation of traditional crude oil and gasoline products. Furthermore, the production of alternative fuels from agro-based feedstocks will inevitably compete with land use and food supply (MacLean et al. 2000). Various studies for estimating impacts of structural changes in the economy have employed extensions to traditional I–O and computable general equilibrium (CGE) models.

The *I–O* model consists of a system of linear equations that show the interindustry relationship between the different sectors of the economy, for which Leontief was awarded the Nobel Prize in Economics in 1973. This model provides a general structure of the economy by considering transaction flows from producers to purchasers of commodities that will either be further processed by a sector or consumed as final goods (Leontief 1936;

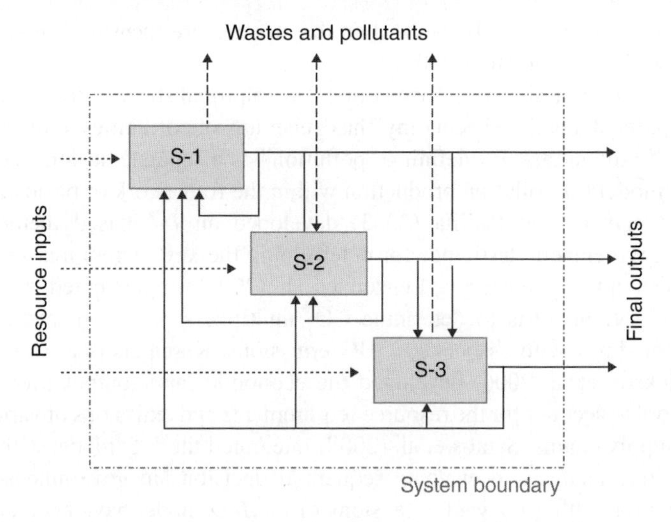

Figure 8.1 *Flow of goods, resources, and wastes in a hypothetical three-sector economy*

Miller and Blair 2009). It assumes that each sector produces a homogenous good and all firms employ the same technology such that they use a fixed proportion of inputs to produce a unit of output, which remains constant over time. Linkages of different sectors may be visualized as shown in Figure 8.1. The flow of goods may be expressed in physical or monetary units, although in practice, the latter approach is used. Given these assumptions, we can write the formulation of the Leontief *I–O* model as shown below:

$$\mathbf{x} = \mathbf{Ax} + \mathbf{c} \tag{8.1}$$

where x is the total output vector (i.e., the element, x_i, denotes the output of sector i); $\mathbf{A}$ is the technical coefficient matrix (i.e., the element, a_{ij}, denotes the input requirement of sector j from sector i, normalized with respect to the total input requirements of sector j); and c is the final demand vector (i.e., the element, c_i, denotes the final demand for the output of sector i). More specifically, the elements of the technical coefficients matrix, a_{ij}, can be derived from the transactions matrix, $\mathbf{Z}$, and the output vector, x. This is done by taking the ratio of each element of the transaction matrix, z_{ij}, which denotes the monetary value of inputs purchased by sector j from sector i, with respect to x_j, which is the total output of sector j:

$$a_{ij} = \frac{z_{ij}}{x_j} \tag{8.2}$$

We further note that Equation 8.1 can be rewritten to its simplified form, where $\mathbf{L} = (\mathbf{I} - \mathbf{A})\text{–}1$ is the Leontief inverse matrix:

$$\mathbf{x} = \mathbf{Lc} \tag{8.3}$$

The Leontief inverse matrix is also called the total requirements matrix since each element, l_{ij}, represents the increase in the output for sector i that is required for every unit of increase in the final demand of sector j. Note that, as a result of convention that has been established among users of *I–O* over the past decades, flows in Equations 8.1–8.3 are expressed in terms of economic value, which facilitates use of *I–O* for economic planning purposes. Nevertheless, it has been clearly established that the same general principles of computation apply even when the streams are measured in physical terms (Leontief 1986, Miller and Blair 2009).

Leontief's (1951) initial vision of executing "an empirical study of interrelations among the different parts of a national economy" has generated opportunities to develop a myriad of extensions to the model. By defining "pollution" as a separate sector, Leontief (1970) formulated a model for pollution production within the framework of basic economic *I–O* accounting. Alcántara and Padilla (2003) developed an *I–O*-based methodology that considers energy demand elasticities for determining the key sectors that are involved in the final consumption of energy. Lenzen et al. (2004) implemented a multiregional environmental *I–O* analysis to determine CO_2 multipliers based on international trade data for commodities with associated GHG emissions. Researchers at Carnegie Mellon (e.g., Hendrickson et al. 2006) developed the economic input–output lifecycle analysis (EIO-LCA) tool to account for the resource requirements and emissions of various products across their supply chains. Santos et al. (2008) integrated the *I–O* model with multiobjective decision tree analysis to evaluate sequential decisions in government-incentivized biofuel production. Other relevant extensions of the *I–O* model have been applied in the fields of energy (Bullard and Herendeen 1975, Casler and Wilbur 1984), environment (Lave et al. 1995, Suh and Nakamura 2007, Wiedmann et al. 2007), climate change (Minx et al. 2009, Huang et al. 2009), and disaster risk management (Okuyama 2007), among others. Methodological extensions such as nonlinear *I–O* models (Sandberg 1973, Chander 1983), multiregional *I–O* models (Hewings et al. 1989), fuzzy *I–O* models (Buckley 1989), and CGE models (Dervis 1975, de Melo 1988) have been introduced to address the limitations of the original model. The *I–O* framework has also been extended for use in the analysis of systems with analogous structures interconnected. Examples include extension models for the analysis of natural ecosystems (Hannon 1973) and industrial networks (Duchin 1992). Furthermore, a dimensionless metric called inoperability, which will be the subject of subsequent sections, has been formulated by normalizing the economic units of production used in traditional *I–O* models.

8.3 Inoperability Input–Output Modeling

The IIM is one of the notable extensions of the Leontief *I–O* model, where inoperability is defined as the "inability of a system to perform its intended function" (Haimes and Jiang 2001). The inoperability metric is a dimensionless number with 0 corresponding to the normal state of a system, 1 being total failure, and intermediate values reflecting partial failure. The initial formulation by Haimes and Jiang (2001), which we will refer to as the physical IIM, focuses on the physical intra- and interconnectivity between infrastructures that require extensive data collection. As a linear metric of system risk, it has been integrated recently into both mathematical programming (Tan 2011) and pinch analysis-based approaches for planning of sustainable energy systems (Tan and Foo 2013). The initial definition of inoperability was later recast by Santos and Haimes (2004) through the demand-reduction IIM, which links the Leontief *I–O* model with the physical IIM. The demand-based IIM can be specified in Equation 8.4 with the model components described below:

$$\mathbf{q} = \mathbf{A}^{*}\mathbf{q} + \mathbf{c}^{*} \tag{8.4}$$

8.3.1 Inoperability

The inoperability vector is denoted as q, where q_i represents the inoperability in sector i. It is derived as

$$q_i = \frac{x_i - \tilde{x}_i}{x_i} \tag{8.5}$$

where x_i is the ideal level of production and $\tilde{x}_i$ is the disrupted level of production. Since the economy is composed of interdependent industries, an increased level of inoperability will result from an initial disruption.

8.3.2 Interdependency Matrix

The interdependency matrix ($\mathbf{A}^*$) is a square matrix representing the pairwise relationships across sectors. In matrix notation, it can be derived from the Leontief *I–O* model as

$$\mathbf{A}^* = \mathbf{diag}(\mathbf{x})^{-1}\,\mathbf{A}\mathbf{diag}(\mathbf{x}) \tag{8.6}$$

where $\mathbf{diag}(\mathbf{x})$ is the diagonalized output vector and $\mathbf{A}$ is the technical coefficients matrix. Its individual elements can be computed in scalar form:

$$a_{ij}^* = a_{ij}\left(\frac{x_j}{x_i}\right) \tag{8.7}$$

The scalar, a_{ij}^*, denotes the additional inoperability contributed by sector j to sector i. Note that the work of Santos and Haimes (2004) allowed IIM models to be calibrated directly from standard economic *I–O* data. On the other hand, Setola and De Porcellinis (2008) proposed methods to estimate $\mathbf{A}^*$ based on physical IIM using the knowledge provided by the experts in assessing the impact of the inoperability of other infrastructure to their infrastructure. The method used fuzzy numbers to handle the uncertainty and vagueness of the subjective judgment of experts and was applied to quantify the degree of interdependency existing among Italian infrastructures (Setola et al. 2009, Oliva et al. 2011).

8.3.3 Perturbation

The normalized degraded demand vector is denoted as $\mathbf{c}^*$. In matrix notation, it can be denoted as

$$\mathbf{c}^* = \mathbf{diag}(\mathbf{x})^{-1}(\mathbf{c} - \tilde{\mathbf{c}}) \tag{8.8}$$

where $\tilde{\mathbf{c}}$ is the degraded final demand vector. Its elements, c_i^*, can be computed in scalar form:

$$c_i^* = \left(\frac{c_i - \tilde{c}_i}{x_i}\right) \tag{8.9}$$

8.3.4 Economic Loss

Economic loss is the product of the inoperability and ideal level output ($q_i x_i$). It is a vector whose elements reflect monetary values associated with a realized inoperability in sector i.

Numerous applications have been done to illustrate the versatility of IIM. These include an ex post analysis of the 9/11 terrorist attacks (Santos and Haimes 2004), cybersecurity (Santos et al. 2007), blackout incidences such as the 2003 Northeast Blackout (Anderson et al. 2007) and the 2003 Italian Power Blackout (Jonkeren et al. 2012), oil shortage (Khanna and Bakshi 2009), loss of pollination in agriculture (Khanna and Bakshi 2009), pandemic outbreaks (Santos et al. 2009, 2013, Orsi and Santos 2010), and natural disasters such as Hurricane Katrina (Crowther et al. 2007) and the 2011 Fukushima earthquake (MacKenzie et al. 2012). Further extensions that build upon the model were later on developed. Leung et al. (2007) explored supply-side changes, incorporating price into IIM. Crowther and Haimes (2010) included spatial explicitness through developing a multiregional IIM. A dynamic inoperability input–output model (DIIM) was later developed by Haimes et al. (2005). Barker and Santos (2010a) incorporated inventories into the DIIM to evaluate resilience of a disrupted system with extensions for identifying critical industries (Resurreccion and Santos 2012) and its impact on the workforce (Akhtar and Santos 2013). In the May 2012 issue of *Risk Analysis*, regional economic impact assessment using tools such as IIM and CGE models has been included in the list of important contributions to risk analysis (Greenberg et al. 2012).

8.4 Illustrative Example

We introduce a hypothetical two-sector economy from Miller and Blair (2009) to illustrate the concepts in the previous section. In addition, the corresponding LINGO code for the example is given in Appendix A. The shaded region in Table 8.3 represents the transactions (**Z**) matrix. The third column represents the monetary value of final goods consumed or the final demand vector (**c**). The fourth column represents the total output of each sector (**x**). The third row represents the amount of value added for each sector. The fourth row represents the total inputs required by each sector. It can be noted that the total inputs of each sector is equal to the total outputs of each sector. This ensures that we have a balanced economy.

Taking the first row, it can be observed that Sector 1 produces \$150 worth of output for its own intermediate consumption, \$500 worth of output which will be used as inputs for Sector 2, and \$350 worth of output is purchased as final goods by end users. The sum of the intermediate demand of each sector for Sector 1's output and final demand for Sector 1's output will yield the total output for the said sector. In addition, the first column shows that Sector 1 uses \$150 worth of input from its own output and \$200 worth of input from Sector 2 and \$650 from value added. Value added can be broken down into employee

Table 8.3 *Flows for a hypothetical two-sector economy (in USD)*

Sector	1	2	Final demand (**c**)	Total output (**x**)
1	150	500	350	1000
2	200	100	1700	2000
Value added	650	1400		
Total input	1000	2000		

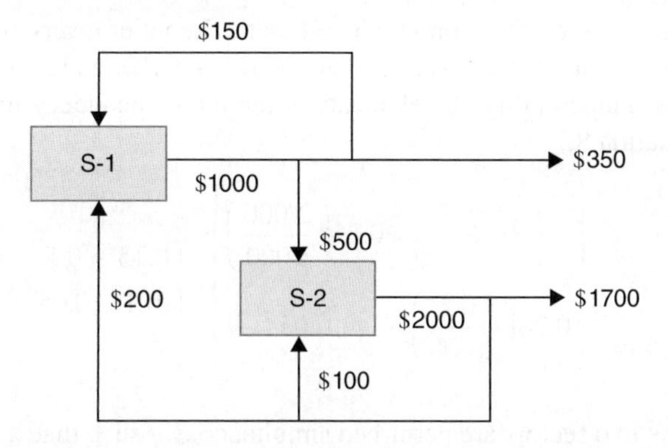

Figure 8.2 *Flow of goods in a hypothetical two-sector economy (in USD)*

compensation, gross operating surplus, and the difference of indirect taxes and subsidies. The flow of goods in this system is also illustrated in Figure 8.2.

The technical coefficients matrix, A, is derived by dividing the elements of transactions matrix (z_{ij}) by the respective column sum (x_j) such that

$$\mathbf{Z} = \begin{bmatrix} 150 & 500 \\ 200 & 100 \end{bmatrix} \tag{8.10}$$

and

$$\mathbf{x} = \begin{bmatrix} 1000 \\ 2000 \end{bmatrix}$$

will yield

$$\mathbf{A} = \left\{ \frac{z_{ij}}{x_j} \right\} \forall i, j = \begin{bmatrix} 150/1000 & 500/2000 \\ 200/1000 & 100/2000 \end{bmatrix} = \begin{bmatrix} 015 & 0.25 \\ 0.20 & 0.05 \end{bmatrix} \tag{8.11}$$

For this example, we can compute for the Leontief inverse matrix, L, as:

$$\mathbf{L} = (I - \mathbf{A})^{-1} = \left(\begin{bmatrix} 1 & 0 \\ 0 & 1 \end{bmatrix} - \begin{bmatrix} 0.15 & 0.25 \\ 0.20 & 0.05 \end{bmatrix} \right)^{-1} = \begin{bmatrix} 0.85 & -0.25 \\ -0.20 & 0.95 \end{bmatrix}^{-1} = \begin{bmatrix} 1.25 & 0.33 \\ 0.26 & 1.12 \end{bmatrix} \tag{8.12}$$

In particular, Matrix **L** allows us to conduct a scenario analysis comparing the impact of a 1-unit worth of monetary increase in the final demand of each sector. Suppose that there is \$1 increase in the final demand for Sector 1. This scenario will require the output of Sector 1 to increase by \$1.25 and the output of Sector 2 to increase by \$0.26. This increases the total output of the economy by \$1.51. On the other hand, a \$1 increase in the final demand of Sector 2 will require the output of Sector 1 to increase by \$0.33 and the output of Sector 2 to increase by \$1.12, which increases total output of the economy by \$1.45. Comparing the impact between a dollar increase in the final demand for output of Sector 1 and Sector 2, the economic benefit is higher if the increase was on Sector 1. Evaluating the impact of each sector on the economy based on the Leontief inverse is one of the most commonly used techniques for key sector identification (Barker and Santos, 2010b).

In the context of inoperability, the elements of the interdependency matrix, **A***, can be derived using Equation 8.7

$$\mathbf{A}^* = \begin{bmatrix} 0.15 & 0.25\left(\dfrac{2{,}000}{1{,}000}\right) \\ 0.20\left(\dfrac{1{,}000}{2{,}000}\right) & 0.05 \end{bmatrix} = \begin{bmatrix} 0.15 & 0.5 \\ 0.1 & 0.05 \end{bmatrix} \tag{8.13}$$

Suppose that the two sectors are perturbed simultaneously such that:

$$\mathbf{c}^* = \begin{bmatrix} c_1^* \\ c_2^* \end{bmatrix} = \begin{bmatrix} 0.1 \\ 0.1 \end{bmatrix} \tag{8.14}$$

The resulting inoperability values are:

$$\begin{aligned} \mathbf{q} &= (I - \mathbf{A}^*)^{-1}\mathbf{c}^* \\ &= \left(\begin{bmatrix} 1 & 0 \\ 0 & 1 \end{bmatrix} - \begin{bmatrix} 0.15 & 0.5 \\ 0.1 & 0.05 \end{bmatrix}\right)^{-1}\begin{bmatrix} 0.1 \\ 0.1 \end{bmatrix} = \begin{bmatrix} 1.25 & 0.66 \\ 0.13 & 1.12 \end{bmatrix}\begin{bmatrix} 0.1 \\ 0.1 \end{bmatrix} = \begin{bmatrix} 0.19 \\ 0.13 \end{bmatrix} \end{aligned} \tag{8.15}$$

while economic losses (**EL**) can be computed as:

$$\mathbf{EL} = \begin{bmatrix} q_1x_1 \\ q_2x_2 \end{bmatrix} = \begin{bmatrix} (0.19)(1{,}000) \\ (0.13)(2{,}000) \end{bmatrix} = \begin{bmatrix} 190 \\ 260 \end{bmatrix} \tag{8.16}$$

This example shows that given a 10% final demand perturbation for both sectors, the resulting total inoperability for Sector 1 will be 19%, which is equivalent to a $190 loss in its total output. By the same token, the total inoperability for Sector 2 will be 13%, which is equivalent to a $260 loss in its total output. From this illustrative example, we have demonstrated that the ranking of key sectors may vary between the two metrics (i.e., economic loss vs. total inoperability). A policymaker that puts more weight on inoperability will choose to prioritize Sector 1. On the other hand, the policymaker will opt to prioritize Sector 2 when economic loss is used as the sole criterion. Hence, a balanced approach that takes into consideration policymaker preferences across economic loss, inoperability, and other appropriate metrics will result in a more holistic method for key sector identification and prioritization.

There are several multiple criteria analysis (MCA) tools that can be used to understand better the trade-offs and improve the decision making process by making the priorities more transparent and analytically robust. Several MCA weighting algorithms are provided in the literature (e.g., see Figueira et al. 2005) to express the criteria or metric's importance either in ordinal or cardinal units. The Analytic Hierarchy Process (Saaty 1980), one of the most widely used weighting technique, is discussed in Appendix D.

8.5 Case Study 1

This case study demonstrates the use of the model for predicting ripple effects from the loss of agricultural output in the Malaysian oil palm sector. The latter crop accounts for about 36% of the economic value of total agricultural output, according to the 2005 I–O data (Malaysia Department of Statistics 2010). Oil palm plantations also account for 4.92 million hectares out of the 7.87 million hectares of total agricultural land in the country (Malaysian Palm Oil Board n.d.). The sector is also seen as a viable source of indigenous energy, with the annual palm biomass energy potential being estimated to have 37.09 million tons of oil equivalent (Ng et al. 2012). However, the vital role played by the palm oil sector in Malaysia also results in some risks in palm-based supply chains, as a result of intersectoral linkages. Thus, there is considerable interest in assessing the vulnerability of the palm oil sector to various disruptive events. Recently, there has been growing concern in Southeast Asia about the effects of "seasonal haze" from forest fires in Indonesia (Quah 2002). For example, the 1997 haze is estimated to have caused a total of US$1.4 billion in total damages (including health, tourism, and other impacts), with 72% occurring in Indonesia and the remaining 28% being incurred in nearby countries, including Malaysia (EEPSEA, 1999). Such forest fires release atmospheric aerosols, which have been shown to cause reduced commercial crop yields due to reduced insolation and photosynthesis (Greenwold et al. 2006). Hence, possible effects of haze incidence also include loss of agricultural output in oil palm plantations in adjoining regions of Malaysia, which may in turn cause further economic damage through sectoral interdependencies.

In this case study, a hypothetical scenario of 5–10 % loss of oil palm output as a result of haze is considered. As much oil palm output is processed into downstream products, in the event of shortage it is assumed that priority is given to satisfying intermediate demand over final demand, which allows the use of "forced demand-reduction" IIM approach (Anderson et al. 2007). The effect of the perturbation is computed within the model using the interval IIM approach of Barker and Rocco (2011), which was proposed as a means of capturing inherent uncertainties in the input data. A brief description of interval arithmetic is given in Appendix C. The Malaysian I–O data has been aggregated into 12 sectors following the procedure outlined in Miller and Blair (2009). Table 8.4 shows the coefficients of matrix A, while Table 8.5 shows the total output (x) and final demand (c) of each sector in Malaysian Ringgit (RM, with approximately RM 3 = US$ 1). Following the procedure described previously, it is then possible to determine the interdependency matrix ($\mathbf{A}^*$), which is shown in Table 8.6. In the scenario considered, the output loss of the oil palm sector is again treated as a "forced demand reduction," as reflected in the perturbation vector, c^*. In this case, the 5–10% loss of supply may be calculated as being equivalent to an absolute estimated loss of RM 1.089–2.178 billion. Thus, the perturbation vector $\mathbf{c}^*$ contains a nonzero interval-valued element corresponding to the perturbation in the oil palm sector (i.e., $c^*_2 = [0.05, 0.1]$), while all the other elements are zero. The corresponding LINGO code for this case study is given in Appendix B. The inoperability vector, $\mathbf{q}$, is presented in Figure 8.3, where we note that six sectors other than the oil palm sector suffer from inoperability values approximating to 0.001. These sectors are agriculture, fishery, and forestry; manufacturing; electricity, gas, and water; trade; transportation, communication and storage; and, finally, finance. Inoperabilities in all other sectors are below 0.001.

Table 8.4 *Coefficients of A in case study 1*

	Agriculture, fishery, and forestry excluding oil palm	Oil palm	Mining and quarrying	Manufacturing	Electricity, gas, and water	Construction	Trade	Transportation, communication, and storage	Finance	Real estate and ownership of dwellings	Private services	Government services
Agriculture, fishery, and forestry excluding oil palm	0.114	0.013	0.002	0.019	0.000	0.001	0.014	0.000	0.001	0.000	0.000	0.003
Oil palm	0.001	0.091	0.001	0.020	0.000	0.000	0.001	0.000	0.000	0.000	0.000	0.001
Mining and quarrying	0.003	0.002	0.005	0.042	0.018	0.032	0.004	0.001	0.002	0.000	0.001	0.003
Manufacturing	0.144	0.108	0.100	0.324	0.223	0.305	0.100	0.106	0.030	0.028	0.063	0.101
Electricity, gas, and water	0.004	0.010	0.002	0.009	0.186	0.004	0.008	0.010	0.007	0.013	0.031	0.020
Construction	0.000	0.000	0.007	0.000	0.043	0.001	0.004	0.022	0.000	0.010	0.001	0.086
Trade	0.024	0.025	0.044	0.047	0.010	0.027	0.043	0.012	0.025	0.139	0.011	0.037
Transportation, communication, and storage	0.014	0.015	0.013	0.021	0.013	0.043	0.044	0.262	0.070	0.017	0.047	0.051
Finance	0.003	0.031	0.000	0.009	0.004	0.032	0.022	0.108	0.300	0.021	0.023	0.005
Real estate and ownership of dwellings	0.000	0.000	0.001	0.000	0.001	0.007	0.014	0.023	0.013	0.081	0.044	0.030
Private services	0.001	0.006	0.006	0.003	0.015	0.026	0.039	0.023	0.029	0.017	0.188	0.037
Government services	0.000	0.000	0.001	0.000	0.004	0.005	0.004	0.015	0.004	0.007	0.009	0.064

Table 8.5 *Total output (x) and final demand (c) case study 1 (in thousand RM)*

	x	c
Agriculture, fishery, and forestry excluding oil palm	38,218,340	14,510,228
Oil palm	21,782,028	1,625,893
Mining and quarrying	95,410,015	53,901,753
Manufacturing	899,165,229	530,734,251
Electricity, gas, and water	35,149,126	13,054,117
Construction	48,051,594	35,400,609
Trade	90,511,083	25,631,177
Transportation, communication, and storage	119,450,597	47,413,486
Finance	74,329,671	23,460,683
Real estate and ownership of dwellings	33,414,975	19,766,907
Private services	69,097,400	38,796,551
Government services	79,326,620	70,027,405

On the other hand, the inoperabilities may be multiplied by the respective baseline output of each sector to estimate economic losses arising from the lost oil palm output. The losses, in thousand RM, are shown in Figure 8.4. There is a notable shift in the vulnerability of the sectors when measured in economic terms, with manufacturing output expected to drop by RM 225–450 million due to indirect (forward and backward) linkages with the oil palm sector. For instance, the latter sector purchases various inputs (e.g., fuels, fertilizers, pesticides) from manufacturing; it also acts as a supplier for the production of vegetable oil, oleochemicals, etc. These results clearly show how economic losses in one sector can cause ripple effects that propagate through an economic system, by virtue of direct and indirect supply chain linkages, to cause economic damage in other sectors. In many cases, the extent of the links are not immediately obvious from intuition but are revealed only through rigorous modeling. For example, the 1997 haze is estimated to have caused RM 393.5 million in industrial production losses (including reduced crop output) in Western Malaysia (EEPSEA, 1999); this damage was geographically limited in scope, but even so, it is unclear if the estimate includes losses from indirect ripple effects. Thus, it is likely that this figure underestimates actual losses. This case study also illustrates that the vulnerability ranking of the different sectors based on absolute economic losses will be different from that derived from inoperability levels.

8.6 Case Study 2

The second case study demonstrates the application of the model in assessing the potential impact of implementing a biofuel plan at a time when climate change-induced disasters are expected. The case study considers the Philippines, an archipelago located in Southeast Asia in the Western Pacific Ocean. It is situated in the typhoon belt of the Pacific; thus, the country experiences an average of 20 typhoons every year causing damage to land and property. With an objective of reducing dependence on imported oil, improving air quality, and increasing economic activity and employment, the Biofuels Act of the Philippines (RA 9367) was ratified in 2006 and signed into law in 2007. RA 9367 mandates the use of

Table 8.6 *Coefficients of* $\mathbf{A}^*$ *in case study*

	Agriculture, fishery and, forestry excluding oil palm	Oil palm	Mining and quarrying	Manufacturing	Electricity, gas, and water	Construction	Trade	Transportation, communication, and storage	Finance	Real estate and ownership of dwellings	Private services	Government services
Agriculture, fishery, and forestry excluding oil palm	0.114	0.008	0.004	0.450	0.000	0.001	0.033	0.001	0.002	0.000	0.000	0.007
Oil palm	0.002	0.091	0.002	0.822	0.000	0.001	0.003	0.001	0.001	0.000	0.000	0.003
Mining and quarrying	0.001	0.001	0.005	0.395	0.007	0.016	0.004	0.001	0.001	0.000	0.000	0.003
Manufacturing	0.006	0.003	0.011	0.324	0.009	0.016	0.010	0.014	0.002	0.001	0.005	0.009
Electricity, gas, and water	0.005	0.006	0.006	0.231	0.186	0.005	0.021	0.035	0.015	0.012	0.062	0.045
Construction	0.000	0.000	0.015	0.003	0.032	0.001	0.007	0.055	0.000	0.007	0.002	0.142
Trade	0.010	0.006	0.046	0.465	0.004	0.014	0.043	0.016	0.020	0.051	0.009	0.033
Transportation, communication, and storage	0.004	0.003	0.011	0.160	0.004	0.017	0.033	0.262	0.044	0.005	0.027	0.034
Finance	0.002	0.009	0.001	0.114	0.002	0.020	0.027	0.173	0.300	0.010	0.022	0.005
Real estate and ownership of dwellings	0.000	0.000	0.002	0.001	0.001	0.010	0.037	0.083	0.028	0.081	0.091	0.072
Private services	0.001	0.002	0.009	0.041	0.007	0.018	0.051	0.040	0.031	0.008	0.188	0.042
Government services	0.000	0.000	0.001	0.005	0.002	0.003	0.005	0.023	0.004	0.003	0.008	0.064

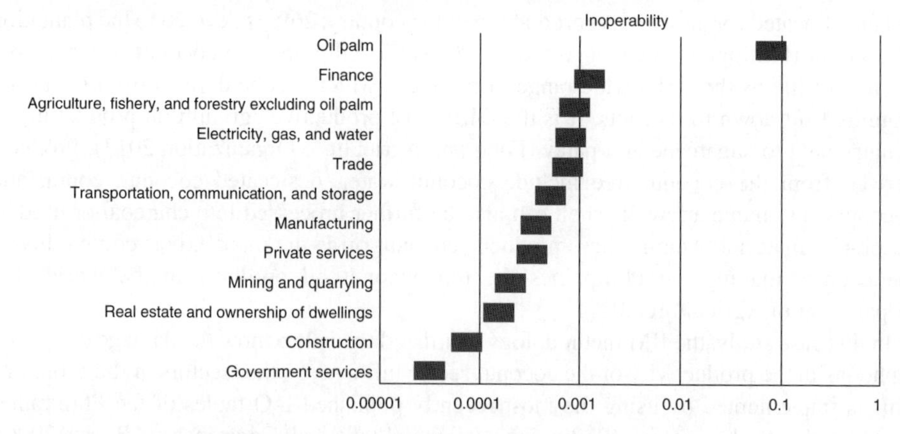

Figure 8.3 *Sector inoperability levels in case study 1*

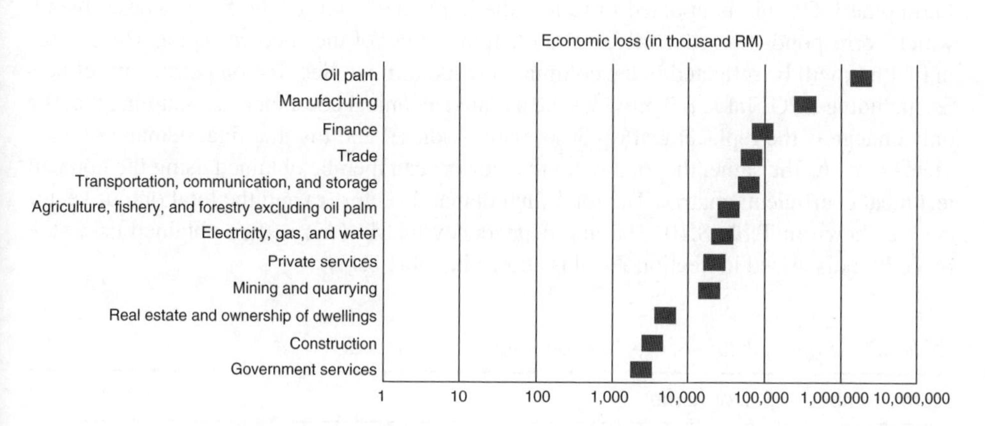

Figure 8.4 *Sector economic losses in case study 1*

a minimum 1% biodiesel blend and 5% bioethanol blend for all diesel and gasoline fuels used in the country by the second quarter of 2007. This made the Philippines the pioneer in implementing the mandatory use of blended fuels in the Asia Pacific Economic Region (APEC) (Milbrandt and Overend 2008). The mandated blend has increased to a minimum of 2% for biodiesel in 2012 and is expected to increase to 5% in the coming years. With coconut and sugarcane as the main feedstock utilized in the production of biodiesel and bioethanol, respectively, it is expected that the implementation of the biofuels program will increase the demand for these energy crops in order to satisfy the demand for biofuels. This increase in dependence on the agricultural sector can potentially make a country more vulnerable to climate change-induced disasters such as storm (Stromberg et al. 2011) and pest infestation.

This case focuses on the coconut sector given its role as primary source for biodiesel feedstock and one of the largest industries in the Philippines. Out of the 13 million hectares of land allocated for agricultural production in the country, 26% is devoted to the plantation of coconut (Philippine Coconut Authority 2013). To Filipinos, the coconut is considered the tree of life as there is a wide range of products, which can be derived from it starting from its fruit down to its roots. It is the third most productive agricultural product in the country next to sugarcane and palay (Food and Agriculture Organization 2013). Products derived from the coconut tree include coconut water, desiccated coconut, copra, and coconut oil to name a few. Its shell can also be further processed into charcoal or used as decorative material. Among these products, coconut oil is the main export commodity of the country making the Philippines the top coconut oil producer in the world (US Department of Agriculture 2013).

In this case study, the IIM methodology is utilized to analyze how the damage caused by typhoons in the productivity of the coconut sector influences other sectors of the economy. This is implemented by using the most recently published I–O tables of the Philippines aggregated into 16 sectors (Philippine National Statistical Coordination Board 2006). The aggregated sector headings are shown in Table 8.7, while the technical coefficients are shown in Table 8.8 (Philippine National Statistical Coordination Board 2006). The 2000 Philippine I–O table is updated to reflect the implementation of the 5% biodiesel blend, which corresponds to a 14% increase in the total output of the coconut sector. The change in fuel mix will be reflected in the column corresponding to Sector 6 on petroleum refineries including LPG. Table 8.9 provides the updated technical coefficients. Assuming that the only change is the replacement of diesel with biodiesel and that the final demand of each sector remains the same, the total output vector (**x**) can then be obtained using the updated technical coefficients matrix. The total final demand vector (**c**) and the total output vector (**x**) are shown in Table 8.10. The interdependency matrix (A^*) is then obtained using the procedure discussed in Section 2 and is shown in Table 8.11.

Table 8.7 *Nomenclature of the 16 economic sectors considered*

Sector number	Sector name
001	Coconut including copra making in the farm
002	Agriculture, fishery, and forestry except coconut and copra making in farm
003	Mining and quarrying
004	Production of crude coconut oil, copra cake and meal
005	Manufacture of refined coconut oil and vegetable oil
006	Petroleum refineries including LPG
007	Manufacture of asphalt, lubricant, and miscellaneous products of petroleum and coal
008	All other manufacturing sectors
009	Construction
010	Electricity, gas, and water
011	Transportation, communication, and storage
012	Trade
013	Finance
014	Real estate and ownership of dwellings
015	Private services
016	Government services

Table 8.8 *Coefficients of A for case study 2*

	001	**002**	**003**	**004**	**005**	**006**	**007**	**008**	**009**	**010**	**011**	**012**	**013**	**014**	**015**	**016**
001	0.0039	0.0000	0.0000	0.5212	0.0000	0.0000	0.0000	0.0017	0.0000	0.0000	0.0000	0.0001	0.0000	0.0000	0.0028	0.0004
002	0.0310	0.0751	0.0012	0.0000	0.0215	0.0000	0.0042	0.1141	0.0001	0.0000	0.0005	0.0175	0.0000	0.0000	0.0232	0.0071
003	0.0000	0.0000	0.0105	0.0000	0.0000	0.6099	0.0053	0.0118	0.0261	0.0675	0.0000	0.0062	0.0000	0.0005	0.0002	0.0000
004	0.0000	0.0000	0.0000	0.0000	0.2932	0.0000	0.0000	0.0014	0.0000	0.0000	0.0000	0.0009	0.0000	0.0000	0.0000	0.0000
005	0.0000	0.0000	0.0000	0.0000	0.1377	0.0000	0.0000	0.0020	0.0000	0.0000	0.0000	0.0013	0.0000	0.0000	0.0023	0.0003
006	0.0034	0.0127	0.0598	0.0233	0.0342	0.0194	0.6682	0.0142	0.0090	0.0951	0.1124	0.0197	0.0168	0.0014	0.0126	0.0078
007	0.0000	0.0002	0.0060	0.0028	0.0135	0.0009	0.0003	0.0034	0.0066	0.0001	0.0061	0.0005	0.0000	0.0013	0.0001	0.0000
008	0.0485	0.1189	0.1008	0.0468	0.1780	0.0523	0.0529	0.3198	0.2659	0.0122	0.2137	0.1170	0.0534	0.0207	0.2417	0.0982
009	0.0002	0.0015	0.0128	0.0001	0.0001	0.0004	0.0000	0.0003	0.0060	0.0013	0.0011	0.0002	0.0034	0.0086	0.0000	0.0251
010	0.0007	0.0098	0.0546	0.0138	0.0241	0.0018	0.0003	0.0222	0.0024	0.0762	0.0044	0.0048	0.0150	0.0015	0.0286	0.0123
011	0.0127	0.0038	0.0136	0.0040	0.0115	0.0069	0.0011	0.0075	0.0913	0.0103	0.0252	0.1130	0.0584	0.0033	0.0144	0.0209
012	0.0050	0.0080	0.0071	0.1205	0.0226	0.0102	0.0006	0.0849	0.0123	0.0199	0.0101	0.0013	0.0012	0.0011	0.0102	0.0046
013	0.0019	0.0086	0.0176	0.0030	0.0015	0.0130	0.0241	0.0065	0.0162	0.0019	0.0315	0.0339	0.0002	0.0315	0.0229	0.0303
014	0.0001	0.0005	0.0015	0.0003	0.0099	0.0001	0.0003	0.0012	0.0068	0.0000	0.0091	0.0061	0.0306	0.0008	0.0086	0.0086
015	0.0029	0.0156	0.0757	0.0007	0.0020	0.0042	0.0006	0.0072	0.0207	0.0225	0.0439	0.0157	0.1653	0.0343	0.0993	0.0612
016	0.0000	0.0000	0.0000	0.0000	0.0000	0.0000	0.0000	0.0000	0.0000	0.0000	0.0000	0.0000	0.0000	0.0000	0.0000	0.0000

Table 8.9 *Updated technical coefficients matrix A due to implementation of 5% biodiesel blend*

	001	**002**	**003**	**004**	**005**	**006**	**007**	**008**	**009**	**010**	**011**	**012**	**013**	**014**	**015**	**016**
001	0.0039	0.0000	0.0000	0.5212	0.0000	0.0000	0.0000	0.0017	0.0000	0.0000	0.0000	0.0001	0.0000	0.0000	0.0028	0.0004
002	0.0310	0.0751	0.0012	0.0000	0.0215	0.0000	0.0042	0.1141	0.0001	0.0000	0.0005	0.0175	0.0000	0.0000	0.0232	0.0071
003	0.0000	0.0000	0.0105	0.0000	0.0000	0.6099	0.0053	0.0118	0.0261	0.0675	0.0000	0.0062	0.0000	0.0005	0.0002	0.0000
004	0.0000	0.0000	0.0000	0.0000	0.2932	0.0000	0.0000	0.0014	0.0000	0.0000	0.0000	0.0009	0.0000	0.0000	0.0000	0.0000
005	0.0000	0.0000	0.0000	0.0000	0.1377	0.0133	0.0000	0.0020	0.0000	0.0000	0.0000	0.0013	0.0000	0.0000	0.0023	0.0003
006	0.0034	0.0127	0.0598	0.0233	0.0342	0.0194	0.6682	0.0142	0.0090	0.0951	0.1124	0.0197	0.0168	0.0014	0.0126	0.0078
007	0.0000	0.0002	0.0060	0.0028	0.0135	0.0009	0.0003	0.0034	0.0066	0.0001	0.0061	0.0005	0.0000	0.0013	0.0001	0.0000
008	0.0485	0.1189	0.1008	0.0468	0.1780	0.0523	0.0529	0.3198	0.2659	0.0122	0.2137	0.1170	0.0534	0.0207	0.2417	0.0982
009	0.0002	0.0015	0.0128	0.0001	0.0001	0.0004	0.0000	0.0003	0.0060	0.0013	0.0011	0.0002	0.0034	0.0086	0.0000	0.0251
010	0.0007	0.0098	0.0546	0.0138	0.0241	0.0018	0.0003	0.0222	0.0024	0.0762	0.0044	0.0048	0.0150	0.0015	0.0286	0.0123
011	0.0127	0.0038	0.0136	0.0040	0.0115	0.0069	0.0011	0.0075	0.0913	0.0103	0.0252	0.1130	0.0584	0.0033	0.0144	0.0209
012	0.0050	0.0080	0.0071	0.1205	0.0226	0.0102	0.0006	0.0849	0.0123	0.0199	0.0101	0.0013	0.0012	0.0011	0.0102	0.0046
013	0.0019	0.0086	0.0176	0.0030	0.0015	0.0130	0.0241	0.0065	0.0162	0.0019	0.0315	0.0339	0.0002	0.0315	0.0229	0.0303
014	0.0001	0.0005	0.0015	0.0003	0.0099	0.0001	0.0003	0.0012	0.0068	0.0000	0.0091	0.0061	0.0306	0.0008	0.0086	0.0086
015	0.0029	0.0156	0.0757	0.0007	0.0020	0.0042	0.0006	0.0072	0.0207	0.0225	0.0439	0.0157	0.1653	0.0343	0.0993	0.0612
016	0.0000	0.0000	0.0000	0.0000	0.0000	0.0000	0.0000	0.0000	0.0000	0.0000	0.0000	0.0000	0.0000	0.0000	0.0000	0.0000

Table 8.10 *Total output and final demand using revised technical coefficients matrix (in thousand PhP)*

Sector	Total output (x)	Final demand (c)
001	27,685,735	5,659,066
002	659,798,402	230,398,096
003	37,958,934	(151,599,611)
004	28,213,327	16,448,098
005	23,059,018	8,423,671
006	208,673,478	26,130,428
007	16,985,665	(278,520)
008	3,027,820,130	1,450,395,866
009	289,364,509	269,486,299
010	196,848,277	68,669,485
011	514,192,903	316,864,932
012	803,816,552	511,748,123
013	300,612,192	185,135,299
014	290,799,136	256,105,187
015	659,993,667	426,180,406
016	442,004,314	442,004,314

The increase in wind speed of typhoons brought by climate change is expected to result in a 2% (Stromberg et al. 2011) direct damage on the coconut sector. This case study thus considers an initial perturbation of 2–5% in Sector 1 corresponding to an estimated loss of 554 M to 1.4 B pesos. The first element (c_1) in the perturbation vector $\mathbf{c}^*$ thus corresponds to the interval [0.02, 0.05], while all other entries are equal to zero. The resulting sector inoperability levels are represented in Figure 8.5. Based on the results, the sectors which are most affected by the initial perturbation are the coconut sector and the mining and quarrying sector. All the other sectors suffer an inoperability level of less than 0.0001. One can clearly see that a perturbation in one sector can easily propagate throughout the other sectors by virtue of their interdependence. Initially, it is quite surprising to find that the mining and quarrying sector ranks second in terms of inoperability, while the sectors on crude coconut oil production, coconut oil production, and petroleum refinery rank 13th, 10th, and 4th, respectively. However, this simply illustrates the inherent structure of the Philippine economy. The significant inoperability in the mining and quarrying sector stems from its large contribution of inputs toward the petroleum sector, which reduces its demand for inputs as a result of the lower levels of coconut production. Furthermore, Figure 8.6 shows the corresponding ranking of sectors based on their economic losses, where the coconut farming sector ranks first followed by the manufacturing sectors. The sectors devoted to crude coconut oil and coconut oil production are in the 15th and 14th place, while the petroleum and refinery sector is in the 6th place. The total economic loss is expected to range between 667 M and 1.67 B pesos. It can be seen that the ranking obtained in consideration of the inoperability is different from the ranking obtained due to absolute economic losses. The reader may modify the code given in Appendix B to reproduce the results in this case study.

Table 8.11 *Coefficients of the matrix* **A*** *for a 5% biodiesel blend*

	001	**002**	**003**	**004**	**005**	**006**	**007**	**008**	**009**	**010**	**011**	**012**	**013**	**014**	**015**	**016**
001	0.0039	0.0000	0.0000	0.5310	0.0000	0.0000	0.0000	0.1859	0.0000	0.0000	0.0000	0.0029	0.0000	0.0000	0.0667	0.0064
002	0.0013	0.0751	0.0001	0.0000	0.0008	0.0000	0.0001	0.5236	0.0000	0.0000	0.0004	0.0213	0.0000	0.0000	0.0232	0.0048
003	0.0000	0.0000	0.0105	0.0000	0.0000	3.3527	0.0024	0.9412	0.1990	0.3500	0.0000	0.1313	0.0000	0.0038	0.0035	0.0000
004	0.0000	0.0000	0.0000	0.0000	0.2397	0.0000	0.0000	0.1503	0.0000	0.0000	0.0000	0.0256	0.0000	0.0000	0.0000	0.0000
005	0.0000	0.0000	0.0000	0.0000	0.1377	0.1204	0.0000	0.2626	0.0000	0.0000	0.0000	0.0453	0.0000	0.0000	0.0658	0.0058
006	0.0005	0.0402	0.0109	0.0031	0.0038	0.0194	0.0544	0.2060	0.0125	0.0897	0.2770	0.0759	0.0242	0.0020	0.0399	0.0165
007	0.0000	0.0078	0.0134	0.0046	0.0183	0.0111	0.0003	0.6059	0.1124	0.0012	0.1846	0.0237	0.0000	0.0223	0.0039	0.0000
008	0.0004	0.0259	0.0013	0.0004	0.0014	0.0036	0.0003	0.3198	0.0254	0.0008	0.0363	0.0311	0.0053	0.0020	0.0527	0.0143
009	0.0000	0.0034	0.0017	0.0000	0.0000	0.0003	0.0000	0.0031	0.0060	0.0009	0.0020	0.0006	0.0035	0.0086	0.0000	0.0383
010	0.0001	0.0328	0.0105	0.0020	0.0028	0.0019	0.0000	0.3415	0.0035	0.0762	0.0115	0.0196	0.0229	0.0022	0.0959	0.0276
011	0.0007	0.0049	0.0010	0.0002	0.0005	0.0028	0.0000	0.0442	0.0514	0.0039	0.0252	0.1767	0.0341	0.0019	0.0185	0.0180
012	0.0002	0.0066	0.0003	0.0042	0.0006	0.0026	0.0000	0.3198	0.0044	0.0049	0.0065	0.0013	0.0004	0.0004	0.0084	0.0025
013	0.0002	0.0189	0.0022	0.0003	0.0001	0.0090	0.0014	0.0655	0.0156	0.0012	0.0539	0.0906	0.0002	0.0305	0.0503	0.0446
014	0.0000	0.0011	0.0002	0.0000	0.0008	0.0001	0.0000	0.0125	0.0068	0.0000	0.0161	0.0169	0.0316	0.0008	0.0195	0.0131
015	0.0001	0.0156	0.0044	0.0000	0.0001	0.0013	0.0000	0.0330	0.0091	0.0067	0.0342	0.0191	0.0753	0.0151	0.0993	0.0410
016	0.0000	0.0000	0.0000	0.0000	0.0000	0.0000	0.0000	0.0000	0.0000	0.0000	0.0000	0.0000	0.0000	0.0000	0.0000	0.0000

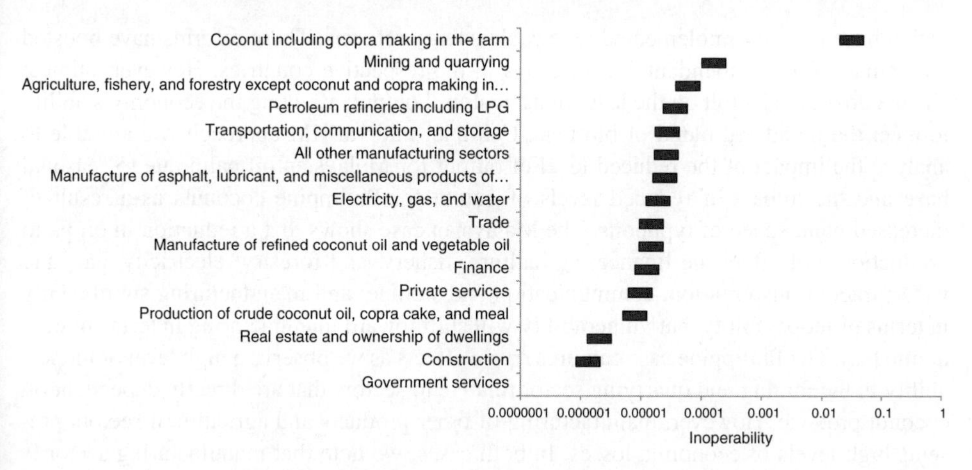

Figure 8.5 *Sector inoperability levels in case study 2*

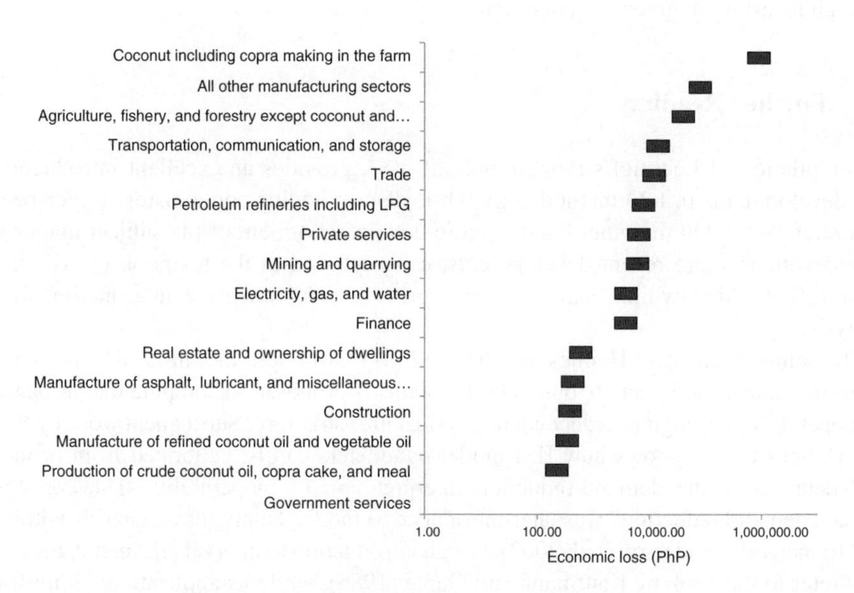

Figure 8.6 *Sector economic losses in case study 2*

8.7 Conclusions

Systematic risk analysis methods are essential to understanding the economic vulnerability of biomass-based supply chains and industrial networks. In particular, IIM methodology is useful for understanding how losses can cascade through such systems as a result of interdependencies across economic sectors. The application of IIM on different forms of natural and man-made disasters has proven its versatility as a tool for economic analysis. This chapter focuses on two Southeast Asian countries, namely, Malaysia and the Philippines, which have recently implemented renewable energy reforms. These reforms have boosted the demand for the abundant energy crops in the respective countries. However, climate change does take its toll on the level of agricultural output, affecting the economy's ability to meet the mandated blend of biofuels. Using an interval IIM approach, we are able to analyze the impact of the reduced level of output for Malaysian oil palm due to seasonal haze and the impact of reduced levels of output for Philippine coconuts as a result of increased wind speed of typhoons. The Malaysian case shows that a reduction in oil palm production will affect the finance; agriculture, fishery, and forestry; electricity, gas, and water; trade; transportation, communication and storage; and manufacturing significantly in terms of inoperability, but vulnerability will shift toward manufacturing in terms of economic loss. The Philippine case captures ripple effects as we observe a high level of inoperability in the mining and quarrying sector relative to sectors that are directly dependent on coconut produce. However, manufacturing of other products and agricultural sectors presents high levels of economic losses. In both cases, we note that manufacturing sector is susceptible to suffering losses as a result of reduced energy crop output. This paves the discussion for an increased stake of the manufacturing sector in environmental protection through investing in greener technologies.

8.8 Further Reading

A compilation of Leontief's most important work provides an excellent introduction to key developments in I–O methodology while also providing some historical perspective (Leontief 1986). On the other hand, a more detailed treatment of the subject matter, that includes much more recent developments, can be found in the textbook by Miller and Blair (2009). Meanwhile, Haimes (2009) gives a comprehensive introduction to risk analysis.

The seminal paper by Haimes and Jiang (2001) introduces the initial IIM framework, wherein a linear model analogous to I–O systems is proposed to compute the propagation of inoperability through interdependent physical infrastructure. Subsequent work by Santos and Haimes (2004) shows how IIM model parameters can be calibrated from economic I–O data, using the demand-reduction interpretation of inoperability. The concept of "forced demand reduction" was later introduced to model supply loss scenarios within the IIM framework (Anderson et al. 2007). For an introduction to interval arithmetic, the reader may refer to the book by Kaufmann and Gupta (1985), while its application within IIM is demonstrated in the work of Barker and Rocco (2011).

Appendix A LINGO Code for Illustrative Example

```
!This section allows the user to define the size of the
transactions matrix, N represents the number of sectors;

Calc:
N = 2;
Endcalc
```

```
!This section defines the properties of the sectors and the
techincal coefficients/interdependency matrix;

Sets:
Sectors: x, q, cstar;
Amatrix(Sectors, Sectors): a, astar, ID, DIFF;
Endsets

!in this section the user is expected to input the required
data/information;

Data:
Sectors = 1..N;

!the technical coefficients matrix should be given, a total
of N x N entries are required sequenced as one goes from
left to right then from top to bottom in the technical
coefficients matrix;

a =   0.15  0.25
      0.20  0.05;

!the total output of each sector should be given, a total of
N entries are required;

x = 1000 2000;

!The initial perturbation for each sector should be
provided, N entries are required;
cstar = 0.1 0.1;
Enddata

@for(Amatrix(i,j): @free(diff(i,j)));

!This section calculates for the (I - Astar) matrix;
Calc:
@for(Amatrix(i,j)|i#NE#j: ID(i,j) = 0);
@for(Amatrix(i,j)|i#EQ#j: ID(i,j) = 1);
@for(Amatrix(i,j): astar(i,j) = a(i,j)*x(j)/x(i));
@for(Amatrix(i,j): DIFF(i,j) = (ID(i,j) - astar(i,j)));
Endcalc

!this section calculates for the total inoperability for
each sector from the initial perturbation;

@for(sectors(i): cstar(i) = @sum(sectors(j): (DIFF(i,j))*q(j)));
```

Appendix B LINGO Code for Case Study 1

```
!This LINGO code solves Case 1 in the Chapter;

!Nomenclature:

  N           - the total number of sectors in the economy
  x(i)        - the total output of sector i
  csmin(i)    - the minimum initial perturbation of sector
                i as a
                result of a disruption
  csmax(i)    - the maximum initial perturbation experienced by
                sector i
  qmin(i)     - the minimum total inoperability of sector
                i as a
                result of the perturbation csmin(i)
  qmax(i)     - the maximum total inoperability of sector
                i as a
                result of the perturbation qmax(i)
  a(i,j)      - the technical coefficient aij
  astar(i,j)- the interdependency matrix coefficient a*ij
  ID(i,j)     - defines the identity matrix
  DIFF(i,j) - corresponds to (I - Astar);

!This section allows the user to define the size of the
transactions matrix, N represents the number of sectors;
Calc:

N = 12;
Endcalc

!This section defines the properties of the sectors and the
techincal coefficients/interdependency matrix;
Sets:

Sectors: x, qmin, qmax, csmin, csmax;
Amatrix(Sectors, Sectors): a, astar, ID, DIFF;

Endsets

!in this section the user is expected to input the required
data/information;
Data:

Sectors = 1..N;
```

```
!the technical coefficients matrix should be given, a total
of N x N entries are required sequence from left to
right and from top to bottom as one goes through the
technical coefficients matrix;

a =
0.114 0.013 0.002 0.019 0.000 0.001 0.014 0.000 0.001 0.000 0.000 0.003
0.001 0.091 0.001 0.020 0.000 0.000 0.001 0.000 0.000 0.000 0.000 0.001
0.003 0.002 0.005 0.042 0.018 0.032 0.004 0.001 0.002 0.000 0.001 0.003
0.144 0.108 0.100 0.324 0.223 0.305 0.100 0.106 0.030 0.028 0.063 0.101
0.004 0.010 0.002 0.009 0.186 0.004 0.008 0.010 0.007 0.013 0.031 0.020
0.000 0.000 0.007 0.000 0.043 0.001 0.004 0.022 0.000 0.010 0.001 0.086
0.024 0.025 0.044 0.047 0.010 0.027 0.043 0.012 0.025 0.139 0.011 0.037
0.014 0.015 0.013 0.021 0.013 0.043 0.044 0.262 0.070 0.017 0.047 0.051
0.003 0.031 0.000 0.009 0.004 0.032 0.022 0.108 0.300 0.021 0.023 0.005
0.000 0.000 0.001 0.000 0.001 0.007 0.014 0.023 0.013 0.081 0.044 0.030
0.001 0.006 0.006 0.003 0.015 0.026 0.039 0.023 0.029 0.017 0.188 0.037
0.000 0.000 0.001 0.000 0.004 0.005 0.004 0.015 0.004 0.007 0.009 0.064
;

!the total output of each sector should be given, a total of
N entries are required;
!x is given in Thousand RM;
x = 38218340, 21782028, 95410015, 899165229, 35149126,
48051594, 90511083, 119450597, 74329671, 33414975,
69097400, 79326620;

!The initial perturbation for each sector should be
provided, N entries are required;
csmin = 0 0.05 0 0 0 0 0 0 0 0 0 0;
csmax = 0 0.10 0 0 0 0 0 0 0 0 0 0;

Enddata

@for(Amatrix(i,j): @free(diff(i,j)));

!This section calculates for the (I - Astar) matrix;
Calc:
@for(Amatrix(i,j)|i#NE#j: ID(i,j) = 0);
@for(Amatrix(i,j)|i#EQ#j: ID(i,j) = 1);
@for(Amatrix(i,j): astar(i,j) = a(i,j)*x(j)/x(i));
@for(Amatrix(i,j): DIFF(i,j) = (ID(i,j) - astar(i,j)));
Endcalc

!this section calculates for the total inoperability for
each sector from the initial perturbation;
```

```
@for(sectors(i): csmin(i) = @sum(sectors(j):
(DIFF(i,j))*qmin(j)));
@for(sectors(i): csmax(i) = @sum(sectors(j):
(DIFF(i,j))*qmax(j)));

Data:
@text() = @table (astar);
@text() = @write (@newline(2));
@text() = @writefor(sectors(i): 'q',(i), ' ', @
format(qmin(i), '12f'), ' to ', @format(qmax(i), '12f'), @
newline(1));
```

Appendix C Interval Arithmetic

Interval numbers provide a simple means of expressing data uncertainties by the specification of lower and upper bounds. An interval thus represents the range of plausible values for any given quantity. Arithmetic operations on interval numbers also provide a means to propagate uncertainties through a mathematical model. Salient aspects of interval arithmetic are given here; further details may also be found in Kaufmann and Gupta (1985).

Given interval quantities $A=\left[a^L,a^U\right]$ and $B=\left[b^L,b^U\right]$, the following operations hold:

$$A+B=\left[\left(a^L+b^L\right),\left(a^U+b^U\right)\right] \tag{C.1}$$

$$A-B=\left[\left(a^L-b^L\right),\left(a^U-b^U\right)\right] \tag{C.2}$$

$$A\times B=\left[min\left(a^Lb^L,a^Ub^L,a^Lb^U,a^Ub^U\right),\ max\left(a^Lb^L,a^Ub^L,a^Lb^U,a^Ub^U\right)\right] \tag{C.3}$$

$$A/B=\left[min\left(\frac{a^L}{b^L},\frac{a^U}{b^L},\frac{a^L}{b^U},\frac{a^U}{b^U}\right),\ max\left(\frac{a^L}{b^L},\frac{a^U}{b^L},\frac{a^L}{b^U},\frac{a^U}{b^U}\right)\right] \tag{C.4}$$

In Equation C.4, the operation becomes defined if b^L is negative and b^U is positive, since that results in the inclusion of zero in the denominator.

Appendix D Analytic Hierarchy Process

Analytic Hierarchy Process (AHP), a multiple criteria analysis tool developed by Prof. Saaty, is a relative measurement theory that provides the objective mathematics to process the subjective and personal judgment of individual or group in decision making. Salient features of AHP weighting process are given here; further details may also be found in Saaty (1980). The priority weight vector (**w**) of n elements measured in ratio scale is derived in AHP from a positive reciprocal square matrix **A**:

$$\mathbf{A} = \begin{bmatrix} 1 & a_{12} & \cdots & a_{1n} \\ \frac{1}{a_{12}} & 1 & \cdots & a_{2n} \\ \vdots & \cdots & \ddots & \\ \frac{1}{a_{1n}} & \frac{1}{a_{2n}} & \cdots & 1 \end{bmatrix} = \begin{bmatrix} \frac{w_1}{w_2} & \frac{w_1}{w_2} & \cdots & \frac{w_1}{w_n} \\ \frac{w_2}{w_1} & \frac{w_2}{w_2} & \cdots & \frac{w_2}{w_n} \\ \vdots & \cdots & \ddots & \vdots \\ \frac{w_n}{w_1} & \frac{w_n}{w_2} & \cdots & \frac{w_n}{w_n} \end{bmatrix} \quad \text{(D.1a)}$$

where $a_{ij} = 1/a_{ji} = w_i / w_j$ (D.1b)

Decision makers are asked to respond to $n(n-1)/2$ pairwise comparisons in which two elements at a time are compared with respect to their parent criteria. The relative dominance values (a_{ij}) are then elicited from the said respondents using the so-called "Saaty's fundamental 9-point scale." After the pairwise comparison matrix **A** is filled up with those value judgments a_{ij}, the relative weights (**w**) of the elements are computed from the normalized right eigenvector associated to the principal eigenvalue (λ_{max}) of **A** as shown in the following equations:

$$\mathbf{A} \cdot \mathbf{w} = \lambda_{max} \cdot \mathbf{w} \quad \text{(D.2a)}$$

$$\sum_{i=1}^{n} w_i = 1 \quad \text{(D.2b)}$$

For a perfectly consistent pairwise comparative judgment matrix, the λ_{max} is equal to n. Note that the eigenvector method also yields a natural measure of consistency as described by the following equation:

$$CI = \frac{\lambda_{max} - n}{n-1} \quad \text{(D.3)}$$

Here is a numerical example to illustrate how weights are derived from a $3x3$ pairwise comparison matrix **A**:

$$\mathbf{A} = \begin{bmatrix} 1 & 3 & 5 \\ 1/3 & 1 & 2 \\ 1/5 & 1/2 & 1 \end{bmatrix} \quad \text{(D.4a)}$$

The priority weight vector $[0.648 \quad 0.230 \quad 0.122]^T$ is computed from the principal right eigenvector method with $\lambda_{max} = 3.004$:

$$\begin{bmatrix} 1 & 3 & 5 \\ 1/3 & 1 & 2 \\ 1/5 & 1/2 & 1 \end{bmatrix} \begin{bmatrix} 0.648 \\ 0.230 \\ 0.122 \end{bmatrix} = 3.004 \begin{bmatrix} 0.648 \\ 0.230 \\ 0.122 \end{bmatrix} \quad \text{(D.4b)}$$

The computed consistency index CI is $(3.004 - 3.0)/(3.0 - 1) = 0.002$. The nearer the CI value to zero, the more consistent the judgments in the pairwise comparison matrix are.

Nomenclature

A	Technical coefficient matrix
A*	Interdependency matrix
c	Final demand vector
c*	Perturbation vector
EL	Economic loss
I	Identity matrix
L	Leontief inverse matrix
q	Inoperability vector
x	Total output vector
Z	Transactions matrix

References

Akhtar, R., Santos, J. R. (2013). Risk-based input–output analysis of hurricane impacts on interdependent regional workforce systems. *Natural Hazards*, **65**(1): 391–405.

Alcántara, V., Padilla, E. (2003). Key sectors in final energy consumption: an input–output application to the Spanish case. *Energy Policy*, **31**(15): 1673–1678.

Anderson, C. W., Santos, J. R., Haimes Y. Y. (2007). A risk-based input–output methodology for measuring the effects of the August 2003 Northeast Blackout. *Economic Systems Research*, **19**(2): 183–204.

Baffes, J. (2013). A framework for analyzing the interplay among food, fuels, and biofuels. *Global Food Security*, **2**(2): 110–116.

Balat, M., Balat, H. (2009). Recent trends in global production and utilization of bio-ethanol fuel. *Applied Energy*, **86**(11): 2273–2282.

Barker, K. A., Rocco, C. S. (2011). Evaluating uncertainty in risk-based interdependency modeling with interval arithmetic. *Economic Systems Research*, **23**(2): 213–232.

Barker, K. A., Santos, J. R. (2010a). Measuring the efficacy of inventory with a dynamic input–output model. *International Journal of Production Economics*, **126**(1): 130–143.

Barker, K. A., Santos, J. R., (2010b). A risk-based approach for identifying key economic and infrastructure systems. *Risk Analysis*, **30**(6): 962–974.

Buckley, J. J. (1989). Fuzzy input output analysis. *European Journal of Operational Research*, **39**(1): 5460.

Bullard, C. W., Herendeen, R. A. (1975). Energy cost of goods and services. *Energy Policy*, **3**(4): 268–278.

Casler, S., Wilbur, S. (1984). Energy input output analysis – a simply guide. *Resources and Energy*, **6**(2): 187–201.

Chander, P. (1983). The non-linear input output model. *Journal of Economic Theory*, **30**(2): 219–229.

Crowther, K. G., Haimes, Y. Y. (2010). Development of the multiregional inoperability input–output model (MRIIM) for spatial explicitness in preparedness of interdependent regions. *Systems Engineering*, **13**(1): 28–46.

Crowther, K. G., Haimes, Y. Y., Taub, G. (2007). Systemic valuation of strategic preparedness through application of the inoperability input–output model with lessons learned from Hurricane Katrina. *Risk Analysis*, **27**(5): 1345–1364.

Dervis, K. (1975). Planning capital-labor substitution and intertemporal equilibrium with a non-linear multi-sector growth model. *European Economic Review*, **6**(1): 77–96.

Duchin, F. (1992). Industrial input–output analysis: implications for industrial ecology. *Proceedings of the National Academy of Sciences*, **89**(3): 851–855.

Economy and Environment Program for Southeast Asia (1999). The Indonesian Fires and Haze of 1997: The Economic Toll. Available at: http://www.eepsea.net/pub/rr/10536124150ACF62.pdf, last accessed September 7, 2013.

Fargione, J., Hill, J., Tilman, D., Polasky, S., Hawthorne, P. (2008). Land clearing and the biofuel carbon debt. *Science*, **319**(5867): 1235–1238.

Figueira, J., Greco, S., Ehrgott, M. (2005). *Multiple Criteria Decision Analysis: State of the Art Surveys*. New York: Springer.

Food and Agriculture Organization (2013). Food and Agricultural Commodities Production. Available at: http://faostat.fao.org/site/339/default.aspx, last accessed August 10, 2013.

Gauderman, W. J., Avol, E., Gilliland, F., Vora, H., Thomas, D., Berhane, K., McConnell, R., Kuenzli, N., Lurmann, F., Rappaport, E., Margolis, H., Bates, D., Peters, J. (2004). The effect of air pollution on lung development from 10 to 18 years of age. *New England Journal of Medicine*, **351**(11): 1057–1067.

Greenberg, M., Haas, C., Cox, A., Lowrie, K., McComas, K., North, W. (2012). Ten most important accomplishments in risk analysis, 1980–2010. *Risk Analysis*, **32**(5): 771–781.

Greenwold, R., Bergin, M. H., Xu, J., Cohan, D., Hoogenboom, G., Chameides, W. L. (2006). The influence of aerosols on crop production: a study using the CERES crop model. *Agricultural Systems*, **89**(2): 390–413.

Haimes, Y. Y. (1991). Total risk management. *Risk Analysis*, **11**(2): 169–171.

Haimes, Y. Y. (2009). *Risk Modeling, Assessment and Management* (3rd edn.). Hoboken, NJ: John Wiley & Sons, Inc.

Haimes, Y. Y., Jiang, P. (2001). Leontief-based model of risk in complex interconnected infrastructures. *Journal of Infrastructure Systems*, **7**(1): 1–12.

Haimes, Y. Y., Horowitz, B. M., Lambert, J. H., Santos, J. R., Crowther, K., Lian, C. (2005). Inoperability input–output model for interdependent infrastructure sectors. I: theory and methodology. *Journal of Infrastructure Systems*, **11**(2): 67–79.

Hannon, B. (1973). The structure of ecosystems. *Journal of Theoretical Biology*, **41**(3): 535–546.

Hendrickson, C. T., Lave, L. B., Matthews, H. S., (2006). *Environmental Life Cycle Assessment of Goods and Services: An Input–Output Approach*. Washington, DC: Resources for the Future Press.

Hewings, G. J. D., Jensen, R. C., West, G. R., Sonis, M., Jackson, R. W. (1989). The spatial organization of production – an input output perspective. *Socio-Economic Planning Sciences*, **23**(1–2): 67–86.

Huang, Y. A., Lenzen, M., Weber, C. L., Murray, J., Matthews, H. S. (2009). The role of input–output analysis for the screening of corporate carbon footprints. *Economic Systems Research*, **21**(3): 217–242.

Jonkeren, O. E., Ward, D., Dorneanu, B., Giannopoulos, G. Economic impact assessment of critical infrastructure failure in the EU: a combined systems engineering–inoperability input–output model. 20th International Input–Output Conference, Bratislava, Slovakia, July 9–12, 2013.

Kaplan, S., Garrick, B. J. (1981). On the quantitative definition of risk. *Risk Analysis*, **1**(1): 11–27.

Kaufmann, A., Gupta, M. M. (1985). *Introduction to Fuzzy Arithmetic: Theory and Applications*. New York: Van Nostrand Reinhold.

Khanna, V., Bakshi, B. R. (2009). Modeling the risks to complex industrial networks due to loss of natural capital. Proceedings of the 2009 IEEE International Symposium Sustainable Systems and Technology, May 18–20, 2009, Tempe, AZ, USA. Curran Associates, Inc., New York.

Lave, L. B., Cobasflores, E., Hendrickson, C. T., McMichael, F. C. (1995). Using Input–output analysis to estimate economy-wide discharges. *Environmental Science & Technology*, **29**(9): A420–A426.

Lenzen, M., Pade, L., Munksgaard, J. (2004). CO_2 multipliers in multi-region input–output models, *Economic Systems Research*, **16**(4): 391–412.

Leontief, W. W. (1936). Quantitative input and output relations in the economic system of the United States, *Review of Economic and Statistics*, **18**(3): 105–125.

Leontief, W. W. (1951). *The Structure of American Economy, 1919–1939: An Empirical Application of Equilibrium Analysis*. New York: Oxford University Press.

Leontief, W. W. (1970). Environmental repercussions and the economic structure: an input–output approach, *Review of Economics and Statistics*, **52**(3) 262–271.
Leontief, W. W. (1986). *Input–Output Economics* (2nd edn.). New York: Oxford University Press.
Leung, M., Haimes, Y. Y., Santos, J. R. (2007). Supply-and output-side extensions to the inoperability input–output model for interdependent infrastructures. *Journal of Infrastructure Systems*, **13**(4): 299–310.
Luo, X., Hu, J., Zhao, J., Zhang, B., Chen, Y., Mo, S. (2014). Multi-objective optimization for the design and synthesis of utility systems with emission abatement technology concerns. *Applied Energy*, **136**: 1110–1131.
MacKenzie, C. A., Santos, J. R., Barker, K. (2012). Measuring changes in international production from a disruption: case study of the Japanese earthquake and tsunami. *International Journal of Production Economics*, **138**(2): 293–302.
MacLean, H. L., Lave, L. B., Lankey R., Joshi, S. (2000). A life-cycle comparison of alternative automobile fuels. *Journal of Air and Waste Management Association*, **50**(10): 1769–1779.
Malaysia Department of Statistics. (2010). Input–Output tables Malaysia 2005. Available at: http://www.statistics.gov.my/portal/index.php?option=com_content&view=article&id=1242&Itemid=111&lang=en, last accessed April 27, 2013.
Malaysian Palm Oil Board (n.d.). Malaysian Palm Oil Industry. Available at: http://www.palmoilworld.org/about_malaysian-industry.html, last accessed April 27, 2013.
de Melo, J. (1988). Computable general equilibrium models for trade policy analysis in developing countries: a survey. *Journal of Policy Modeling*, **10**(4): 469–503.
Milbrandt, A., Overend, R. P. (2008). The future of liquid biofuels for APEC economies. APEC Energy Working Group. APEC#208-RE-1.8. Golden, CO: National Renewable Energy Laboratory.
Miller, R. E., Blair, P. D. (2009). *Input–Output Analysis: Foundations and Extensions* (2nd edn.). New York: Cambridge University Press.
Minx, J. C., Wiedmann, T., Wood, R., Peters, G. P., Lenzen, M., Owen, A., …, Ackerman, F. (2009). Input–output analysis and carbon footprinting: an overview of applications. *Economic Systems Research*, **21**(3): 187–216.
National Pollution Prevention Roundtable (2009). Road to Sustainability: Pollution Prevention Progress from 2004 to 2006 Results from the National Pollution Prevention Data Management System. Available at: http://www.epa.gov/ppic/pubs/ppicdist.html#new, last accessed August 10, 2013.
Ng, W. P. Q., Lam, H. L., Ng, F. Y., Kamal, M., Lim, J. H. E. (2012). Waste-to-wealth: green potential from palm biomass in Malaysia. *Journal of Cleaner Production*, **34**: 57–65.
Philippine National Statistics Coordination Board (2006). *The 2000 Input–Output Accounts of the Philippines*. Makati: NSCB.
Okuyama, Y. (2007). Economic modeling for disaster impact analysis: past, present, and future. *Economic Systems Research*, **19**(2): 115–124.
Oliva, G. Panzieri, S., Setola, R., (2011) Fuzzy dynamic input–output inoperability model. *International Journal of Critical Infrastructure Protection*, (4)**3**: 165–172.
Orsi, M. J., Santos, J. R. (2010). Probabilistic modeling of workforce-based disruptions and input–output analysis of interdependent ripple effects. *Economic Systems Research*, **22**(1): 3–18.
Perlack, R. D., Wright, L. L., Turhollow, A. F., Graham, R. L. Stokes, B. J., Erback, D. C. (2005). *Biomass as Feedstock for a Bioenergy and Bioproducts Industry: The Technical Feasibility of a Billion-Ton Annual Supply.* ORNL/TM-2011/224. Springfield, VA: U.S. Department of Commerce National Technical Information Service.
Philippine Coconut Authority – Department of Agriculture (2013). Coconut Statistics. Available at: http://www.pca.da.gov.ph/cocostat.php, last accessed 16 August 2013.
Quah, E. (2002). Transboundary pollution in Southeast Asia: the Indonesian fires. *World Development*, **30**(3): 429–441.
Resurreccion, J., Santos, J. R. (2012). Multiobjective prioritization methodology and decision support system for evaluating inventory enhancement strategies for disrupted interdependent sectors. *Risk Analysis*, **32**(10): 1673–1692.

Rockström, J., Steffen, W., Noone, K., Persson, Å., Chapin III, F. S., Lambin, E., …, Liverman, D. (2009). Planetary boundaries: exploring the safe operating space for humanity. *Ecology and Society*, **14**(2): 472–475.

Saaty, T. L. (1980). *The Analytic Hierarchy Process*. New York: McGraw-Hill.

Sandberg, I. W. (1973). Nonlinear input–output model of a multisectored economy. *Econometrica*, **41**(6): 1167–1182.

Santos, J. R., Haimes, Y. Y. (2004). Modeling the demand reduction input–output (I–O) inoperability due to terrorism of interconnected infrastructures. *Risk Analysis*, **24**(6): 1437–1451.

Santos, J. R., Haimes, Y. Y., Lian, C. (2007). A framework for linking cybersecurity metrics to the modeling of macroeconomic interdependencies. *Risk Analysis*, **27**(5): 1283–1297.

Santos, J. R., Barker, K. A., Zelinke, P. J. (2008). Sequential decision-making in interdependent sectors with multiobjective inoperability decision trees. *Economic Systems Research*, **20**(1): 29–56.

Santos, J. R., Orsi, M. J., Bond, E. J. (2009). Pandemic recovery analysis using the dynamic inoperability input-output model. *Risk Analysis*, **29**(12): 1743–1758.

Santos, J. R., May, L., El Haimar, A. (2013). Risk-Based Input–Output Analysis of Influenza Epidemic Consequences on Interdependent Workforce Sectors. *Risk Analysis*, **33**(9): 1620–1635.

Schaeffer, R., Szklo, A. S., Pereira de Lucena, A. F., Moreira Cesar Borba, B. S., PupoNogueira, L. P., Fleming, F. P., …, Boulahya, M. S. (2012). Energy sector vulnerability to climate change: a review. *Energy*, **38**(1): 1–12.

Setola R., De Porcellinis, S. (2008). A methodology to estimate input–output inoperability model parameters. In: J. Lopez, B. Hamerli (Eds), *Critical Information Infrastructures Security*, Berlin: Springer Verlag, pp. 149–160.

Setola R., De Porcellinis, S., Sforna, M. (2009). Critical infrastructure dependency assessment using the input–output inoperability model. *International Journal of Critical Infrastructure Protection*, **2**(4): 170–178.

Stromberg, P. M., Esteban, M., Gasparatos, A. (2011). Climate change effects on mitigation measures: the case of extreme wind events and the Philippines' biofuel plan. *Environmental Science & Policy*, **14**(8): 1079–1090.

Suh, S., Nakamura, S. (2007). Five years in the area of input–output and hybrid LCA. *International Journal of Life Cycle Assessment*, **12**(6): 351–352.

Tan, R. R. (2011). A general source-sink model with inoperability constraints for robust energy sector planning. *Applied Energy*, **88**(11): 3759–3764.

Tan, R. R., Foo, D. C. Y. (2013). Pinch analysis for sustainable energy planning using diverse quality measures. In J. J. Klemes (Ed.), *Handbook of Process Integration*. Cambridge, UK, Woodhead Publishing Limited, pp. 505–523.

United States Department of Agriculture (2013). Coconut Oil Production by Country. Available at: http://www.indexmundi.com/agriculture/?commodity=coconut-oil&graph=production, last accessed September 4, 2013.

United States Department of Energy. (2013). Energy Efficiency & Renewable Energy Bioenergy Technologies Office Multi-Year Program Plan: May 2013 Update. Available at: http://www1.eere.energy.gov/bioenergy/pdfs/mypp_may_2013.pdf, last accessed August 10, 2013.

Wiedmann, T., Lenzen, M., Turner, K., Barrett, J. (2007). Examining the global environmental impact of regional consumption activities — Part 2: review of input–output models for the assessment of environmental impacts embodied in trade. *Ecological Economics*, **61**(1): 15–26.

Zhang, R., Zhou, J., Wang, Y. (2012). Multi-objective optimization of hydrothermal energy system considering economic and environmental aspects. *Electrical Power and Energy Systems*, **42**(1):384–395.

Zhang, Q., Mclellan, B. C., Tezuka, T., Ishihara, K. N. (2014). Economic and environmental analysis of power generation expansion in Japan considering Fukushima nuclear accident using a multi-objective optimization model. *Energy*, **44**(1): 986–995.

Part 3
Other Applications of Biomass Conversion Systems

9

Process Systems Engineering Tools for Biomass Polygeneration Systems with Carbon Capture and Reuse

Jhuma Sadhukhan[1], Kok Siew Ng[1], and Elias Martinez-Hernandez[2]

[1] *Centre for Environmental Strategy, University of Surrey, Guildford, UK*
[2] *Department of Engineering Science, University of Oxford, Oxford, UK*

9.1 Introduction

A polygeneration system is economic when one or two main biofuels and commodity chemicals are produced, alongside a number of high-value, low-volume chemicals. The economic and environmental sustainability can be enhanced by polygeneration of added value chemicals (for which small volume is required) and materials through process integration opportunities. Reuse and recycle of CO_2 rich as well as lean streams (post-combustion and pre-combustion, respectively) can generate syngas, hydrogen, formic acid, methane, ethylene, methanol, dimethyl ether, urea, Fischer–Tropsch liquid, succinic acid, etc. as well as materials such as epoxides as precursor to polymers and carbonates through mineralisation reactions for construction applications. As such, materials like carbon black can also be produced from biomass, that is, lignin fraction of lignocellulose materials. There are three main ways to capture carbon dioxide directly or indirectly by biomass:

1. During biomass growth: By the process of photosynthesis, biomass directly sequestrates biogenic carbon dioxide during growth. After accounting for all emissions across a whole system, a lowest of 50% carbon dioxide reduction is possible from biofuel

Process Design Strategies for Biomass Conversion Systems, First Edition. Edited by
Denny K. S. Ng, Raymond R. Tan, Dominic C. Y. Foo, and Mahmoud M. El-Halwagi.
© 2016 John Wiley & Sons, Ltd. Published 2016 by John Wiley & Sons, Ltd.

production from crops. From lignocellulose feedstocks, higher percentage of carbon dioxide equivalent global warming potential reduction is possible alongside reduction in land use and avoidance of food versus fuel debate.
2. Chemical production: Carbon dioxide becomes one of the main reactants to produce biomass-derived chemicals.
3. Material production: Insertion of carbon dioxide in biomass-derived epoxides, acetals and orthoesters generates important precursors to many polymers.

Descriptions of reactions or processes for chemical production and material production using examples are shown in the following section. After that, process systems engineering tools that are proved to be useful for complex biorefinery process integration are shown with specific applications. Main findings and recommendations are discussed at the end of each case study.

9.2 Production Using Carbon Dioxide

9.2.1 Chemical Production from Carbon Dioxide

Gaseous phase reactions of carbon dioxide: Examples of gaseous phase reactions of carbon dioxide are methane production in Sabatier's reaction and syngas (carbon monoxide and hydrogen) production in tri-reforming reaction, shown in Equations 9.1 and 9.2, respectively:

$$CO_2 + 4H_2 \leftrightarrow CH_4 + 2H_2O \quad \Delta H_r^o = -167 \text{ kJ mol}^{-1} \tag{9.1}$$

The temperature and pressure conditions of Sabatier's reaction are 300°C and 2 bar. Complete conversion of carbon dioxide can be achieved in the presence of hydrogen-rich feed gas.

The tri-reforming reaction occurs when methane gas is reacted with a gas mixture containing carbon dioxide, steam and oxygen. Ng et al. investigated the tri-reforming process fed with CH_4, CO_2, H_2O and O_2 with a molar ratio of 1:0.475:0.475:0.1, operating at 1 bar and 850°C (Ng et al. 2013). In their study, the product gas contains H_2 to CO at a molar ratio of 1.68. The reactions are shown in Equation 9.2. The overall reaction performance is endothermic as expected from the reforming reaction:

$$\begin{aligned} &CH_4 + CO_2 \leftrightarrow 2CO + 2H_2 \quad \Delta H_r^o = 247.3 \text{ kJ mol}^{-1} \\ &CH_4 + H_2O \leftrightarrow CO + 3H_2 \quad \Delta H_r^o = 206.3 \text{ kJ mol}^{-1} \\ &CH_4 + 0.5O_2 \leftrightarrow CO + 2H_2 \quad \Delta H_r^o = -35.6 \text{ kJ mol}^{-1} \end{aligned} \tag{9.2}$$

Carbon dioxide in carbonate formations: Solid oxides are used to capture carbon dioxide to form carbonates that are useful chemicals. The technology is used to capture carbon dioxide in chemical looping combustion processes. In an integrated CO_2 capture process, calcium oxide (CaO) is employed as a high-temperature CO_2 sorbent and carrier between two reactors: a steam gasifier and an oxygen-fired regenerator (Sadhukhan et al. 2014). CO_2 is captured in situ by CaO sorbent in the gasifier; thereby, a hydrogen-rich gas is produced. The adsorbed CO_2 (calcium carbonate, $CaCO_3$, or calcite or limestone, used in cement) is transferred to the regenerator where the sorbent is calcined generating a CO_2-rich stream suitable for storage. The adsorption (exothermic) and calcination (endothermic) reactions are shown in Equation 9.3:

$$\begin{aligned} &\mathrm{CaO} + \mathrm{CO_2} \rightarrow \mathrm{CaCO_3} \quad \Delta H_r^o = -178.4 \text{ kJ mol}^{-1} \\ &\mathrm{CaCO_3} \rightarrow \mathrm{CaO} + \mathrm{CO_2} \quad \Delta H_r^o = 178.4 \text{ kJ mol}^{-1} \end{aligned} \tag{9.3}$$

Copper(II) oxide also acts similar way to produce copper(II) carbonate. The reverse reaction can be easily performed and has been widely adopted for demonstration in school laboratory. Copper(II) carbonate (green) is heated up to produce copper(II) oxide (black). The change in colour is easily visible. Similarly, magnesium carbonate, calcium magnesium carbonate, iron carbonate, sodium carbonate and potassium carbonate can be formed by carbon dioxide adsorption (exothermic reaction) on corresponding oxides. These processes occur naturally in mineral formation. Carbonates are useful chemicals for smelting, glazing, glassmaking, ceramic and construction industries.

Carbon dioxide in metabolic products: Succinic acid can be produced from glucose, as shown in Equation 9.4:

$$\mathrm{C_6H_{12}O_6} + \mathrm{CO_2} \rightarrow \mathrm{C_4H_6O_4} + \mathrm{CH_3COOH} + \mathrm{HCOOH} \tag{9.4}$$

A theoretical yield of 2 moles of succinic acid per mole glucose and per mole of carbon dioxide captured can be obtained. Carbon feedstock, pH and carbon dioxide are critical for the production of succinic acid and any succinate derivative. A proper combination of these parameters must be selected for each microorganism, as they use different pathways for succinate production and tolerate different levels of CO_2, pH and H_2. CO_2 is an electron acceptor that diverts metabolism to pyruvate and lactate/ethanol when present at low levels but to succinate when present at high levels (Sadhukhan et al. 2014). CO_2 can be supplied from an external gas stream and carbonates added to the medium (e.g. $MgCO_3$, $NaCO_3$, $NaHCO_3$ or $CaCO_3$) or from a combination of these sources.

9.2.2 Material Production from Carbon Dioxide

Coupling of epoxides and carbon dioxide produces polycarbonates with enhanced biodegradable thermoplastic properties. Polycarbonates are polymers containing carbonate groups (–O–CO–O) in the main chain. They are mainly synthesised by condensation of phosgene and aromatic or aliphatic diols. Bisphenol A (BPA) polycarbonate, shown in Figure 9.1, used to be the most important polycarbonate in the market.

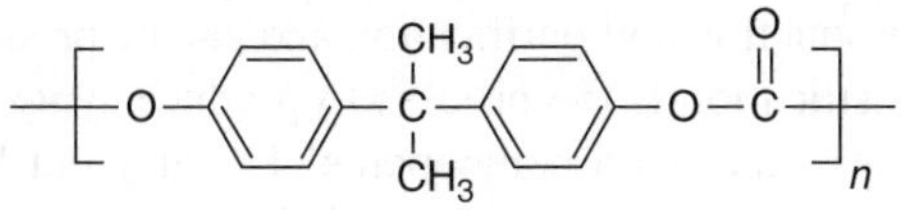

Figure 9.1 *Structure of bisphenol A polycarbonate*

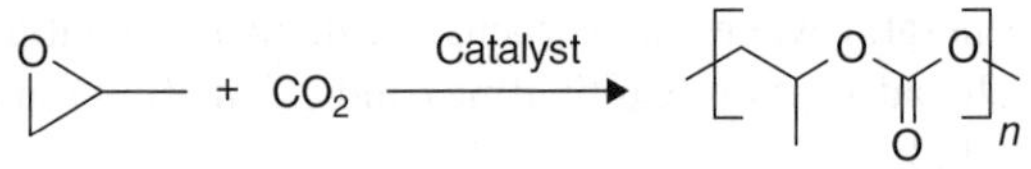

Figure 9.2 Synthesis of poly(propylene carbonate)

The potential release of the toxic BPA from the polymer is driving the search for an alternative monomer. Diphenolic acid (DPA) is a potential substitute for BPA in polymer synthesis. The synthesis of polycarbonates involves a two-step process. First, DPA is treated with NaOH to produce a sodium diphenoxide. This intermediate reacts with phosgene ($COCl_2$) to start the polymerisation. A disadvantage of this process is the toxicity of phosgene (phosgene was used as a chemical weapon during World War I). An alternative route to polycarbonates consists of the transesterification of DPA with diphenyl carbonates. DPA can also be copolymerised using the same reaction route.

Despite the potential replacement of the toxic BPA by the less harmful and bio-based DPA, the synthesis route uses the harmful reactant phosgene. Thus, alternatives are being explored for the production of novel polycarbonates (Sadhukhan et al. 2014). A route promising various environmental advantages comprises the use of CO_2 for carbonate product. This route eliminates the need for phosgene and BPA. In addition, CO_2 is captured in a durable polymer, and CO_2 is less expensive than phosgene.

CO_2 and epoxides are polymerised into polycarbonates in the presence of zinc-based catalysts (e.g. $ZnEt_2$) combined with water, *tert*-butylcatechol or pyrogallol (benzene with three –OH groups). For example, the polymerisation of propylene oxide with CO_2 produces poly(propylene carbonate) (PPC), as shown in Figure 9.2.

9.3 Process Systems Engineering Tools for Carbon Dioxide Capture and Reuse

This section discusses the techno-economic analysis tools in Section 9.3.1 and CO_2 pinch analysis, targeting and exchange network design tools in Section 9.3.2 using suitable examples.

9.3.1 Techno-economic Analysis Tools for Carbon Dioxide Capture and Reuse in Integrated Flowsheet

Conceptual process synthesis using heuristics, process simulation, heat integration and economic analysis are applied in sequence for feasible techno-economics of the integrated system. A process flowsheet can be developed logically starting from the core reaction–separation–heat recovery processes to fully developed heat integrated flowsheet.

The first example is a biomass gasification-based process integrated with Sabatier's reaction via hydrogen production and purification and calcite production via carbon dioxide capture. Biomass gasification is the process to produce syngas, consisting of carbon monoxide and hydrogen as the main components. To carry out biomethane production using Sabatier's reaction shown in Equation 9.1, allothermal gasification (Sadhukhan et al. 2010) and water gas shift (WGS) reaction (Equation 9.5; undertaken at a temperature of

250–300°C) can be carried out at the core of the process flowsheet to be synthesised. This will increase the proportion of hydrogen to the highest, equal to the hydrogen chemically bonded in the biomass and that available in the form of moisture in the biomass. Hence, the preprocessing step is a wet milling process to retain the moisture in the biomass feedstock:

$$CO + H_2O \leftrightarrow CO_2 + H_2 \quad \Delta H_r^o = -41.2 \text{ kJ mol}^{-1} \tag{9.5}$$

An allothermal gasifier consists of two interconnected fluidised beds, a char combustor, combusting char in the presence of air and a steam gasifier, gasifying the balance of the biomass, that is, volatilised gases and tars. As the steam gasification occurs separately from the char combustor, the char combustor does not need pure oxygen for combustion, as it can run on air. The heat required by the steam gasification is supplied from the char combustor (i.e. exothermic heat of combustion) by means of the circulation of bed particles, generally sand. It is in a loop with end-to-end configuration composed of a combustor, a cyclone and a gasifier. The scheme avoids dilution of the product gas with nitrogen whilst avoiding the use of an oxygen plant (air separation unit) for supplying pure oxygen to the gasifier. In view of the thoroughness of mixing and good gas–solid contact, the use of fluidised bed reactors for both is a commonplace. Gasification occurs at around 800–1100°C, more commonly at 950°C, and combustion needs to be undertaken at a slightly higher temperature to maintain the temperature gradient between the two reactors. The steam gasifier and the char combustor can be operated in perfect heat balance giving an overall heat neutral operation of the allothermal gasifier. In between the three main reactors, gasifier, combustor and WGS reactor, there are heat and water recovery schemes to ensure that there is no loss of useful work and thermodynamically highest recoveries are achieved within the limits of process operations.

A Rectisol process needs to be integrated to the gasification product gas after heat recovery in order to purify the gas into syngas production (Sadhukhan et al., 2010). The Rectisol technology uses refrigerated methanol as the solvent for physical absorption/removal of undesired contaminants producing ultra-clean syngas and is widely used in gasification plants. Rectisol provides an excellent option for co-removal of a number of contaminants including H_2S, COS, HCN, NH_3, nickel and iron carbonyls, mercaptans, naphthalene, organic sulphides, etc. to a trace level (e.g. H_2S to less than 0.1 ppm by volume) using one integrated plant. The clean syngas is suitable for WGS reaction, giving rise to hydrogen and carbon dioxide. Hydrogen from the WGS product gas mixture can be separated using pressure swing adsorption. The other outlet stream from the pressure swing adsorption, carbon dioxide-rich gas, is combusted in the char combustor for unreacted carbon monoxide from the WGS reaction. Pure hydrogen and a part of the carbon dioxide-rich gas (in the molar ratio of hydrogen to carbon dioxide 4:1) from the pressure swing adsorption are routed to Sabatier's reactor for biomethane formation. The exhaust gas from the char combustor are cooled and then reacted with calcium oxide for carbon dioxide capture into the production of calcite.

The heat sources are the char combustor exhaust gas cooler and steam gasifier product gas coolers. Hence, steam can be generated recovering this heat. The gasifier product gas has higher heat content due to latent heat of the steam contained in biomass. The first gasifier product gas cooler is suitable for superheating high-pressure (HP) steam, generated from the second gasifier product gas cooler called heat recovery steam generator (HRSG, used to saturate boiler feed water (BFW)). The char combustor exhaust gas cooler is also suitable to generate superheated HP steam. Some steam would be used on-site for the WGS reactor and steam gasifier. Excess superheated steam can be exported or used for power generation to export. The clean syngas from the Rectisol process needs preheating up to the WGS reaction temperature, 250–300°C, and HRSG is suitable to heat this stream up. Therefore, a multiple stream exchanger can be designed with one hot stream, gasifier product gas (from 450°C to the dew point), and two cold streams, clean syngas from the Rectisol process and BFW, in the countercurrent arrangement. Therefore, two exchangers in series to cool down the gasifier product gas up to the dew point are recommended. After this point, the gas can be directly quenched with cooling water to clean the gas and remove the waste water before entering the Rectisol process for final contaminant removal and ultra-purification of the gas. The heat recovery system design carried out follows the heat integration exercise, shown in our various works (Sadhukhan 2009; Ng et al. 2010, 2013; Ng and Sadhukhan 2011a, 2011b; Martinez-Hernandez et al. 2013). The integrated scheme is shown in Figure 9.3. The net result is the production of biomethane, calcite and superheated steam, from biomass and the other main raw material, calcium oxide, further discussed with the consideration of constraints.

The following *generic heuristics* can be applied to develop a conceptual process flowsheet:

1. Select appropriate reactors and essential separation processes for a given production objective.
   Here, allothermal gasifier, WGS reactor and Sabatier's reactor are selected for biomethane production by carbon capture. The strategy was to separate the carbon or char fraction from the rest of the volatile components, so that the theoretical highest amount of hydrogen can be produced to facilitate Sabatier's reaction. The essential separation process involved is the Rectisol process, which is required to obtain high-purity syngas to avoid catalyst poisoning in the downstream processes. While biomethane production by carbon capture is the key objective, carbon dioxide needed in Sabatier's reaction is

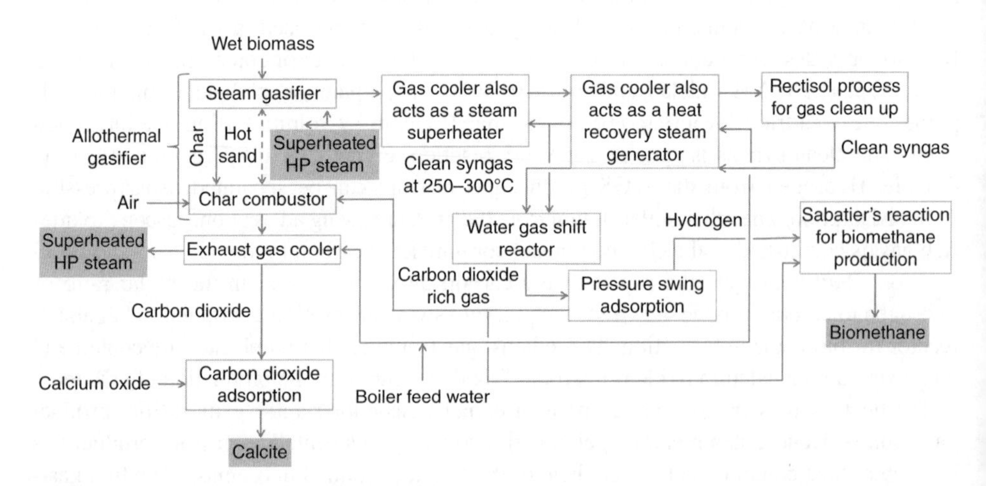

Figure 9.3 Conceptual process flowsheet for carbon dioxide capture in Sabatier's reaction and by adsorbent. The shaded texts indicate products for export to the market

only one fourth of the total hydrogen production capacity from the biomass. Thus, it makes economic sense to produce a carbonate product from the balance of the carbon dioxide that has a market value, such as calcite, by the addition of sorbent calcium oxide. The theoretical highest amount of hydrogen production capacity from a biomass is given by the amount of hydrogen chemically bonded within the biomass and hydrogen present in the form of moisture contained in the biomass and a part of steam generated by heat recovery within the site to react (WGS reaction) with remaining carbon monoxide sourced from volatile carbon present in the biomass.

2. Synthesise flowsheet to maximise high-value product.
   The objective of biomass utilisation for high-value production in this case is methane production, which is limited by hydrogen availability as shown in Equation 9.1. The second priority product is the calcite for the remaining carbon dioxide capture on-site. This carbon dioxide results from all of char and a part of tar fractions of the biomass pyrolysis products. The balance of heat recovered in the form of HP steam can be exported or used for power generation using condensing turbine. Hence, the products from the site in the order of priority are (i) methane, (ii) calcite and (iii) HP steam or power. To maximise methane production effectively, that is, hydrogen production on-site (as hydrogen is the limiting reactant in the Sabatier's reaction), we can rationalise the following strategies:
   a. Release chemically bonded hydrogen present in the biomass by primary pyrolysis process. Primary pyrolysis or devolatilisation occurs as soon as biomass comes in contact with hot steam gasifier reactor bed. Steam atomisation further helps to dissociate volatile gases of biomass, including hydrogen. Some hydrogen could attach to methane and ethane (low molecular weight alkanes) that could be minimised by keeping the gasifier temperature above 900°C and maintaining oxygen lean (no external oxygen input) steam gasification medium. The latter would minimise full combustion of volatile carbon into carbon dioxide formation and instead partially oxidise carbon to produce carbon monoxide. Steam gasification is essential to reform tar to consequently release the remaining chemically bonded hydrogen. The tar, rich in phenol, can be reformed catalytically in the steam gasifier. Steam is also known to reduce the concentration of other forms of oxygenates including condensable.
   b. The second source of hydrogen is the moisture in biomass. This could be released by steam reforming and WGS reactions, mostly in steam gasifier and balance in WGS reactor. Remember, the overall thermal performance of steam gasification is endothermic because chemical bond breakage, devolatilisation and reforming reactions require heat. Also, biomass needs to be heated up to the gasifier temperature. The net heat required by the steam gasification is met by the heat from the char combustor operated at slightly higher temperature than the gasifier to maintain a temperature gradient by the circulation of hot sand bed.
   c. It would reach to a point when entire steam and hydrogen present in the biomass could be released by the process of steam gasification and WGS reaction. However, there would still be some remaining carbon monoxide in the syngas that could be used in WGS reaction to produce hydrogen. At this point, the overall site requires external water (or make-up BFW) supply. The site is already a heat source, giving an opportunity for heat recovery in the form of steam generation. The steam generated through heat recovery can be used in WGS reactor to completely convert the

remaining carbon monoxide into the production of hydrogen and carbon dioxide. At this point, all carbon present in biomass except char would be used up to react with steam to produce hydrogen. This will determine the highest theoretical amount of methane to be generated from the site.

3. Carry out heat integration and any other process integration opportunities identified to save capital and to recover heat. Simplify the flowsheet and save capital by combined units and functionalities.

   Saving in capital is evident from the combined combustion and its downstream processes for biomass char and off-gas from the pressure swing adsorption process. Also, multiple heat exchangers, stream splitting (generally suggested by heat exchanger network design by pinch analysis), etc. are avoided for better controllability and ease of operation of the plant.

9.3.1.1 Heat integration heuristics and application to the case study

There are two grades of heat available on-site. The grade of heat is determined based on the temperature level, that is, high grade implies high temperature of the heat source and whether the heat is latent heat (high grade) or sensible heat with high capacity of the medium (low grade):

1. High-grade heat from steam superheater and the high-temperature heat of the char combustor exhaust gas cooler
2. Medium-grade heat from the HRSG and from the remaining char combustor exhaust gas cooler (low temperature)

For tighter heat integration (highest heat recovery), the heat available can be extracted in series from low through medium to high grade. However, for improved flexibility and controllability in operation and lesser piping and instrumentation or complexity, the heat recovery strategy shown in Figure 9.3 should be adopted. The gasifier gas coolers' heat can be extracted in series, first into HP saturated steam generation in HRSG, followed by HP steam superheating in steam superheater, using the same BFW stream. The heat from the char combustor exhaust gas cooler is separately extracted into another HP superheated steam using BFW.

The heat sinks are the WGS reactor and steam gasifier. Steam is the reactant to produce hydrogen in WGS reactor and also used to heat up the clean syngas from the room temperature (Rectisol process outlet) to WGS reactor temperature. For this, HP saturated steam from the HRSG can be used. Hence, the balance of the HP saturated steam is routed to steam superheater. The steam gasifier requires HP superheated steam for steam atomising and as heating medium, reforming reaction and hydrogen production. This HP superheated steam is supplied from the steam superheater.

After the heat source and sink integration as shown in Figure 9.3 is performed, excess HP superheated steam (the balance of heat from the steam superheater and the heat from the char combustor exhaust gas cooler) can be exported or used to generate power for export using back-pressure and condensing steam turbines. There would be some low-grade heat recovery in the form of hot water generation on-site that could be used for space heating (lower ends or grades of heat from the char combustor exhaust gas cooler and gasifier product gas cooler).

9.3.1.2 Data for the case study

The ultimate analysis of an example biomass and its primary pyrolysis product compositions is shown in Table 9.1. The results of stream analysis in terms of compositions and temperature for the example biomass processing into proposed products through the flowsheet shown in Figure 9.3 are provided in Table 9.2. Table 9.3 lists the key process units for capital cost evaluation. The energy balance shows 239 kW net output heat generations in the form of 13.7 kmol per hour HP superheated steam at 550°C and 50 bar. Hence, from Table 9.3, the main products are biomethane, 65 kg per hour; calcite, 1204 kg per hour; and HP steam, 13.7 kmol per hour from 314 kg per hour biomass of given ultimate analysis shown in Table 9.1.

9.3.1.3 Economic Analysis Methodologies and Application to the Case Study

The economics of the site must be analysed for technical feasibility. The netback is calculated by subtracting the annualised capital and operating costs from the annualised products' sales price to set the upper bound of the annualised feedstock cost. If the annualised feedstock market price is less than the netback estimated, the plant will make money. The capital cost is evaluated in terms of the direct (inside and outside battery limits, i.e. ISBL and OSBL) and indirect capital costs. The ISBL comprises the cost of equipment which can be estimated using

Table 9.1 *Biomass ultimate analysis and primary pyrolysis product compositions*

Biomass	wt%
C	36.57
H	4.91
N	0.57
O	40.70
S	0.14
Ash	8.61
Moisture	8.50
Lower heating value (LHV) (MJ kg^{-1})	14.60
Gas from biomass primary pyrolysis	
Molar composition	
CO	0.3954
H_2	0.0406
CO_2	0.0358
H_2O	0.2284
CH_4	0.0769
H_2S	0.0025
N_2	0.0115
C_2H_6	0.2089
Tar from biomass primary pyrolysis	
Molar composition	
C_6H_6O	0.181365
H_2O	0.055735
O_2	0.7629
Char from biomass primary pyrolysis is assumed to be only carbon	

Table 9.2 *Streams' compositions and temperature in the flowsheet shown in Figure 9.3*

Product gas from steam gasifier	
Molar composition	
CO	0.4899
H_2	0.4220
CO_2	0.0316
H_2O	0.0396
CH_4	0.0121
H_2S	0.0008
N_2	0.0040
Temperature (°C)	950
Pressure (bar)	10–30
Product gas from steam superheater	
Temperature (°C)	450
Product gas from HRSG	
Temperature (°C)	<60
(Below the dew point of the condensates)	
Clean syngas	
H_2S	<1 ppm
Product gas from WGS	
Molar composition	
CO	0.0480
H_2	0.5846
CO_2	0.3138
H_2O	0.0424
CH_4	0.0084
N_2	0.0027
Temperature (°C)	250
Carbon dioxide-rich gas	
Molar composition	
CO	0.1155
CO_2	0.7555
H_2O	0.1021
CH_4	0.0201
N_2	0.0066
Exhaust gas from char combustor	
Molar composition	
CO_2	0.5311
H_2O	0.0190
N_2	0.4499
Temperature (°C)	970
(The gas is mostly dry and hence has a large heat capacity)	
Char combustor exhaust gas from cooler	
Temperature (°C)	<60
(To use cooling water for cooling)	

Table 9.3 *Basis streams in the key process units for mass and energy balance and capital cost evaluation*

Name of the process	Representative stream	Quantity	Unit
Total biomass	Biomass (gas + tar + char)	152.26 + 149.2209 + 12.59	kg per hour
Steam gasifier	Biomass (gas + tar)	152.26 + 149.2209	kg per hour
Steam gasifier	Enthalpy (endothermic)	70	kW
Steam superheater	Heat (exothermic)	82.25	kW
HRSG	Heat (exothermic)	61.2	kW
Rectisol	Gas flow/impurity	300.65/<1 ppm	kg per hour
Rectisol	Gas volumetric flow	448	m^3 per hour
Rectisol	Gas molar mass	17	
WGS reactor	Syngas	300.16	kg per hour
WGS reactor	Enthalpy (endothermic)	33.4	kW
Pressure swing adsorption	Feed gas/H_2 molar recovery	444.85/99%	kg per hour
Pressure swing adsorption	Purge gas molar flow rate	10.67	kmol per hour
Sabatier reactor	Hydrogen/biomethane	32.51/65	kg per hour
Sabatier reactor	Feed gas molar flow rate	20.32	kmol per hour
Char combustor	Char (carbon)	12.59	kg per hour
Char combustor	Enthalpy (exothermic)	76.85	kW
Exhaust gas cooler	Exhaust gas	1001.78	kg per hour
Exhaust gas cooler	Enthalpy (exothermic)	129	kW
Carbon dioxide adsorption	Calcite	1204	kg per hour
Carbon dioxide adsorption	Purge gas molar flow rate	10.63	kmol per hour

cost and size correlation, shown in Equation 9.6. The parameters such as base cost, base scale and scale factor θ are from Hamelinck and Faaij (2002), Larson et al. (2005) and Dutta et al. (2011). Each cost is levelised according to Equation 9.7, where chemical engineering plant cost index (CEPCI) is applied. The discounted cash flow method is applied for determining the annual charge for the capital investment, that is, 13% using the following assumptions:

- Discount rate: 10%
- Plant life: 15 years
- Start-up period: 2 years (25, 75%)
- 8000 operating hours per annum:

$$\frac{\mathrm{COST}_{\mathrm{size2}}}{\mathrm{COST}_{\mathrm{size1}}} = \left(\frac{\mathrm{SIZE}_2}{\mathrm{SIZE}_1}\right)^{\theta} \tag{9.6}$$

SIZE_1 and $\mathrm{COST}_{\mathrm{size1}}$ represent the capacity and the cost of a base unit, whilst SIZE_2 and $\mathrm{COST}_{\mathrm{size2}}$ represent the capacity and the cost of the unit after scaling up/down, respectively:

$$\text{Present cost} = \text{original cost} \times \left(\frac{\text{index at present}}{\text{index when original cost was obtained}} \right) \quad (9.7)$$

The operating cost consisting of the fixed and variable costs is evaluated. The parameters for estimating the operating costs are from Tijmensen et al. (2002) and Sinnott (2006). Fixed operating cost is estimated based on a percentage of indirect capital cost, except for the cost for personnel. Other associated costs such as laboratory, supervision and plant overheads are estimated based on a certain percentage of cost of personnel. Variable operating costs such as fuel, feedstock and electricity are estimated using the latest available price data. The current market prices/estimated cost of production are identified for evaluating the total value of the products.

The ISBL calculation is shown in Table 9.4. The annualised capital and operating costs are shown in Table 9.5 and Table 9.6, respectively. The product values and the estimated netback are shown in Table 9.7 of this section. The netback is far more acceptable than many usual biorefinery cases without carbon capture and reuse to produce added value products. As can be seen, the highest product value is generated from the calcite product capturing carbon released from the biomass during energy generation. The other two products for primarily energy generation, biomethane followed by HP steam, have lesser value generations.

9.4 CO_2 Pinch Analysis Tool for Carbon Dioxide Capture and Reuse in Integrated Flowsheet

This section presents the conceptualisation of biorefineries based on the utilisation of CO_2 streams and a systematic methodology for integration of CO_2 streams based on the mass pinch analysis technique (El-Halwagi and Manousiouthakis 1989). To design highly integrated biorefineries for enhanced sustainability, any residue stream (if cannot be prevented or reduced any further) must not be treated as wastes or emission but as a stream with potential for energy generation or added value production. This broadens the possibilities for CO_2 utilisation within a biorefinery system, thus avoiding emissions and improving environmental performance of processes and products. In order to achieve highest CO_2 utilisation and recycling within the biorefinery, process integration tools such as mass pinch analysis can be used to find the targets for design. The value of mass pinch analysis tools, either graphical, mathematical or a combination thereof, has been demonstrated in biorefineries for integration of carbon (Ng 2010), syngas (Tay and Ng 2012) and bioethanol (Martinez-Hernandez et al. 2013). A pinch analysis approach forms the basis for an automated targeting procedure to find the highest biofuel production rate and revenue without detailed design of biorefineries (Pham and El-Halwagi 2012). Mass pinch analysis has also been applied for the planning of carbon-constrained energy sector (Tan and Foo 2007) and eco-industrial parks (Manan and Wan-Alwi 2012) to minimise CO_2 emissions.

A biorefinery design that aims to make optimal use of available streams to minimise inputs and emissions will form mass exchange networks. Highly integrated biorefineries will not only exchange intermediate streams but also product streams that can be used as in-process raw materials such as CO_2, hydrogen or bioethanol (Martinez-Hernandez et al. 2013). CO_2

Table 9.4 Parameters and calculation of ISBL costs

Process unit	Base size	Unit of capacity	Base cost (million $)	Scaling factor	Year	Reference	Quantity	Cost (million $)	CEPCI (corresponding year)	Levelised cost (million $, 2013)
Allothermal gasifier and auxiliaries	500	MTPD biomass	6.5	0.6	2010	1	3.65	0.3396	550.8	0.3478
Steam superheater	77	MW_{th} heat duty	25.4	0.6	2003	2	0.08225	0.4188	402	0.5877
HRSG	355	MW_{th} heat duty	41.2	1	2003	2	0.06377	0.0074	402	0.0104
Rectisol	200000	N m^3/h gas feed	20	0.65	2003	2	448	0.3790	402	0.5319
WGS reactor	1377	MWLHV biomass input	30.6	0.67	2003	2	1.395	0.3016	402	0.4233
Pressure swing adsorption	0.294	kmol/s purge gas flow	5.46	0.74	2003	2	0.00296	0.1817	402	0.2550
Sabatier reactor	1390	kmol/h feed	9.4	0.6	2001	3	20.32	0.7449	394.3	1.0656
Exhaust gas cooler	77	MW_{th} heat duty	25.4	0.6	2003	2	0.129	0.5486	402	0.7699
Carbon dioxide adsorption	0.294	kmol/s purge gas flow	5.46	0.74	2003	2	0.00295	0.1813	402	0.2543
									Total ISBL	**4.2458**

Table 9.5 Annualised capital cost estimation

Item	Cost estimation	Cost (million $, 2013)
Instrumentation and control	5% of ISBL	0.2123
Buildings	1.5% of ISBL	0.0637
Grid connections	5% of ISBL	0.2123
Site preparation	0.5% of ISBL	0.0212
Civil works	10% of ISBL	0.4246
Electronics	7% of ISBL	0.2972
Piping	4% of ISBL	0.1698
	Total OSBL	1.4011
	Total direct capital (TDC)	5.6469
Indirect capital cost		
	Cost estimation	**Cost (million $, 2013)**
Engineering	15% of TDC	0.8470
Contingency	10% of TDC	0.5647
Fees/overheads/profits	10% of TDC	0.5647
Start-up	5% of TDC	0.2823
	Total indirect capital (TIC)	2.2588
	Total CAPEX	7.9057
	Annual capital cost	1.0394

Table 9.6 Annualised operating cost estimation

Specification	Cost estimation	Cost (million $, 2013)
Fixed OPEX		
Maintenance	10% of indirect capital cost	0.2259
Personnel	0.595 million Euro/100 MW_{th} LHV	0.0120
Laboratory costs	20% of Personnel	0.0024
Supervision	20% of Personnel	0.0024
Plant overheads	50% of Personnel	0.0060
Capital charges	10% of indirect capital cost	0.2259
Insurance	1% of indirect capital cost	0.0226
Local taxes	2% of indirect capital cost	0.0452
Royalties	1% of indirect capital cost	0.0226
Total fixed OPEX		0.5649
Total fixed OPEX per year		0.0743
Total variable OPEX per year		0
Direct production cost (DPC) = variable + fixed operating costs		
DPC per year		0.0743
Miscellaneous		
Sales expense	30% of DPC	0.0223
General overheads		
Research and development		
Total OPEX per year		0.0965

Table 9.7 *Product prices and netback*

Product	Unit cost ($ kg^{-1})	Flow rate (kg h^{-1})	Amount (kg y^{-1})	Value (million $ y^{-1})
Biomethane	0.75	65	520,000	0.39
Calcite	8.1	1204	9,632,000	78.0192
Steam	0.0112	246.6	1,972,800	0.02209536
			Total	78.43
			Netback	$ kg^{-1} 30.70

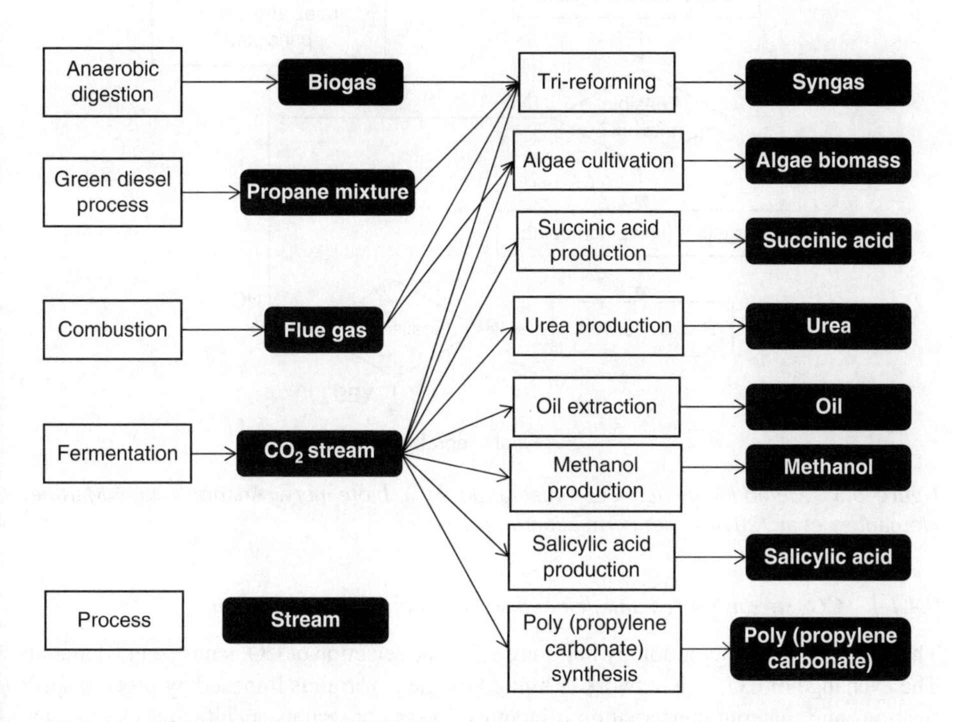

Figure 9.4 *Potential CO_2 sources and demands in a biorefinery*

can be contained in a feed, an intermediate or a product stream. CO_2 streams are consumed and produced by the various biorefinery processes at different purities and flow rates resulting in a CO_2 exchange network. This opens the opportunity for the application of mass pinch analysis for process integration in order to achieve the highest CO_2 utilisation and recycling within the biorefinery. Figure 9.4 gives an overview of potential processes producing CO_2 streams (the CO_2 sources) and processes consuming or utilising CO_2 streams (the CO_2 demands).

9.4.1 Overview of the Methodology for CO_2 Integration

The generic methodology for CO_2 integration in a biorefinery is shown in Figure 9.5. The various steps are discussed as follows and applied later in a biorefinery case study.

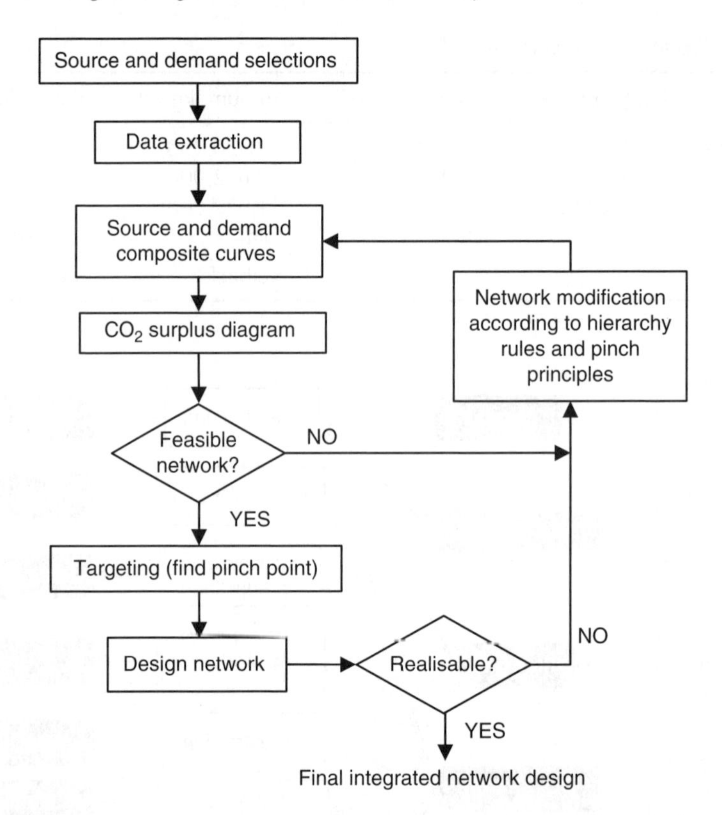

Figure 9.5 *Methodology for CO_2 integration in a biorefinery. Adapted from Martinez-Hernandez et al. (2014), with permission from Elsevier*

9.4.1.1 CO_2 stream source and demand selection and data extraction

The first step of the methodology in Figure 9.5 is the selection of CO_2 sources and demands. The exchange of CO_2 in a network is limited by the constraints imposed by process specifications and material conservation principles. These constraints specifically include lowest flow rate, lowest or highest purity of supply to a demand (e.g. lowest 98% for chemical production), limited CO_2 content for process unit operations (e.g. CO_2 concentration in the feed to algae cultivation is limited by the CO_2 tolerance of the algae), the nature and content of impurities, etc. The purpose and specification of products are also important. If the chemical species involved in the various streams are not compatible for all the processes or products, the CO_2 streams might not be exchangeable without purification. Once the sources and demands are selected, the next step is data extraction consisting of collecting flow rates and purities of the source and demand streams.

9.4.1.2 CO_2 purity profile and CO_2 surplus diagram

To visualise the mass balances of the CO_2 exchange network and identify any bottleneck or constraint for feasibility, the purity profile and the surplus diagram are constructed as shown in Figure 9.6. Figure 9.6a shows a generic purity profile diagram for a network

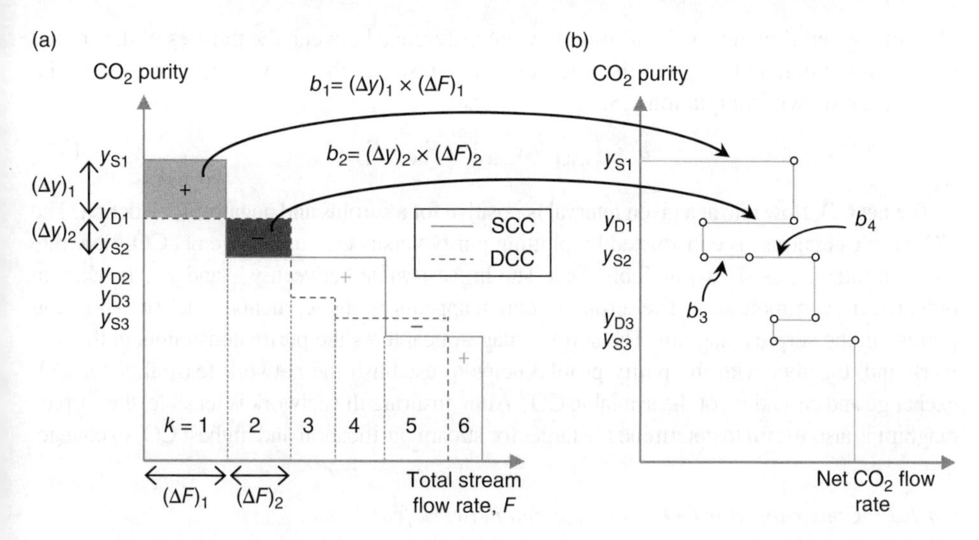

Figure 9.6 *(a) CO_2 purity profile formed by source composite curve (SCC) and demand composite curve (DCC). The areas enclosed between the composite curves correspond to the net CO_2 flow rate shown as horizontal segments in the corresponding (b) CO_2 surplus diagram. Adapted from Martinez-Hernandez et al. (2013), with permission from Elsevier*

comprising three source streams and three demand streams. The CO_2 purity profile is formed by the source composite curve (SCC) and the demand composite curve (DCC). To construct the SCC, sources are first arranged from high to low purities. Then, the CO_2 purities of the source streams are plotted against their cumulative flow rate. The DCC is formed following the same procedure. Each composite curve forms a cascade of steps. The step length in the SCC indicates the total stream flow rate available at the corresponding purity level. In the DCC, the step length indicates the total stream flow rate required at the corresponding purity level.

The areas enclosed between the SCC and DCC represent the CO_2 pockets in the network (Figure 9.6a). If the SCC is above the DCC for a given purity range, then the sources provide more CO_2 than is required by the demands, indicating a CO_2 excess or surplus (+). If the SCC is below the DCC, the CO_2 from the sources is not enough to cover the demands, producing a deficit (–) of CO_2. If the area covered by the SCC is larger than that of the DCC, then there is an excess of CO_2 streams in the network. The excess CO_2 may come from a source stream and is not exchangeable due to purity constraints (e.g. purity lower than that required by the demands). In such case, the CO_2 is lost as emission unless options for stream purification or utilisation by a suitable process are introduced into the network. In summary, the purity profile helps to identify where in the system CO_2 is being wasted and where the CO_2 supply is deficient.

After the areas of CO_2 excess and deficit are identified from the purity profile, the net CO_2 flow rate can be determined for each flow interval. The number of intervals is equal to the total number of flow rate segments in the two composite curves. The purity profile in Figure 9.6a is divided into six flow intervals (I–VI). The area between the SCC and DCC in a particular flow rate interval k represents the material balance on CO_2. This area is calculated from the flow rate difference between the upper bound (F_{Uk}) and the lower bound

(F_{Lk}) of a given flow interval multiplied by the difference between the purities of the source ($y_{Si,k}$) and the demand ($y_{Dj,k}$) in the interval k. The result is the net CO_2 flow rate b_k in the interval, as shown in Equation 9.8:

$$b_k = \left(y_{Si,k} - y_{Dj,k}\right) \times \left(F_{Uk} - F_{Lk}\right) \tag{9.8}$$

The net CO_2 flow rate at a given interval is positive for a surplus and negative for a deficit. The CO_2 surplus diagram is constructed by plotting purity versus the cumulative net CO_2 flow rate for each interval, as shown in Table 9.6b. The highest value between $y_{Si,k}$ and $y_{Dj,k}$ is taken in order to set a common scale. The surplus or deficit appears as line segments at the corresponding purities in the surplus diagram. The surplus diagram captures the purity constraints in the network and, together with the purity profile, helps to establish the network feasibility for CO_2 exchange and cascading of the available CO_2. After ensuring the network is feasible, the surplus diagram is also useful to determine the target for stream purification and highest CO_2 exchange.

9.4.1.3 Constraints for CO_2 exchange and network feasibility

The purity and flow rate constraints imposed by the CO_2 demand streams are captured by formulating a material balance on the total streams and a material balance on CO_2. The total flow rate of CO_2-containing streams available from the sources must be in excess or equal to the total flow rate of CO_2-containing streams required by the demands as the first necessary condition for the exchange network to be feasible. This is established from the overall mass balance of the network as shown in Equation 9.9:

$$\sum_{i=1}^{n_s} F_{Si} \geq \sum_{j=1}^{n_D} F_{Dj} \tag{9.9}$$

where F_{Si} is the CO_2 flow rate available from source i, F_{Dj} is the CO_2 flow rate required by demand j, and n_S and n_D are the numbers of sources and demands, respectively. Since the overall mass balance is captured in the purity profile, the constraint can be verified from this diagram. If the SCC is shorter than the DCC, then the overall material balance on the total streams is violated, and the network is not feasible.

The surplus diagram captures the purity constraints in the network. If the cumulative net CO_2 is negative at any purity level (i.e. surplus curve crosses y-axis), then the network is not receiving the required amount of CO_2 at the adequate purity. In that case, at least one of the constraints imposed by the demands cannot be satisfied by the sources, rendering the network unfeasible. To make the network feasible, a purified CO_2 stream is required. Then, the options are (i) to import a high-purity stream or (ii) introduce a purification unit to increase the purity of a low-purity stream. Therefore, the second necessary condition for network feasibility is that the balance of CO_2 in the overall system (i.e. the cumulative net CO_2 flow rate) must always be positive. This means that the entire CO_2 surplus curve must lie at or above zero flow rate for the network to be feasible.

9.4.1.4 CO_2 pinch analysis and targeting

If both the first and second necessary conditions are met, then the CO_2 integration problem has at least one feasible solution. One possible solution is when the system is, at some point, constrained in CO_2 supply, that is, there is no surplus or deficit. This condition sets the initial target for network design and indicates the highest exchange that leads to lowest CO_2 emissions. To determine the values for the exchange and emission flow rates, the CO_2 surplus diagram is used to find the CO_2 pinch. The CO_2 pinch corresponds to the point at which the CO_2 exchange network has neither excess nor deficit. Figure 9.7a shows how the flow rate of the first source is varied until a pinch occurs in the surplus diagram in Figure 9.7b. The purity of the CO_2 source stream at the pinch corresponds to the CO_2 pinch purity (y_p). The pinch appears in the surplus diagram by a discontinuity segment at surplus equal to zero between y_p and the corresponding y_D. In the pinched surplus diagram (Figure 9.7b), the surplus curve is shifted towards the *y*-axis showing an increase in CO_2 exchange and reducing the need for high-purity CO_2, thus setting the target for highest exchange and lowest CO_2 emission.

The CO_2 pinch divides the network into a subsystem with net zero CO_2 surplus (region above the pinch) and a subsystem with net positive CO_2 surplus (region below the pinch). Above the pinch, there is a portion of the flow rate from the source stream at the pinch purity indicated as F_{PR} in Figure 9.7a. This flow rate corresponds to the amount that must be reused by the demand streams above the pinch to meet the exchange and emission target. In intervals where a net flow rate surplus exists, the net flow rate can be cascaded to lower purity intervals. In intervals where a deficit of net CO_2 flow rate exists, the excess CO_2 from higher purity intervals must be used first. Only after exhausting flow rate surpluses from higher purity intervals, CO_2 purification or other modifications can be introduced. The balance of CO_2 surplus and deficits at the various purity levels by the cascading procedure described earlier is the key to achieve an efficient integration in a network design. Other helpful guidelines from the CO_2 pinch point are described as follows.

As discussed, the network is divided into a region above and below the pinch. Since the subsystem above the pinch is balanced, reusing a CO_2 stream from below the pinch implies the transference of the same amount from a source above the pinch (at higher purity) across

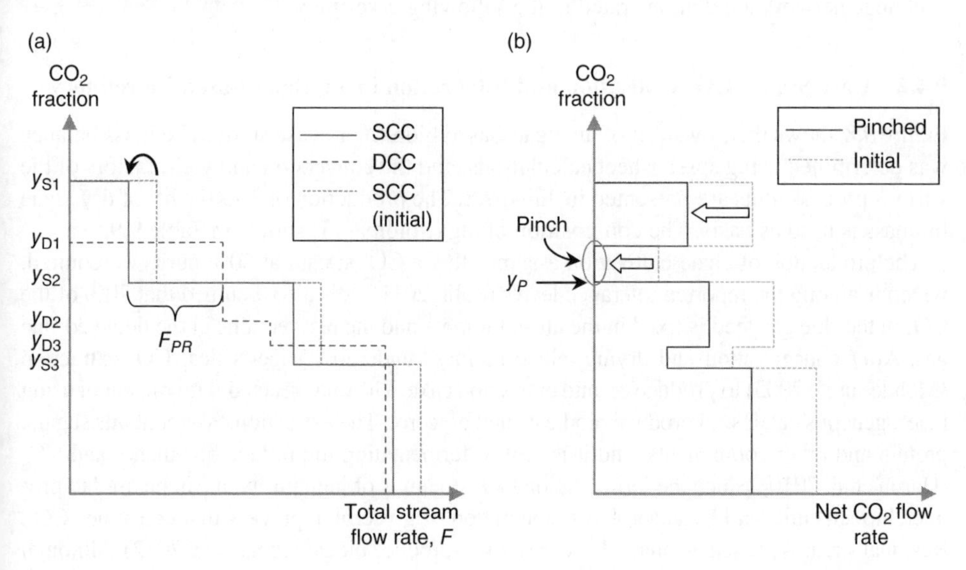

Figure 9.7 *(a) The change in flow rate displaces the SCC producing a (b) shift in the surplus diagram until pinched with the y-axis. Adapted from Martinez-Hernandez et al. (2013), with permission from Elsevier*

the pinch in order to preserve the material balance. This produces a reduction in CO_2 surplus above the pinch, and additional high-purity CO_2 stream must be supplied as a penalty. Then, the requirement of high-purity CO_2 stream would exceed the lowest target identified by the pinch method. Therefore, the CO_2 streams must never be directly exchanged across the pinch. As with pinch analysis in other contexts, this is the first fundamental principle for the design of an exchange network at lowest supply of high-quality streams.

If a network has excess CO_2 even after highest reuse indicated by the pinch point, opportunities for system improvement can be further explored by network modifications. Modifications can consist of introducing demands to increase utilisation of existing source streams, removal of sources to decrease emissions or introduction of purification units. The general hierarchical decision making for the implementation of modifications to achieve targets in the initial network and for further CO_2 emissions reduction is summarised as follows (Manan and Wan-Alwi 2012):

1. Direct reuse of CO_2 streams by maximising exchange between existing processes. This option is applied first, after the target is known from the CO_2 pinch analysis.
2. Source and demand manipulation. Either modification of process conditions/flow rates or introduction/removal of processes.
3. Purification of CO_2 streams across the pinch. A purifier placed below the pinch purity would make purer CO_2 in a region of surplus that will end up as waste stream since it cannot be exchanged to supply a demand above the pinch. Thus, a purifier should always be placed across the pinch purity in order to exchange CO_2 from a region of surplus to a region of limited supply. This leads to further CO_2 exchange, keeping the CO_2 within the biorefinery.
4. CO_2 capture and storage.

The application of the CO_2 pinch targeting method to minimise the CO_2 emissions by maximising CO_2 exchange and the use of the integration principles for the design of a CO_2 exchange network are demonstrated in the following case study.

9.4.2 Case Study: CO_2 Utilisation and Integration in an Algae-Based Biorefinery

Figure 9.8 shows the flowsheet of an algae-based biorefinery case study. The mass balance was determined using spreadsheet calculations, and the conversion and yield factors of the various process units are presented in Table 9.8. The production of 1000 kg h^{-1} of dry algae biomass is used as basis. The composition of algae biomass is shown in Table 9.9.

The production of algae cultivation assumes that a CO_2 stream at 30% purity is required, which is among the reported tolerance levels (Salih 2011). It is also assumed that 20% of the CO_2 in the flue gas feed is fixed in the algae biomass and the rest remains in the depleted flue gas. After concentration and drying, algae biomass undergoes supercritical CO_2 extraction (Mendes et al. 2003) to produce oil and extraction cake. The oil is reacted with methanol using heterogeneous catalyst to produce biodiesel and glycerol. The extraction cake contains sugars, protein and other components, and it is sent to fermentation to produce bioethanol and CO_2 (Harun et al. 2010). Since the fermentation CO_2 stream is of high purity, it can be used to produce bio-succinic acid by anaerobic fermentation of glycerol, a process that consumes CO_2. Residual streams are sent to anaerobic digestion to produce biogas (Frear et al. 2012). Although biogas contains about 40% CO_2 by volume, this methane-rich stream is used in a combined heat and power (CHP) plant to provide energy to the biorefinery processes. Combustion of biogas in the CHP plant generates a flue gas with CO_2 which might be combined with the main flue gas feed as an effective way to recycle the CO_2 within the biorefinery.

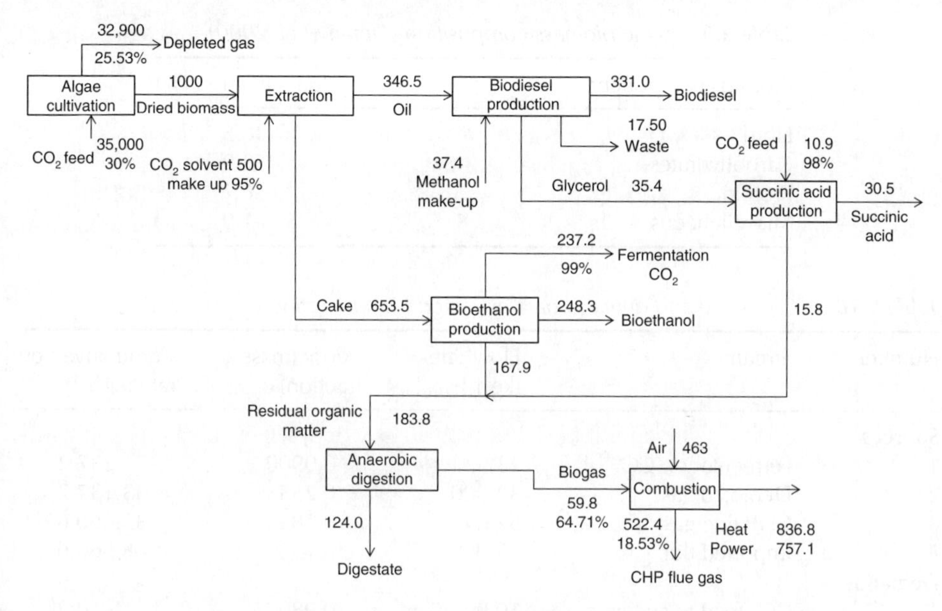

Figure 9.8 *Algae-based biorefinery process flowsheet. Mass flow rates are in $kg\,h^{-1}$ and energy in $MJ\,h^{-1}$. The percentage CO_2 purity on mass basis is shown*

Table 9.8 *Mass balance factors for the algae-based biorefinery*

Process unit	Mass balance factors	CO_2 purity (mass basis)
Algae cultivation	CO_2 fixation: 2.1 $kg\,kg^{-1}$ algae	30% in feed 25.53% in depleted gas (from mass balance)
Oil extraction	CO_2 solvent: 50 $kg\,kg^{-1}$ algae CO_2 solvent recycle: 99% Efficiency: 0.9 kg oil extracted kg^{-1} oil in feed	95% lowest in feed
Biodiesel process	Methanol feed: 0.1079 $kg\,kg^{-1}$ oil Biodiesel yield: 0.9552 $kg\,kg^{-1}$ oil Glycerol yield: 0.1022 $kg\,kg^{-1}$ oil	–
Bioethanol production	Ethanol yield: 0.38 $kg\,kg^{-1}$ sugars CO_2 yield: 0.363 $kg\,kg^{-1}$ sugars	99% in CO_2 product
Succinic acid (SA) production	SA yield: 0.86 $kg\,kg^{-1}$ glycerol CO_2 feed: 0.301 $kg\,kg^{-1}$ glycerol	98% lowest in feed
Anaerobic digestion	Biogas yield: 0.3255 $kg\,kg^{-1}$ organic matter	40% in biogas
CHP plant	Air feed: 15% excess Electric efficiency: 0.42 Heat efficiency: 0.38 Biogas heating value: 33.3 $MJ\,kg^{-1}$	18.52% in CHP flue gas (from mass balance)

Table 9.9 *Algae biomass composition (Illman et al. 2000)*

Biomass component	Mass fraction (%)
Lipid	38.5
Carbohydrates	52.9
Protein	6.7
Miscellaneous	1.9

Table 9.10 *CO_2 source and demand data extracted from biorefinery flowsheet*

Number	Stream	Flow rate ($kg\,h^{-1}$)	Purity (mass fraction)	Cumulative flow rate ($kg\,h^{-1}$)
Sources				
1	Fermentation CO_2	237.2	0.9900	237.2
2	Depleted gas	32,900	0.2553	33,137.2
3	CHP flue gas	523.4	0.1853	33,660.6
4	Imported flue gas	35,000	0.1500	68,660.6
Demands				
1	CO_2 feed to succinic acid production	10.9	0.9800	10.9
2	CO_2 solvent make-up to oil extraction	500	0.9500	510.9
3	CO_2 feed to algae	35,000	0.3000	35,510.9

The objective is to design an integrated biorefinery for the efficient utilisation of CO_2 streams to produce valuable products with the highest exchange within the biorefinery and the lowest CO_2 emission. The aim is also to minimise the flow rate of CO_2 streams that require purification. A flue gas stream is available to the biorefinery from an adjacent power plant with a total flow rate of 30,000 $kg\,h^{-1}$ containing 15% mass content of CO_2 and is to be used in the biorefinery. It is assumed that the stream is previously treated in the power plant to remove undesirable impurities such as nitrogen and sulphur oxides and thus only contains CO_2, oxygen and nitrogen. Table 9.10 shows the data extracted from the sources and demands in the biorefinery in Figure 9.8. The data is arranged in decreasing order of purity, and the cumulative flow rate is calculated for the sources and demands in order to produce the purity profile.

9.4.2.1 Targeting Lowest Purification and Emission Flow Rates Using CO_2 Pinch Analysis for the Case Study

The purity profile and CO_2 surplus diagrams are shown in Figure 9.9. A small excess area can be observed at the beginning of the purity profile compared to the deficit area. It can be observed that there is a large excess of CO_2 as the SCC is larger than the DCC. This means the first condition for network feasibility is met. However, the excess is not at the purity required by the network, and the effect of this is clear on the surplus diagram. The surplus

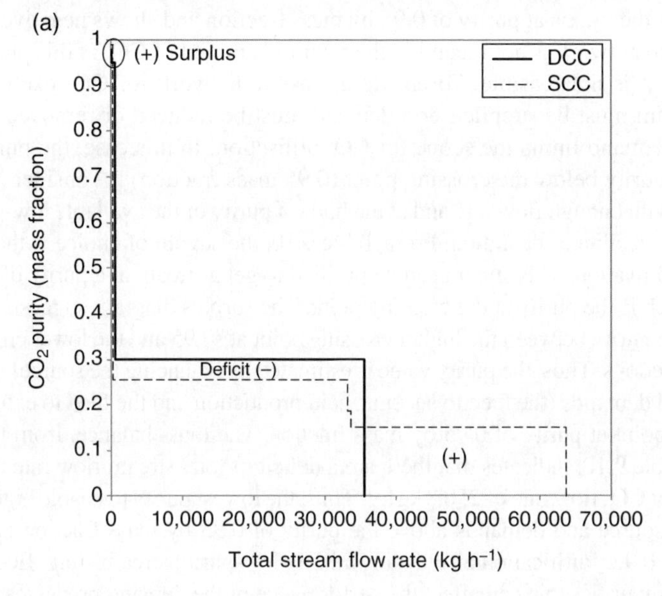

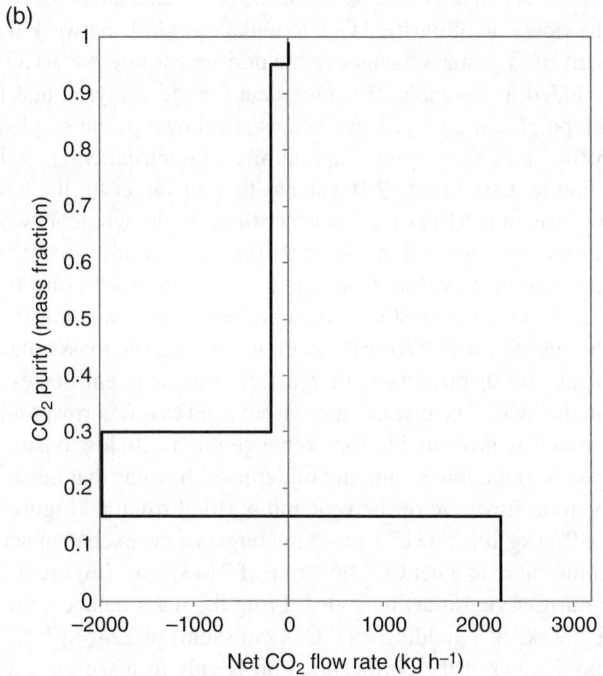

Figure 9.9 *(a) Purity profile showing the demand composite curve (DCC) and the source composite curve (SCC) and (b) surplus diagram for the case study*

curve crosses the y-axis at purity of 0.95 by mass fraction and shows negative surplus. This means that the network is not receiving the required amount of CO_2 at this purity. Thus, the initial network is not feasible. To enable a feasible network for CO_2 exchange, either a purified stream must be supplied or a demand must be reduced or removed. Reducing or removing a demand limits the scope for CO_2 utilisation. In this case, the purification of a stream with purity below the crossing point (0.95 mass fraction) is considered.

A stream with enough flow rate and at the highest purity of the available low-purity streams is preferred to facilitate separation. From Table 9.10, the stream of choice is the depleted gas from algae cultivation. This stream can be purified to get a stream at a purity of 0.95 by mass fraction, which is the purity at the crossing point. The surplus diagram suggests that there is a pinch at some purity between the higher crossing point at s0.95 and the lower crossing point at 0.15 mass fractions. Thus, the purity value is estimated by balancing the sources (the fermentation CO_2) and demands (the feed to succinic acid production and the feed to extraction) above the crossing point at purity of 0.95 by mass fraction. The mass balance, from flow rates and purities in Table 9.10, indicates that there are a deficit of total stream flow rate of 273.7 kg h^{-1} and a deficit of CO_2 flow rate of 250.8 kg h^{-1}. Thus, the lowest purity to enable highest exchange between the source and demands above the purity of 0.95 by mass fraction is estimated as 250.8/273.7 = 0.92. Purification also avoids mixing low-purity streams (e.g. flue gas streams) that contain impurities that can affect the performance of the demand processes.

The target for the flow rate of purified CO_2 to make the whole network feasible is found from CO_2 pinch analysis. The mass balance of the purifier assumes 98% CO_2 recovery. The purified stream is added to the table of source data for the analysis, and its flow rate is varied until a pinch point and only positive values are shown in the surplus diagram. The pinched purity profile and CO_2 surplus diagram after the introduction of the purification unit are shown in Figure 9.10. Figure 9.10a shows that the target for the lowest amount of 92% (by mass) CO_2 stream is 5164 kg h^{-1}, in order to make the whole network feasible.

Two pinch points can be observed in Figure 9.10b, pinch A at $y_{p,A} = 0.92$ and pinch B at $y_{p,B} = 0.15$. The exchange or recycling flow rate above each pinch point is the difference between the DCC and the pinched SCC at the corresponding pinch purity. This is shown as $F_{R,A} = 273.7$ kg h^{-1} and $F_{R,B} = 15{,}676$ kg h^{-1} in Figure 9.10a. The mass balance of the purifier is represented by the displacement of the flow rate segment corresponding to the depleted flue gas in the SCC. The displacement from right to left corresponds to the vented gas stream. This stream is not suitable for exchange due to its low purity. The total displacement corresponds to the lowest amount of depleted flue gas that needs to be purified. This is equal to the total flow rate of the vent and purified streams (Figure 9.10a), that is, 13,826 + 5,164 = 18,990 kg h^{-1}. The CO_2 emission target of the exchange network is shown in the pinched surplus curve at a net CO_2 flow rate of 2148 kg h^{-1} (Figure 9.10b). However, the amount in the purifier vent must be added. From the mass balance, the CO_2 flow rate in the vent stream is 97 kg h^{-1}, yielding total CO_2 emissions of 2245 kg h^{-1}.

Note that the introduction of the purification unit is only to make the exchange network feasible. Other options need to be explored for lowering CO_2 emissions further. Purification is also not useful, unless a new demand process is integrated such as production of polycarbonate, urea, salicylic acid, etc. Alternatively, the low-purity flue gas could be used for tri-reforming, but CO_2 conversion is relatively low, and economic feasibility is yet to be analysed. Although highest flue gas utilisation is desirable, Figure 9.10 shows that its flow rate can be reduced. Imported flue gas is the stream with the lowest purity and is producing the last excess pocket,

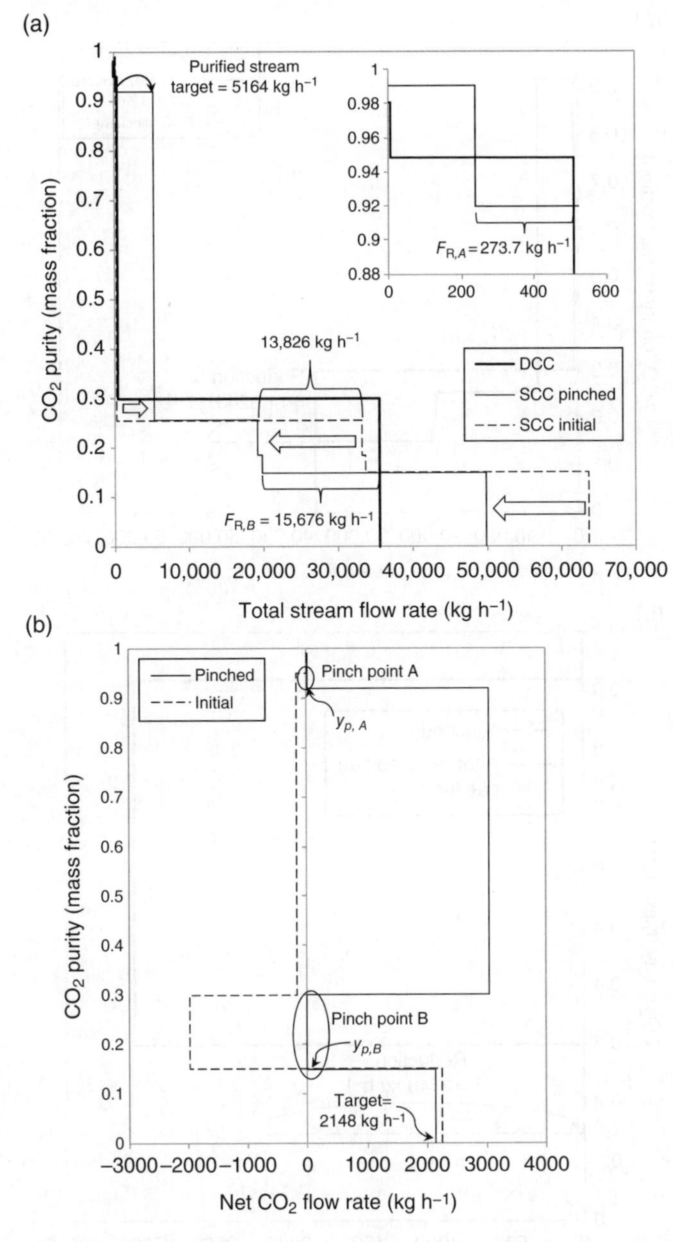

Figure 9.10 *(a) Purity profile and (b) surplus diagram as in the initial network and pinched after the introduction of purification unit*

and it is not being exchanged. Thus, the flow rate is reduced until the SCC ends at the same point as the DCC. Figure 9.11a shows that the amount of imported flue gas can be reduced by 14,324 kg h^{-1}. Thus, the highest amount of imported flue gas that the network can process in an efficient manner is 30,000 − 14,324 = 15,676 kg h^{-1}. In this case, this is equal to the exchange

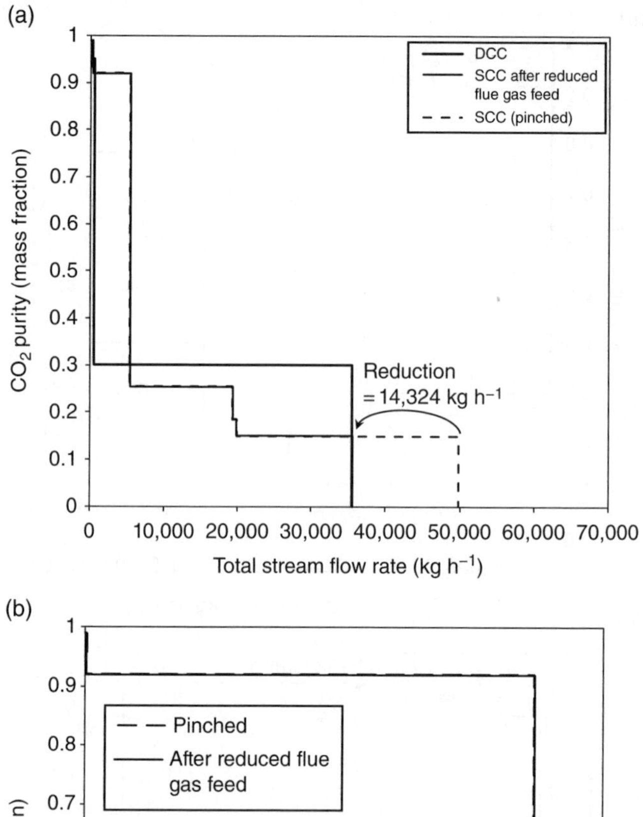

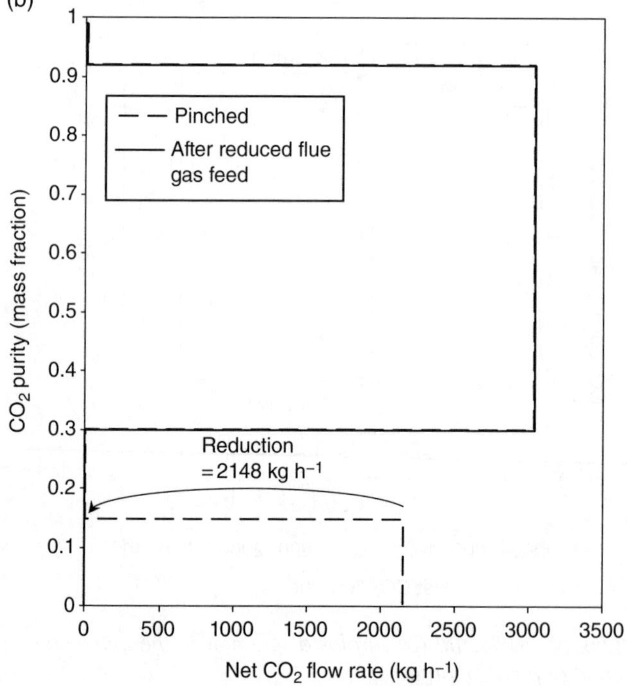

Figure 9.11 *(a) Purity profile and (b) surplus diagram pinched after the introduction of purification unit and after reducing the amount of imported flue gas*

flow rate of the imported flue gas ($F_{R,B}$) above the second pinch, as this is the source at this point. The remaining of the imported flue gas could be used in a similar algae-based biorefinery or processed for carbon capture and storage (CCS). Alternatively, it could be purified if other demand processes are introduced into the exchange network. Figure 9.11b shows that CO_2 emissions (net CO_2 flow rate) in the exchange network are reduced to zero when the flue gas is reduced, indicating an efficient exchange network. The only emission is the vent gas from the purifier, not shown in the diagrams as this stream does not participate in the exchange network. Thus, the net CO_2 emission is reduced from 2148 to 97 kg h^{-1}, that is, 96% reduction.

9.4.2.2 CO_2 Exchange Network Design for the Case Study

Once the targets for purification and exchange are known, a CO_2 exchange network can be designed to achieve the lowest CO_2 emissions. The final diagram shows a threshold problem in which one end of the curve is at zero surplus. This reduces the design problem to only two areas, above and below the pinch point A. The design above pinch point A must use 273.7 kg h^{-1} of the purified stream at 0.92 purity by mass fraction to meet the exchange target. The design below the pinch point must use the highest flow rate of imported flue gas (15,676 kg h^{-1}) and the remaining flow rate of the purified stream. The mass balances for both subproblems were solved in spreadsheet. The integrated network design is shown in Figure 9.12. Note that to meet the target above the pinch and avoid sending the fermentation CO_2 stream to below the pinch, the purities of the stream feeds to the oil extraction and succinic acid production are slightly above the values specified in Table 9.10. Those values are the lowest required and therefore the process is not expected to be affected.

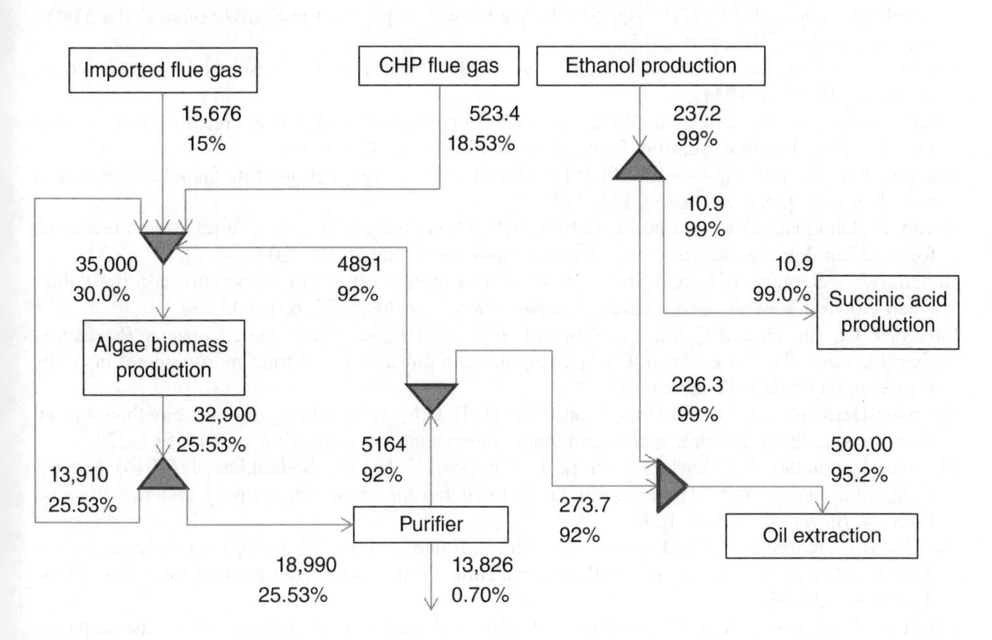

Figure 9.12 Integrated CO_2 exchange network design for the case study. Flow rates in kg h^{-1} and purities in mass percentage

Although the current design has been strategically developed, other factors may need to be considered to make sure the network is still feasible and realisable. Examples of such factors are the capital and operating costs (which can be estimated as shown in Section 9.3.1.3), the difference in pressure among the various streams that are mixed, the technical feasibility of piping connections, etc.

9.5 Conclusions

Reuse of carbon dioxide can not only generate main added value products – such as biomethane using Sabatier's reaction, calcite, succinic acid and algae-based biofuel productions, for example, bioethanol and biodiesel – but also improves economic and environmental sustainability of the biorefining systems compared to those emitting carbon dioxide to the atmosphere. This chapter shows the two most effective tools to develop biorefinery flowsheets incorporating CO_2 capture and reuse: techno-economic analysis tools, by process synthesis, simulation, heat integration and economic analysis and CO_2 pinch analysis, targeting and exchange network design. The systematic tools have been demonstrated to develop integration configurations for CO_2 utilisation and exchange that leads to lowest or zero CO_2 emission.

References

Dutta, A., Talmadge, M., Hensley, J. et al. (2011) Process Design and Economics for Conversion of Lignocellulosic Biomass to Ethanol: Thermochemical Pathway by Indirect Gasification and Mixed Alcohol Synthesis, NREL/TP-5100-51400. Available at: http://www.nrel.gov/biomass/pdfs/51400.pdf (accessed on 29 August 2015).

El-Halwagi, M.M. and Manousiouthakis, V. (1989) Synthesis of mass exchange networks, *AIChE J.*, **35**, 1233–1244.

Frear, C., Zhao, B., Zhao, Q. et al. (2012) Anaerobic digestion of algal biomass residues with nutrient recycle, Algae Biomass Summit, September 24–27, Denver, CO, USA.

Hamelinck, C.N. and Faaij, A.P.C. (2002) Future prospects for production of methanol and hydrogen from biomass, *J Power Sources*, **111**, 1–22.

Harun, R., Danquah, M.K. and Forde, G.M. (2010) Microalgal biomass as a fermentation feedstock for bioethanol production, *J. Chem. Technol. Biotechnol.*, **85**(2), 199–203.

Illman, A.M., Scragg, A.H., and Shales, S.W. (2000) Increase in *Chlorella* strains calorific values when grown in low nitrogen medium. *Enzyme Microb. Technol.*, **27**, 631–635.

Larson, E.D., Jin, H. and Celik, F.E. (2005) *Gasification-Based Fuels and Electricity Production from Biomass, Without and with Carbon Capture and Storage*. Princeton Environmental Institute, Princeton University, Princeton, NJ.

Martinez-Hernandez, E., Sadhukhan, J. and Campbell, G.M. (2013) Integration of bioethanol as an in-process material in biorefineries using mass pinch analysis. *Appl. Energy*, **104**, 517–526.

Martinez-Hernandez, E., Martinez-Herrera, J., Campbell, G.M., and Sadhukhan, J. (2014). Process integration, energy and GHG emission analyses of Jatropha-based biorefinery systems. *Biomass Convers. Bioref.*, **4**(2), 105–124.

Mendes, R.L., Nobre, B.P., Cardoso, M.T., Pereira, A.P. and Palavra, A.F. (2003) Supercritical carbon dioxide extraction of compounds with pharmaceutical importance from microalgae, *Inorg. Chim. Acta*, **356**, 328–334.

Munir, S.M., Manan, Z.A. and Wan-Alwi, S.R. (2012) Holistic carbon planning for industrial parks: a waste-to-resources process integration approach, *J. Clean. Prod.*, **33**, 74–85.

Ng, D.K.S. (2010) Automated targeting for the synthesis of an integrated biorefinery, *Chem. Eng. J.*, **162**(1), 67–74.

Ng, K.S. and Sadhukhan, J. (2011a) Techno-economic performance analysis of bio-oil based Fischer–Tropsch and CHP synthesis platform, *Biomass Bioenergy*, **35**, 3218–3234.

Ng, K.S. and Sadhukhan, J. (2011b) Process integration and economic analysis of bio-oil platform for the production of methanol and combined heat and power, *Biomass Bioenergy*, **35**, 1153–1169.

Ng, K.S., Lopez, Y., Campbell, G.M. and Sadhukhan, J. (2010) Heat integration and analysis of decarbonised IGCC sites, *Chem. Eng. Res. Des.*, **88**, 170–188.

Ng, K.S., Zhang, N., Sadhukhan, J. (2013) Techno-economic analysis of polygeneration systems with carbon capture and storage and CO_2 reuse, *Chem. Eng. J.*, **219**, 96–108.

Pham, V. and El-Halwagi, M.M. (2012) Process synthesis and optimization of biorefinery configurations, *AIChE J.* **58**(4), 1212–1221.

Sadhukhan, J., Ng, K.S., Shah, N. and Simons, H.J. (2009) Heat integration strategy for economic production of CHP from biomass waste, *Energy Fuels*, **23**, 5106–5120.

Sadhukhan, J., Zhao, Y., Shah, N., Brandon, N.P. (2010) Performance analysis of integrated biomass gasification fuel cell (BGFC) and biomass gasification combined cycle (BGCC) systems, *Chem. Eng. Sci.*, **65**(6), 1942–1954.

Sadhukhan, J., Ng, K.S., Martinez-Hernandez, E. (2014) *Biorefineries and Chemical Processes: Design, Integration and Sustainability Analysis*, John Wiley & Sons, Ltd, Chichester.

Salih, F.M. (2011) Microalgae tolerance to high concentrations of carbon dioxide: a review, *J. Environ. Protect.*, **2**(5), 648–654.

Sinnott, R.K. (2006) *Coulson & Richardson's Chemical Engineering Design, Volume 6*, 4th edition, Butterworth-Heinemann, Oxford.

Tan, R.R. and Foo, D.C.Y. (2007) Pinch analysis approach to carbon-constrained energy sector planning, *Energy*, **32**(8), 1422–1429.

Tay, D.H.S. and Ng, D.K.S. (2012) Multiple-cascade automated targeting for synthesis of a gasification-based integrated biorefinery, *J. Clean. Prod.*, **34**, 38–48.

Tijmensen, M.J.A., Faaij, A.P.C., Hamelinck, C.N. and van Hardeveld, M.R.M. (2002) Exploration of the possibilities for production of Fischer–Tropsch liquids and power via biomass gasification, *Biomass Bioenergy* **23**, 129–152.

10

Biomass-Fueled Organic Rankine Cycle-Based Cogeneration System

Nishith B. Desai and Santanu Bandyopadhyay

Department of Energy Science Engineering, Indian Institute of Technology Bombay, Mumbai, India

10.1 Introduction

An organic Rankine cycle (ORC) is a Rankine cycle which uses an organic fluid as a working fluid. It offers significant advantages over steam Rankine cycle in efficiently converting power from low- and medium-temperature heat sources up to 370°C (Hung et al., 1997). Water is not a good working fluid for such heat sources because the lower pressure requires large volumes and costly equipments. Organic working fluids have the advantage that they can use smaller and more efficient turbines. Growing energy demand and greenhouse gas emission make energy conservation imperative. In order to ensure the electricity generation without environmental pollution, the new energy conversion technologies are also required. In recent years, the interest for low- and medium-grade heat recovery using an ORC has grown considerably. ORC has been applied for generating power from different low- and medium-temperature heat sources such as waste heat, solar thermal, biomass, geothermal, etc. (Tchanche et al., 2011).

The key technical benefits of the ORC are (Bini and Manciana, 1996; Obernberga et al., 2002):

- High turbine efficiency (up to 85%)
- High cycle efficiency
- Improved part-load characteristics

Process Design Strategies for Biomass Conversion Systems, First Edition. Edited by Denny K. S. Ng, Raymond R. Tan, Dominic C. Y. Foo, and Mahmoud M. El-Halwagi.
© 2016 John Wiley & Sons, Ltd. Published 2016 by John Wiley & Sons, Ltd.

- Low RPM of the turbine which will allow the direct drive of the electric generator without reduction gear
- No erosion of blades due to absence of moisture
- Long service life
- Low maintenance costs

Biomass is the world's fourth largest energy source (after coal, oil, and natural gas), contributing to about 10% of the world's primary energy demand. In developing countries, its contribution to the national primary energy demand is higher and usually used in unsustainable ways. The power production and cogeneration from biomass are the most effective solutions for reliable and sustainable energy supply in small-scale applications, where conventional power plants are technologically and economically unfeasible. ORC-based cogeneration system increases potential due to its uncomplicated and efficient operation under lower temperatures and pressures compared to conventional steam Rankine cycle at comparably low investment and operating costs. Biomass-fueled ORC-based cogeneration plants at medium scale (100 kW to few MW) have been successfully demonstrated and are now commercially available. Apart from these, the small-scale systems of few kW are also started coming into the market. The number of installed plants, mostly in Europe, is increasing rapidly as the technology is becoming mature and cost effective. The benefits of the low-capacity distributed generation system using biomass-fueled ORC are as follows:

- Rural regions of many countries lack centralized grid power, and in that case, the distributed small- to medium-scale units may be economically competitive with alternative off-grid power technologies.
- Heat and electricity demand is usually available on-site and that makes a biomass plant suitable in the case of off-grid or unreliable grid connection.
- Local generation requires small-scale power plants (few kWe to MWe range) where the conventional steam Rankine cycle is not cost effective.

10.2 Working Fluids for ORC

The selection of proper working fluid for an ORC plays a vital role in efficient and economical utilization of any available heat source. The criteria for selection of the working fluids can be given as follows (Maizza and Maizza, 2001; Drescher and Bruggemann, 2007):

- Safety, health, and environmental aspects
- Good material compatibility and fluid stability limits
- Good thermodynamic properties: high heat of vaporization and acceptable pressures
- Noncorrosive to common engineering materials
- Availability and cost

Working fluids, based on the saturation vapor curve on the *T–s* diagram, can be categorized into three types: (i) dry fluids, such as n-pentane, n-hexane, etc., having a positive slope as shown in Figure 10.1a; (ii) isentropic fluids, such as R-11, R-12, etc., having an infinite slope (the saturation vapor curve is almost vertical on the *T–s* diagram, as shown in

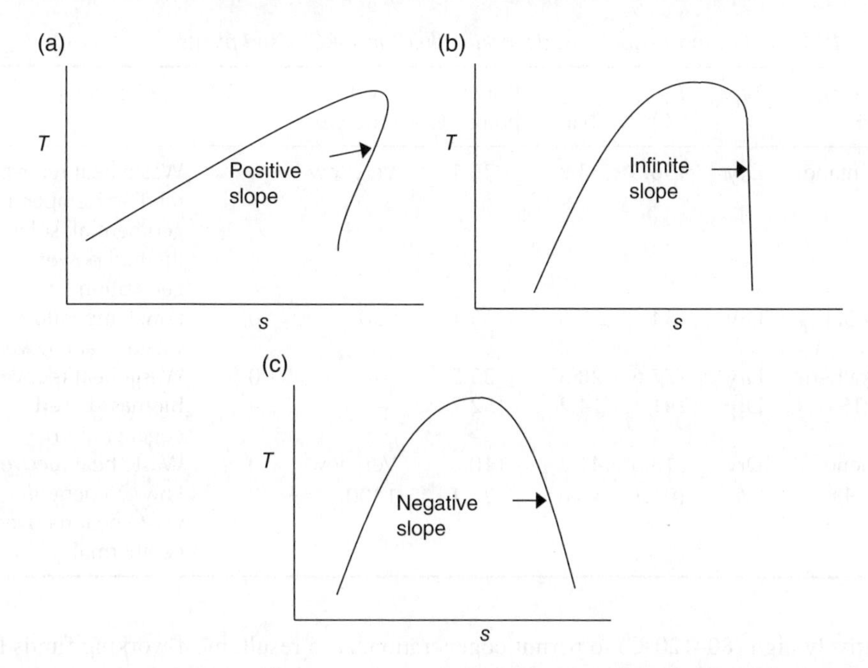

Figure 10.1 *Typical saturation curves on* T–s *diagram for (a) a dry fluid, (b) an isentropic fluid, and (c) a wet fluid. Reproduced with permission from Desai and Bandyopadhyay (2009), © 2009, Elsevier*

Figure 10.1b); and (iii) wet fluids, such as water, R-134a, etc., having a negative slope as shown in Figure 10.1c. In an ideal steam Rankine cycle, fluid expands isentropically through a turbine to generate power. Due to the negative slope of the saturation vapor curve for water (wet fluid), the outlet stream of a turbine contains lot of saturated liquid. The presence of liquid inside the turbine damages its blades and also reduces the isentropic efficiency of the turbine. Therefore, the minimum dryness fraction, at the outlet of a turbine, is typically kept around 15%. To satisfy the minimum dryness fraction at the outlet of turbine, the wet fluid should be superheated before entering into the turbine. Due to reduction in heat transfer coefficient in the gas phase, the heat transfer area requirement and the cost of the superheater go up significantly. Apart from these, there are other operational issues related to the superheater (Desai and Bandyopadhyay, 2009). On the other hand, the dry and isentropic fluids do not offer such disadvantages as the condition of expanded stream, at the outlet of turbine, is always either saturated or superheated. Therefore, the dry fluids and the isentropic fluids are the most preferred working fluids for the ORC system which utilizes low- and medium-grade heat sources (Hung, 2001; Lui et al., 2004).

In the case of biomass-based systems, the flame temperature of combustion process is very high (about 900°C) compared to all other ORC applications. Therefore, an indirect heating of organic working fluid, using a thermal oil circuit, is necessary to avoid local overheating and preventing the working fluid to become chemically unstable. The heat absorbed by the thermal oil is used to heat the working fluid of ORC and converted to electricity. In the case of cogeneration, the condensation temperature is usually

Table 10.1 *List of most widely used working fluids in ORC-based plants*

Working fluid	Type	$T_{critical}$ (°C)	$P_{critical}$ (bar)	Boiling point (°C)	GWP (100 years)	ODP	Applications
n-Pentane	Dry	196.6	33.7	36.1	Very low	0	Waste heat recovery, medium-temperature geothermal, solar thermal power generation
R-245fa	Dry	154	36.5	15.1	950	0	Low-temperature waste heat recovery
Solkatherm	Dry	177.6	28.5	35.5	—	0	Waste heat recovery
OMTS	Dry	291	14.2	152.7	—	—	Biomass-based cogeneration
Toluene	Dry	318.6	41.3	110.6	Very low	0	Waste heat recovery
R-134a	Wet	101.1	40.6	–26.1	1300	0	Low-temperature waste heat recovery, geothermal

relatively high (80–120°C) to permit cogeneration. As a result, most working fluids for low- to medium-temperature heat source cannot be used due to high vapor pressure at these condensation temperatures (Tchanche et al., 2011). Octamethyltrisiloxane (OMTS) is the most widely used working fluid for biomass-based ORC systems. However, Drescher and Brüggemann (2007) reported that the thermal and the global efficiency of the system are comparatively low when OMTS is used as a working fluid. In the literature, working fluid selection studies for an ORC system cover a range of fluids, but only few are used in commercial ORC-based plants. The list of most widely used working fluids in ORC-based plants with their important properties is given in Table 10.1 (Quoilin and Lemort, 2009).

10.3 Expanders for ORC

An expander is the most important equipment for an efficient and cost-effective ORC system. Expanders can be categorized into two types: (i) velocity type such as axial turbine and (ii) volume type such as scroll expanders, reciprocating piston expanders, screw expanders, rotary vane expanders, etc. The turbine, which uses an organic fluid as a working fluid, can reach very high isentropic efficiency with just one or two stages and rotating at a much lower speed. In the systems with high flow rates and low pressure ratios, one-stage axial turbines are most widely used. On the other hand, one-stage radial-inflow turbines are suitable for the systems with lower flow rates and higher pressure ratios. Therefore, one-stage radial-inflow turbines are most widely used in the ORC systems (Bao and Zhao, 2013). Radial-inflow turbine is suitable for the large-size system.

The screw expanders are widely used for waste heat recovery and geothermal applications. Reciprocating piston expanders have been used for the waste heat recovery of internal combustion engine exhaust. Screw and reciprocating piston expanders can be used

Table 10.2 *The comparison between various types of expanders used in the ORC-based system*

Type	Speed (rpm)	Cost	Capacity	Advantages	Disadvantages
Radial-inflow turbine	8,000–80,000	High	High–med.	Lightweight, mature technology, high efficiency	High cost, low efficiency at off-design conditions, intolerable two phase
Screw expander	<6,000	Med.	Med.–small	Low rotating speed, high efficiency at off-design conditions, tolerable two phase	Difficult manufacturing, lubrication requirement
Reciprocating piston expander	—	Med.	Med.–small	High pressure ratio, mature technology, tolerable two phase	Many moving parts, heavy weight, valves as well as torque impulse
Scroll expander	<6,000	Low	Small–micro	High efficiency, easy manufacturing, lightweight, low rotating speed, tolerable two phase	Low capacity, lubrication requirement
Rotary vane expander	<6,000	Low	Small–micro	Stable torque, simple structure, low cost, low noise, tolerable two phase	Low capacity, lubrication requirement

for a small–medium capacity system. Scroll expanders have a fixed volumetric ratio, and it can be used in a very small-scale system. Compared to other expanders, rotary vane expanders have simpler structure, easier manufacturing, and lower cost. The scroll and rotary vane expanders are used in a small or a micro ORC system. The comparison between various types of expanders, used in the ORC-based system, is given in Table 10.2 (Bao and Zhao, 2013).

10.4 Existing Biomass-Fueled ORC-Based Cogeneration Plants

ORC-based system manufacturers have been present in the market since the beginning of 1980s. They give ORC-based system in a wide range of power and temperature levels, as shown in Table 10.3 (Quoilin and Lemort, 2009; Qui et al., 2011; Vélez et al., 2012). The list mentioned in the table is not exhaustive. There are other manufacturers also available like Enertime, Barber-Nichols, GE/Calnetix Technologies, Enex, etc. (Enertime, 2008).

Table 10.3 *List of the ORC-based system manufacturers*

Manufacturer	Applications	Power range (kWe)	Heat source temperature (°C)	Working fluids and/or technology
Turboden, Italy	Geothermal, cogeneration	200–2000	100–300	Fluid: OMTS, Solkatherm Turbines: two-stage axial
ORMAT, United States	Geothermal, waste heat recovery, solar thermal power generation	50–72,000	150–300	Fluid: n-pentane Turbine: two-stage axial
ElectraTherm, United States	Waste heat recovery	50	>93	Fluid: R-245fa Expander: twin screw
UTC, United States	Geothermal, waste heat recovery	280	>93	—
Infinity Turbine, United States	Waster heat recovery	10–250	>80	Fluid: R-134a, R-245fa Expander: radial
Adoratec, Germany	Cogeneration	315–1600	300	Fluid: OMTS
GMK, Germany	Geothermal, waste heat recovery, cogeneration	50–2000	120– 350	Fluid: GL 160 (GMK patented) Turbine: multistage axial
Köhler & Ziegler, Germany	Cogeneration	70–200	150–270	Fluid: hydrocarbons Expander: screw
Eneftech, Switzerland	Waste heat recovery, cogeneration	10–30	125–150	Expander: scroll
Cryostar, France	Geothermal, waste heat recovery	—	100–400	Fluids: R-245fa, R-134a Turbine: radial inflow
Freepower, United Kingdom	Waste heat recovery	6–120	180–225	Scroll expander
Triogen, Netherlands	Waste heat recovery	160	>350	Fluid: toluene Expander: radial turbo
Green Energy, Australia	Geothermal, waste heat recovery, cogeneration	10–1000	>80	R-245fa

There are many ORC-based plants installed worldwide, which use different heat sources such as waste heat, biomass, geothermal, solar thermal, etc. Biomass is a very important renewable energy source, available almost everywhere. It can be stored for a longer period. Biomass can be utilized in the best way in cogeneration plants. It is most suitable for small-scale power systems (a few hundred kWe to MWe), built near the heat consumer. The ORC manufacturer Turboden, through their installations, demonstrated that ORC is a well-established technology for small-scale biomass-based cogeneration plants (lesser than 2.5 MWe). They have built about 220 such plants, mostly in Europe (Turbodan, 2013). Turboden uses OMTS as a working fluid for the heat source temperature range of 100–300°C. On the other hand, the ORC manufacturer ORMAT uses n-pentane as a working fluid for the heat source temperature range of 150–300°C. The working fluids of other ORC system manufacturers include hydrocarbons, R-245fa, R-134a, etc. (Quoilin and Lemort, 2009).

10.5 Different Configurations of ORC

The schematic of a basic ORC with dry organic working fluid and corresponding *T–s* diagram is shown in Figure 10.2a and b, respectively. A basic ORC consists of four processes: increasing pressure of the working fluid through a pump (from state 1 to 2 in Figure 10.2a), high-temperature heat addition through an evaporator (from state 2 to 3 in Figure 10.2a), expansion of the high-temperature and the high-pressure working fluid through a turbine (from state 3 to 4 in Figure 10.2a), and low-temperature heat rejection through a condenser (from state 4 to 1 in Figure 10.2a). The need of efficient and economical utilization of any available heat source leads to changes in the basic cycle. The various configurations are conceivable for the ORC, depending on the nature of the heat source and conditions in which the heat recovery is performed.

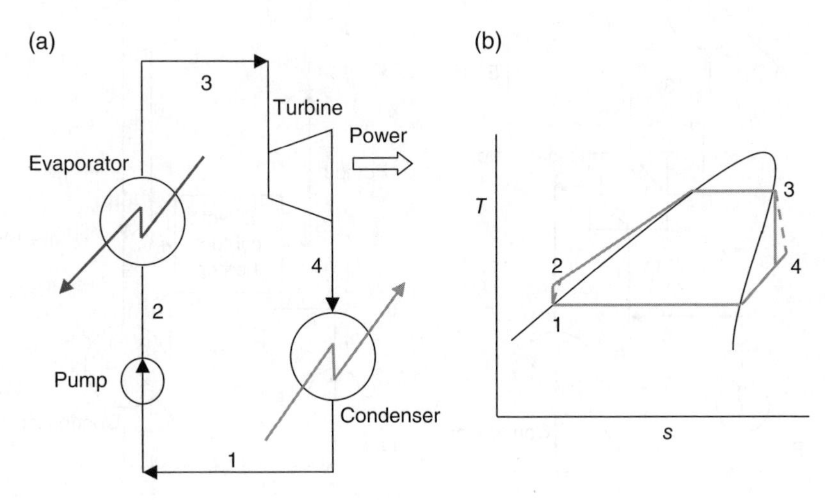

Figure 10.2 *Schematic diagram of an ORC: (a) flow diagram of a basic ORC and (b)* T–s *diagram of a basic ORC working with a dry fluid*

10.5.1 Regeneration Using an Internal Heat Exchanger

In the case of dry fluids, the state point after the expansion in the turbine (state 4 in Figure 10.2b) lies in the superheated vapor region. As the temperature of the superheated vapor at the turbine outlet (state 4 in Figure 10.2b) is more than that of the liquid at the inlet of the evaporator (state 2 in Figure 10.2b), it is possible to improve the thermal efficiency of the cycle through regeneration (Angelino and Piliano, 1998). The schematic of ORC with regeneration is shown in Figure 10.3a.

10.5.2 Turbine Bleeding

A small amount of the working fluid can be extracted from the turbine and mixed with the working fluid entering into the evaporator. This process is known as turbine bleeding. Through turbine bleeding, the mean temperature of heat addition increases, and as a result, the thermodynamic efficiency of the power generating cycle increases. However, it may be noted that the net shaft work reduces due to turbine bleeding. The schematic of the ORC with turbine bleeding is shown in Figure 10.3b. A direct contact heater is basically a mixing chamber, where the extracted fluid from the turbine (state 6) mixes with the liquid and the hot mixture (state 3) enters into the second pump. The mixture leaves the heater as a saturated liquid at the intermediate pressure. Mago et al. (2008) have demonstrated that the thermal efficiency of an ORC can be improved by incorporating turbine bleeding.

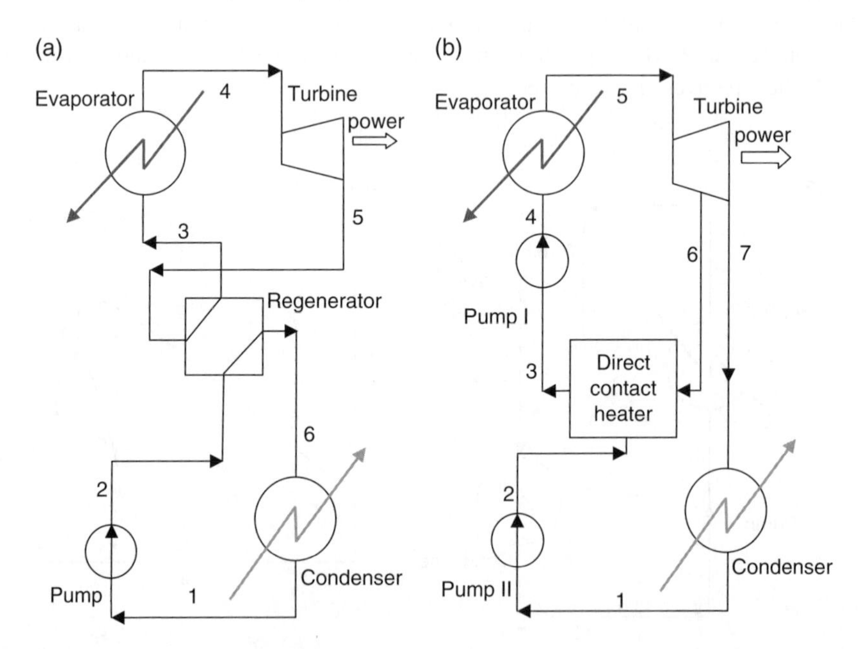

Figure 10.3 *Schematic diagram of an ORC: (a) flow diagram of an ORC incorporating regeneration and (b) flow diagram of an ORC incorporating turbine bleeding*

10.5.3 Turbine Bleeding and Regeneration

Simultaneous regeneration and turbine bleeding improve the thermal efficiency of the ORC significantly (Desai and Bandyopadhyay, 2009). For an ORC with turbine bleeding and regeneration, an internal heat exchanger (regenerator) and a direct contact heater are incorporated into the basic ORC. The schematic of a modified ORC incorporating both regeneration and turbine bleeding is shown in Figure 10.4a. A corresponding *T–s* diagram is shown in Figure 10.4b.

10.5.4 Thermodynamic Analysis of the ORC with Turbine Bleeding and Regeneration

To perform thermodynamic analysis of the ORC with turbine bleeding and regeneration, steady-state condition and no pressure drop in any equipment are assumed. The analysis of modified ORC can be given as follows (Desai and Bandyopadhyay, 2009):

Pumping process: The pump I increases the pressure of working fluid to an intermediate pressure (from state 1 to 2 in Figure 10.4a), while the pump II increases the pressure

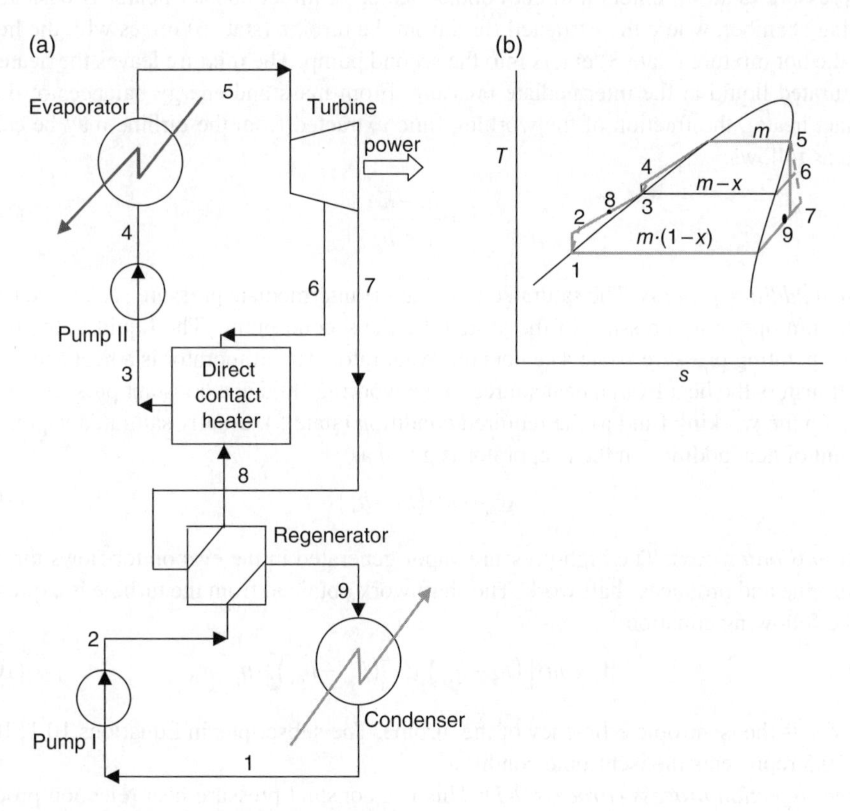

Figure 10.4 *Schematic diagram of an ORC: (a) flow diagram of an ORC incorporating turbine bleeding and regeneration and (b)* T–s *diagram of an ORC incorporating turbine bleeding and regeneration working with a dry fluid. Reproduced with permission from Desai and Bandyopadhyay (2009), © 2009, Elsevier*

of working fluid to the maximum operating pressure of the cycle (from state 3 to 4 in Figure 10.4a). The work done by the pumps can be given as follows:

$$W_{P1} = \frac{(1-x)\cdot m\cdot(h_{2s}-h_1)}{\eta_{P1}} \approx \frac{(1-x)\cdot m\cdot v_1\cdot(P_2-P_1)}{\eta_{P1}} \quad (10.1)$$

$$W_{P2} = \frac{m\cdot(h_{4s}-h_3)}{\eta_{P2}} \approx \frac{m\cdot v_3\cdot(P_4-P_3)}{\eta_{P2}} \quad (10.2)$$

where m is the mass flow rate of working fluid, η_P is the isentropic efficiency of pump, and v is the specific volume of working fluid. It may be noted that P_i and h_i denote the pressure and specific enthalpy, respectively, at ith state point. In Equation 10.1, x denotes the fraction of working fluid extracted from the turbine. As the enthalpy of a subcooled liquid is not readily available, an approximate expression for subcooled liquid is used in calculating pump work.

Direct contact heater: The intermediate liquid at state 2 enters a regenerator where the low-pressure vapor from the turbine (state 7) supplies heat. Heated liquid at the intermediate pressure (state 8) enters a direct contact heater. A direct contact heater is basically a mixing chamber, where the extracted fluid from the turbine (state 6) mixes with the liquid and the hot mixture (state 3) enters into the second pump. The mixture leaves the heater as a saturated liquid at the intermediate pressure. From mass and energy balances of direct contact heater, the fraction of the working fluid extracted from the turbine may be calculated as follows:

$$x = \frac{h_3 - h_8}{h_6 - h_8} \quad (10.3)$$

Heat addition process: The saturated liquid at the intermediate pressure is elevated to the maximum operating pressure of the system by the second pump. The liquid at the maximum operating pressure (state 4) enters the evaporator. The evaporator is a heat exchanger that transfers the heat from a heat source to the working fluid at a constant pressure, and it heats up the working fluid to the required condition (state 5), usually saturated vapor. The amount of heat addition in the evaporator is given as

$$Q_{\text{in}} = m\cdot(h_5 - h_4) \quad (10.4)$$

Expansion process: The high-pressure vapor generated in the evaporator flows through the turbine and produces shaft work. The shaft work obtained from the turbine is expressed by the following equation:

$$W_t = m\cdot\left[(h_5 - h_{7s}) + x\cdot(h_{7s} - h_{6s})\right]\cdot\eta_t \quad (10.5)$$

where η_t is the isentropic efficiency of the turbine. The subscript s in Equations 10.1, 10.2, and 10.5 represents the isentropic condition.

Heat rejection process (Process 9-1): This is a constant pressure heat rejection process in the condenser. The turbine exhaust is condensed in a condenser after it transfers a portion of heat through the regenerator. The condenser heat rejection rate is

$$Q_{\text{out}} = m\cdot(1-x)\cdot(h_9 - h_1) \quad (10.6)$$

Regeneration process: Regenerator is assumed to be a countercurrent heat exchanger with a given minimum temperature driving force. The intermediate liquid at state 2 enters a regenerator where the low-pressure vapor from the turbine (state 7) supplies heat.

The overall thermal efficiency (η_{th}) and net electrical power output of the ORC (P_{ORC}) may be calculated as follows:

$$\eta_{th} = \frac{W_t - W_{P1} - W_{P2}}{Q_{in}} = \frac{W_{net}}{Q_{in}} \tag{10.7}$$

$$P_{ORC} = \frac{W_{net}}{\eta_{em}} \tag{10.8}$$

where η_{em} is the electromechanical efficiency, which accounts for the mechanical and electrical losses.

10.6 Process Description

A typical biomass-fueled ORC-based cogeneration system is consist of a biomass feed boiler and an ORC linked through a thermal oil loop as shown in Figure 10.5 (Algieri and Morrone, 2012). Biomass fuel is burned through a process similar to a conventional steam boiler. Use of thermal oil as a heat transfer medium provides advantages like low pressure in the biomass boiler, no sensitivity to change in load, and simple as well as safe control and operation. Apart from these, it avoids local overheating of the organic working fluid and preventing it to become chemically unstable. The heat absorbed by the thermal oil is transferred to the ORC and converted into electricity. Typically, the regeneration using an internal heat exchanger is used for the ORC with dry organic working fluid. The condensation heat is used to produce hot water at a temperature range between 80 and 120°C, which is suitable for district heating and other thermal processes such as absorption cooling, wood drying, etc. (Dong et al., 2009). The exhaust gases coming out of the biomass boiler can be used for preheating the thermal oil, air, as well as water for heat consumer (Rentizelas et al., 2009). The heat input to the biomass boiler is given by

$$Q_{in,b} = m_{bf} \cdot \mathrm{LHV}_{bf} \tag{10.9}$$

where m_{bf} is the mass flow rate of the biomass fuel and LHV_{bf} is the lower heating value of the biomass fuel.

The energy utilization factor (EUF) and the cogeneration efficiency are given by

$$\mathrm{EUF} = \frac{P_{ORC} + Q_{cogen}}{Q_{in,b}} \tag{10.10}$$

$$\eta_{cogen} = \frac{P_{ORC}}{Q_{in,b} - \dfrac{Q_{cogen}}{\eta_{sb}}} \tag{10.11}$$

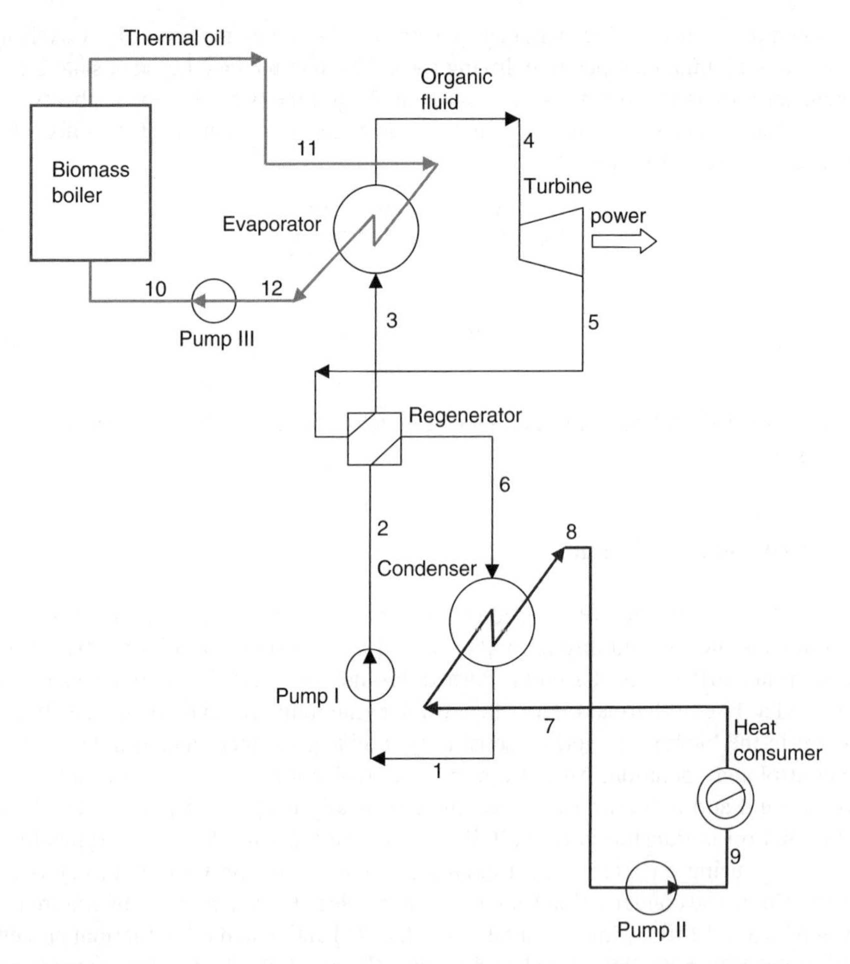

Figure 10.5 *Typical flow diagram of a biomass-fueled organic Rankine cycle-based cogeneration system*

where P_{ORC} is the net power output of the ORC, Q_{cogen} is the heat requirement of the cogeneration process, and η_{sb} is the efficiency of a second boiler that should be added to produce Q_{cogen} separately.

10.7 Illustrative Example

The data used for example are given in Table 10.4. Thermodynamic properties of organic fluid are calculated using the REFPROP software (Lemmon et al., 2002). The results are given in the Table 10.5 and Table 10.6. In the case of ORC with regeneration, the values of thermal efficiency, EUF, and cogeneration efficiency are 22.4%, 84.1%, and 67.7%, respectively. It may be noted that the thermal efficiency and cogeneration efficiency, for ORC with turbine bleeding and regeneration, are 23.8% and 68.6%, respectively.

Table 10.4 *Data used for example*

Input parameter	Value/type
Working fluid for the ORC	Toluene
Evaporation pressure/temperature of the ORC	32.76 bar/300°C (saturated)
Condensation pressure/temperature of the ORC	0.7425 bar/100°C
Isentropic efficiency of the turbine (η_t)	80%
Isentropic efficiency of the pump (η_p)	80%
Efficiency of the biomass boiler (η_b)	85%
Efficiency of the secondary boiler (η_{sb})	90%
Electromechanical efficiency (η_{em})	95%
Net power output (P_{ORC})	1000 kW
Water temperature at condenser inlet (T_7)	80°C
Lower heating value of the biomass (LHV_{bf})	14,000 kJ/kg
Temperature driving force (ΔT_{min})	For evaporator = 10°C; For condenser = 5°C

Table 10.5 *Results, for example, pressure, temperature, and enthalpy for different state points of the ORC (cycle with regeneration)*

State point no.	*P* (bar)	*T* (°C)	*h* (kJ/kg)	*x*
1	0.743	100	−21.02	0
2	32.76	101.5	−15.96	Sub.
3	32.76	160.7	106.6	Sub.
4	32.76	300	602.1	1
5	0.743	187.4	486	Sup.
6	0.743	111.5	363.4	Sup.

Table 10.6 *Results, for example, main performance parameters*

Parameter	Cycle with regeneration	Cycle with turbine bleeding and regeneration
Thermal efficiency (η_{th})	22.4%	23.8%
Organic working fluid mass flow rate (m)	9.48 kg/s	11 kg/s
Cooling water mass flow rate (m_{cw})	55.3 kg/s	64.2 kg/s
Temperature of the water for heat consumer (T_9)	95.7°C	92.5°C
Biomass fuel mass flow rate (m_{bf})	0.395 kg/s	0.372 kg/s
Heat for cogeneration (Q_{cogen})	3644 kW	3370 kW
Energy utilization factor (EUF)	84.1%	84%
Cogeneration efficiency (η_{cogen})	67.7%	68.6%

10.8 Conclusions

Water is not a good working fluid for low- and medium-temperature heat sources because the lower pressure requires large volumes and costly equipments. Organic working fluids have the advantage that they can use smaller and more efficient turbines. The power production and cogeneration from biomass are the most effective solutions for reliable and sustainable energy supply in small-scale applications, where conventional power plants are technologically and economically unfeasible. Biomass-fueled ORC-based cogeneration plants at medium scale (100 kW to few MW) have been successfully demonstrated and are now commercially available. Apart from these, the small-scale systems of few kW are also started coming into the market. Heat and electricity demand is usually available on-site and that makes a biomass plant suitable in the case of off-grid or unreliable grid connection. The dry fluids and the isentropic fluids are the most preferred working fluids for the ORC system which utilizes low- and medium-grade heat sources. In the case of biomass-based systems, the flame temperature of combustion process is very high (about 900°C) compared to all other ORC applications. Therefore, an indirect heating of organic working fluid, using a thermal oil circuit, is necessary to avoid local overheating and preventing the working fluid to become chemically unstable. Apart from these, in the case of cogeneration, the condensation temperature is usually relatively high (80–120°C) to permit cogeneration. As a result, most working fluids for low- to medium-temperature heat source cannot be used due to high vapor pressure at these condensation temperatures. OMTS is the most widely used working fluid for biomass-based ORC systems.

In the case of dry fluids, the state point after the expansion in the turbine lies in the superheated vapor region enabling regeneration to improve thermal efficiency. Simultaneous regeneration and turbine bleeding improve the thermal efficiency of the ORC significantly. An expander is the most important equipment for an efficient and cost-effective ORC system. One-stage radial-inflow turbines are most widely used in the large-size ORC systems. The screw expanders are used for waste heat recovery and geothermal applications. Reciprocating piston expanders have been used for the waste heat recovery of internal combustion engine exhaust. Screw and reciprocating piston expanders are used for a small–medium capacity system. The scroll and rotary vane expanders are used in a small or a micro ORC system.

References

Algieri, A., Morrone, P. (2012). Comparative energetic analysis of high-temperature subcritical and transcritical Organic Rankine Cycle (ORC). A biomass application in the Sibari district. *Applied Thermal Engineering*, **36**: 236–244.

Angelino, G., Paliano, P. (1998). Multicomponent working fluids for Organic Rankine Cycle (ORCs). *Energy*, **23**(6): 449–463.

Bao, J., Zhao, L. (2013) A review of working fluid and expander selections for organic Rankine cycle. *Renewable and Sustainable Energy Reviews*, **24**: 325–342.

Bini, R., Manciana, E. (1996). Organic Rankine cycle turbogenerators for combined heat and power production from biomass. Third Munich Discussion Meeting on Energy Conversion from Biomass Fuels Current Trends and Future Systems, 22–23 October 1996, Munich, Germany.

Desai, N.B., Bandyopadhyay S. (2009). Process integration of organic Rankine cycle. *Energy*, **34**: 1674–1686.

Dong, L., Liu, H., Riffat, S. (2009). Development of small-scale and micro-scale biomass-fuelled CHP systems: A literature review. *Applied Thermal Engineering*, **29**: 2119–2126.
Drescher, U., Bruggemann, D. (2007). Fluid selection for the organic Rankine cycle (ORC) in biomass power and heat plants. *Applied Thermal Engineering*, **27**(1): 223–228.
Enertime (2008). http://www.cycle-organique-rankine.com/market-markers.php (accessed July 11, 2013).
Hung, T.C., Shai, T.Y., Wang, S.K. (1997). A review of organic Rankine cycles (ORCs) for the recovery of low-grade waste heat. *Energy*, **22**(7): 661–667.
Hung, T.C. (2001). Waste heat recovery of organic Rankine cycle using dry fluids. *Energy Conversion and Management*, **42** :539–553.
Lemmon, E., McLinden, M., Huber, M. (2002). NIST reference fluid thermodynamic and transport properties-REFPROP. NIST standard reference database 23-Version 7.0.
Liu, B.T., Chien, K.H., Wang, C.C. (2004). Effect of working fluids on organic Rankine cycle for waste heat recovery. *Energy*, **29**(8): 1207–1217.
Mago, P.J., Chamra, L.M., Srinivasan, K., Somayaji, C. (2008). An examination of regenerative organic Rankine cycles using dry fluids. *Applied Thermal Engineering*, **28**: 998–1007.
Maizza, V., Maizza, A. (2001). Unconventional working fluids in organic Rankine cycles for waste energy recovery systems. *Applied Thermal Engineering*, **21**(3): 381–390.
Obernberger, I., Thonhofer, P., Reisenhofer, E. (2002). Description and evaluation of the new 1,000 kWel organic Rankine cycle process integrated in the biomass CHP plant in Lienz, Austria. *Euroheat & Power*, **10**: 1–9.
Qiu, G., Liu, H., Riffat, S. (2011). Expanders for micro-CHP systems with organic Rankine cycle. *Applied Thermal Engineering*, **31**: 3301–3307.
Quoilin, S., Lemort, V. (2009). Technological and Economical Survey of Organic Rankine Cycle systems. Fifth European Conference on Economics and Management of Energy in Industry, Vilamoura, Portugal, 14–17 April.
Rentizelas, A., Karellas, S., Kakaras, E., Tatsiopoulos, I. (2009). Comparative techno-economic analysis of ORC and gasification for bioenergy applications. *Energy Conversion and Management*, **50**: 674–681.
Tchanche, B.F., Lambrinos, G., Frangoudakis, A., Papadakis, G. (2011). Low-grade heat conversion into power using organic Rankine cycles – A review of various applications. *Renewable and Sustainable Energy Reviews*, **15**: 3963–3979.
Turboden (2013). http://www.turboden.eu/en/references/references.php (accessed July 11, 2013).
Vélez, F., Segovia, J.J., Martín, M.C., Antolín, G., Chejne, F., Quijano, A. (2012). A technical, economical and market review of organic Rankine cycles for the conversion of low-grade heat for power generation. *Renewable and Sustainable Energy Reviews*, **16**, 4175–4189.

11

Novel Methodologies for Optimal Product Design from Biomass

Lik Yin Ng, Nishanth G. Chemmangattuvalappil, and Denny K. S. Ng

Department of Chemical and Environmental Engineering/Centre of Sustainable Palm Oil Research (CESPOR), The University of Nottingham, Malaysia Campus, Selangor, Malaysia

11.1 Introduction

An integrated biorefinery is a processing facility that integrates multiple chemical reaction pathways to convert biomass into value-added products along with heat and power (Fernando et al., 2006). To date, there are a large number of established biomass conversion pathways available for implementation in an integrated biorefinery. Despite this, considering a large number of possible pathways prior to designing an integrated biorefinery would be a highly complex task. In this respect, Kokossis and Yang (2010) highlighted the important role of systematic screening tools in designing an integrated biorefinery. Systematic screening tools are required to identify the optimum conversion pathway based on different production objectives. In view of this, several systematic approaches have been developed in the past. These systematic approaches include (but not limited to) insight-based approaches and mathematical optimisation approaches. Insight-based approaches are used to integrate physical insights for acquiring a better understanding of a process. For instance, Halasz et al. (2005) adapted the P-graph method, which is based on combinatorial principles using branched and bound optimisation method in developing a green biorefinery by considering processing technology as well as raw material utilisation. Ng et al. (2009) presented a hierarchical approach which utilises two screening tools

Process Design Strategies for Biomass Conversion Systems, First Edition. Edited by Denny K. S. Ng, Raymond R. Tan, Dominic C. Y. Foo, and Mahmoud M. El-Halwagi.
© 2016 John Wiley & Sons, Ltd. Published 2016 by John Wiley & Sons, Ltd.

(evolutionary technique and forward–reverse synthesis tree) to synthesise and screen the potential alternatives for an integrated biorefinery. On the other hand, Tay et al. (2010) extended the utilisation of C–H–O ternary diagram in determining the overall performance of the synthesised integrated biorefineries. The work proposed by Tay et al. (2010) acts as a quick targeting tool that aids in the evaluation and analysis of integrated biorefinery. Svensson and Harvey (2011) adapted pinch analysis in identifying biorefinery pathway with increased and diversified revenues as well as reduced carbon dioxide reduction. Martinez-Hernandez et al. (2013) developed an integration approach based on mass pinch analysis for the analysis and design of product exchange networks formed in biorefinery pathways featuring a set of processing units.

Apart from insight-based approaches, numerous mathematical optimisation approaches have been developed for the synthesis of integrated biorefineries. For example, Sammons et al. (2007, 2008) introduced a flexible framework for optimal biorefinery product allocation by utilising mathematical optimisation techniques in evaluating and identifying optimal combination of production routes and product portfolios. Bao et al. (2009) presented a systematic approach based on technology pathway to determine the optimum pathway that achieves the highest conversion of the desired products. Later, Pham and El-halwagi (2012) presented a systematic two-stage approach in synthesising and optimising biorefinery configurations. The presented approach is based on the concept of 'forward and backward' approach which involves forward synthesis of biomass that leads to possible intermediates and backward synthesis that starts with the desired products and identifies potential pathways that produces the products. Ng and Ng (2013) adapted the concept of industrial symbiosis to develop palm oil processing complex (POPC). The developed POPC integrates the entire palm oil processing industry in maximising material recovery between processing technologies to achieve maximum economic performance. Murillo-Alvarado et al. (2013) adapted the disjunctive programming approach to identify optimal reaction pathways of a biorefinery while taking into account the maximisation of the net profit and minimisation of greenhouse gas emissions. Later, an optimum biorefinery via fast pyrolysis, hydrotreating and hydrocracking is designed by Gebreslassie et al. (2013) by developing a bi-criteria non-linear programming (NLP) model that seeks to maximise net present value (NPV) and minimise global warming potential (GWP). The optimum design is identified by solving the bi-criteria NLP model with ε-constraint method.

Besides economic performance, design aspects such as raw material allocation, environmental, safety and health impacts are considered during the synthesis of integrated biorefineries. Santibañez-Aguilar et al. (2011) presented a multi-objective optimisation model that simultaneously maximises the economic performance and minimises environmental impact while considering different feedstock, processing technology as well as end products. Zondervan et al. (2011) proposed a superstructure-based optimisation model for the design of optimal processing routes for multi-product biorefinery system by considering different feedstock, processing steps, final products and optimisation objectives. Later, Ponce-Ortega et al. (2012) presented a disjunctive programming approach in designing optimal integrated biorefinery. The proposed approach decomposes and solves a complex biorefinery design problem as set of sub-problems to identify the optimal pathway configuration for a given criterion. Meanwhile, Ng et al. (2013) extended fuzzy optimisation to develop a systematic multi-objective optimisation approach for the synthesis of integrated biorefinery which takes into consideration economic performance, environmental, safety and health impacts. El-halwagi et al. (2013) introduced an approach that considers the techno-economic factors as well as the effects of associated risk into the selection, sizing and supply chain network development of a biorefinery. Wang et al. (2013) proposed a superstructure-based multi-objective mixed-integer non-linear programming (MINLP) optimisation model for a biorefinery via gasification pathway that simultaneously maximises economic objective and minimises environmental concern. By solving the MINLP model with the ε-constraint method, the optimal solution is identified in terms of maximised economic objective measured by NPV and minimised environmental impact measured by GWP. Recently, Gong and You (2014) developed a detailed superstructure-based optimisation model for algae-based biorefinery that considers carbon sequestration and utilisation. Based on the developed superstructure, the optimal design of algae-based biorefinery is determined by minimising the unit carbon sequestration and utilisation costs.

It can be seen that the aforementioned works have focused on process design aspects of designing an integrated biorefinery where the focus is mainly on identifying and designing the optimal processing routes that leads to the product without integrating the product design aspects of the biorefinery. In order to synthesise an optimal integrated biorefinery, the product design has to be considered. This can be achieved by integrating the design of integrated biorefinery and chemical product design. According to Cussler and Moggridge (2001), chemical product design is a process of choosing the optimal product to be made for a specific application. Conventionally, the optimal chemical product is designed using generate-and-test techniques (Odele and Macchietto, 1993). However, these techniques are often very expensive and time consuming (Venkatasubramanian et al., 1994). In addition, as these approaches are largely dependent on expert knowledge and design heuristics, it is difficult and challenging to search for new chemical products with optimal performance (Churi and Achenie, 1996). Alternatively, chemical product design process can be done through reverse engineering approaches, where the design process searches for molecule that possesses properties which meet the product needs (Gani et al., 1991). Generally, functionality and performance of a product are defined in terms of physical properties instead of chemical structure of the product. For example, to design an effective refrigerant, the volumetric heat capacity for the designed refrigerant should be high so that the amount of refrigerant required is reduced for the same refrigeration duty. Besides, the designed refrigerant should have a low viscosity to achieve a low pumping power requirement. Hence, as long as the product possesses high volumetric heat capacity and low viscosity and fulfils other product needs, it is suitable to be used as an effective refrigerant regardless of its chemical structure. Therefore, chemical product design can be considered as an inverse property prediction problem where the desired product attributes are represented in terms of physical properties of the molecule (Gani and O'Connell, 2001). As the desired product attributes are usually extracted from customer requirements, it is required to translate descriptive customer requirements into physical properties of the product (Achenie et al., 2003). For example, in order to design a product which is non-hazardous, toxic limit concentration of the product should be measured; to design a product which will not cool easily, the heat capacity of the product is measured and taken into consideration during the design process. In addition, there are situations where a product need is required to be fulfilled by measuring and taking more than one physical property into account. For instance, density and viscosity of a fluid have to be considered in order to measure the consistency of a fluid flow; octane rating and heating value should be taken into account to design a fuel which provides high engine efficiency. The process of representing product attributes by using measurable product properties is often done by computer-aided molecular design (CAMD) techniques. Therefore, by utilising CAMD techniques in designing chemical products, it is made sure that the designed chemical products possess properties that satisfy customer requirements.

Throughout the years, CAMD techniques are being developed as powerful techniques in the field of chemical product design as they are able to predict, estimate and design molecules with a set of predefined target properties (Harper and Gani, 2000). Applications of CAMD techniques have included the design of different chemical products. For example, Camarda and Maranas (1999) developed an algorithm which includes structure–property correlation of polymer repeat unit in an optimal polymer design problem. Sahinidis, Tawarmalani and Yu (2003) proposed an MINLP model which can handle large number of preselected molecular groups and search for global optimal solution in an alternative refrigerant design problem. A two-step method in designing novel pharmaceutical products is introduced by developing structure-based correlations for physical properties and identifying the molecules having the desired properties (Siddhaye et al., 2004). Karunanithi, Achenie and Gani (2005) developed a framework in designing crystallisation solvents by solving the problem with decomposition-based solution approach. Samudra and Sahinidis (2013) utilised CAMD approach to identify efficient refrigerant components that fulfil process targets, environmental regulations as well as safety guidelines.

In this chapter, the aspect of product design of integrated biorefinery is discussed by integrating process and product design of integrated biorefinery.

11.2 CAMD

CAMD techniques are important for chemical product design for their ability in predicting, estimating and designing molecules with a set of predefined target properties (Harper and Gani, 2000). Property estimation is normally done by utilising property prediction models in predicting molecular properties from structural descriptors (Gani and Pistikopoulos, 2002). Some of the commonly used structural descriptors to quantify molecular structure include chemical bonds and molecular geometry (Randić et al., 1994). Most of the CAMD techniques utilise property prediction models based on group contribution (GC) methods to verify that the generated molecules possess the specified set of target properties (Harper et al., 1999). By utilising molecular groups as structural descriptors, GC methods estimate the property of the molecule by summing up the contributions from the molecular groups in the molecule according to their appearance frequency (Ambrose, 1978). Throughout the years, GC methods are extended and improved to include the property estimation of isomers and polyfunctional and structural molecular groups (Constantinou and Gani, 1994; Marrero and Gani, 2001). A general representation of property prediction model based on GC methods can be shown with the following equation:

$$f(X) = \sum_i N_i C_i + w \sum_j M_j D_j + z \sum_k O_k E_k \tag{11.1}$$

where $f(X)$ is a function of the property X; w and z are binary coefficients depending on the levels of estimation; N_i, M_j, O_k are the number of occurrence of first-, second- and third-order molecular GC correspondingly; and C_i, D_j, E_k are contribution of first-, second- and third-order molecular group subsequently.

In addition to GC methods, established method in developing property prediction models include the application of topological indices (TIs). TIs are molecular descriptors calculated based on principles in chemical graph theory (Trinajstić, 1992). In chemical graph theory which considers the molecules as the vertices and edges in a graph, atoms in the graph are named vertices, while the bonds used to connect them are called edges (Wilson, 1986). This method allows the capture of molecular information such as types of atoms and bonds, total number of atoms and bonding between the atoms. Hence, interactions among different atoms/molecular groups and their effects are known and utilised in describing a molecular graph as an index. This index is used to correlate the chemical structure to physical properties of a molecule. The correlated relationships are called quantitative structure–property/activity relationships (QSPR/QSAR) (Kier & Hall, 1986). Currently, a variety of properties can be predicted by using these QSPR/QSAR models which utilise their own property prediction methods. Wiener indices (Wiener, 1947), Randić's molecular connectivity indices (CI) (Randić, 1975), Kier's shape indices (Kier, 1985) and edge adjacency indices (Estrada, 1995) are among the well-known TIs which correlate the chemical structure to physical properties of a molecule. For illustration purpose, property prediction by using CI is shown. CI is calculated by using a molecular descriptor called delta (δ) value, which is the count of all bonded carbon atoms in a molecular structure. In order to take into consideration the effect and presence of heteroatoms and multiple bonds, valence δ value, δ^v is introduced, as shown by using the following equations:

$$\delta^{v} = \frac{Z^{v} - H}{Z - Z^{v} - 1} \tag{11.2}$$

Here, δ^v is the valence delta of an atom, Z is the total number of electrons in the atom, Z^v is the number of valence electrons in the atom, and H is the number of hydrogen atoms attached to the atom. The nth-order CI (${}^{v}\chi^{n}$) can then be determined by using the following equation:

$${}^{v}\chi^{n} = \frac{1}{\sqrt{\delta_i^{v} \ldots \delta_n^{v}}} \tag{11.3}$$

A schematic representation of the CI can be shown by Figure 11.1.

11.2.1 Signature-Based Molecular Design

In some chemical product design problems, the desired target properties could not be estimated by using a single class of property prediction model. Hence, different classes of property prediction models are required for the estimation of different target properties in the design problem. Although property prediction models are useful in estimating target product properties, applying different classes of property prediction models together in an inverse molecular design problem is a computationally challenging task (Camarda and Maranas, 1999). As mathematical formulations are exclusive for different property prediction models, it is difficult to utilise these different models by using a similar calculation method. This difficulty is addressed by utilising molecular signature descriptors as structural descriptors (Visco et al., 2002). Signature is a systematic coding system to represent

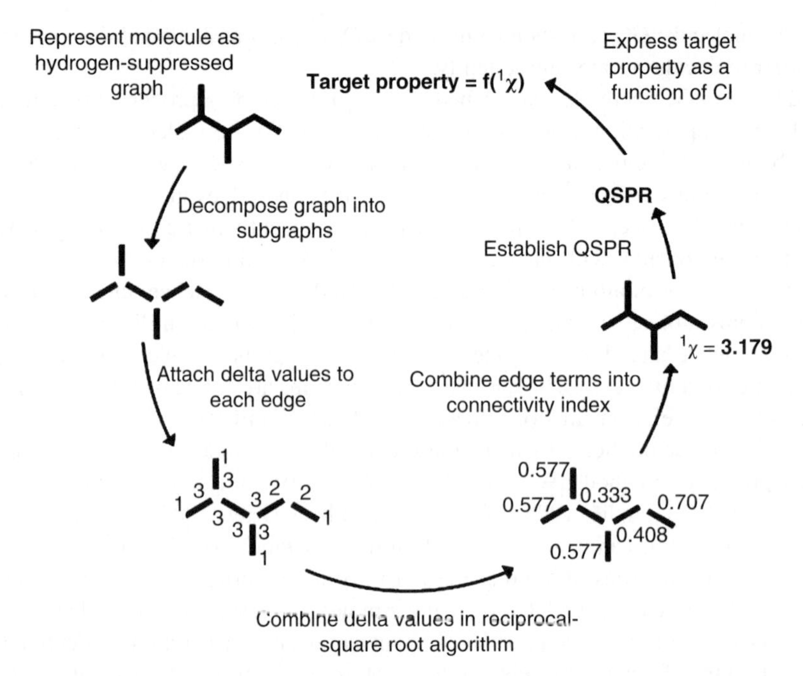

Figure 11.1 *Property prediction by using QSPR model. Reproduced with permission from Ng et al. (2014), © 2014, Elsevier*

the atoms in a molecule by using the extended valencies to a predefined height. Equation 11.4 represents the relationship between a TI and its signature:

$$\mathrm{TI}(G) = k\,{}^{h}\alpha_{G} \cdot \mathrm{TI}\left(\mathrm{root}\left({}^{h}\Sigma\right)\right) \tag{11.4}$$

Here, ${}^{h}\alpha_{G}$ is the occurrence number of each signature of height h, TI(root (${}^{h}\Sigma$)) is the TI values for each signature root, and k is a constant specific to TI. Signature of a molecule can be obtained as a linear combination of its atomic signatures by representing a molecule with atomic signature. One of the significance of signature descriptors is its ability to take into consideration the contributions of second- and third-order molecular groups in property prediction model developed based on GC methods (Chemmangattuvalappil and Eden, 2013). While signature descriptors represent individual building blocks for a complete molecule, they are related to the rest of the building blocks in the molecule as they carry information of their neighbouring atoms. Therefore, TIs can be described by using molecular signature (Faulon et al., 2003). Hence, by writing a molecule in terms of signature, GC methods and TIs with different mathematical formulations can now be expressed and utilised on a common platform. The application of molecular signature is important for chemical product design problem which involves multiple property targets which are required to be estimated by different classes of property prediction models (Chemmangattuvalappil et al., 2010). In order to utilise molecular signature descriptors in a molecular design problem, Chemmangattuvalappil et al. (2010) developed a signature-based algorithm for molecular design.

11.2.2 Multi-objective Chemical Product Design with Consideration of Property Prediction Uncertainty

To date, most of the developed and available product/molecular design algorithms emphasise on designing optimal products/molecules. In most cases, the optimal product is designed in terms of optimum target property(s). This optimality of product properties (termed as property superiority throughout the chapter) is the main factor that defines the quality of a product. Hence, in inverse property prediction product design problem, the product with optimal predicted target property(s) will be regarded as the optimal product. As mentioned earlier, most of the CAMD frameworks involve property predictions, which are done by using property prediction models. It is noted that effectiveness and usefulness of these property prediction models in estimating a property and eventually identifying the optimum molecule rely heavily on the accuracy of the property prediction models. Property prediction models are usually developed from regression analysis over a set of compounds. In the context of chemical product property prediction, regression analysis is a process of estimating the relationships between the property and the TI/molecular groups from GC methods that correlate with each other. According to Kontogeorgis and Gani (2004), the development of property models is an iterative process of theory/hypothesis definition, model equations solving, validation of model against experimental data, and modification of theory/model parameters if required. While providing relatively simple and accurate methods in property predictions, it is noted that these property prediction models are an approximation in reality, and there are always some discrepancies between experimental measurements and models predictions. The disagreement between the prediction and experimental values applies to all property estimation methods such as factor analysis, pattern recognition, molecular similarity, different TIs and GC methods (Maranas, 1997). From the cyclic process of property prediction models, it can be said that the accuracy of a model is affected by the uncertainties, which can arise from deficiency in theories or models and their parameters, and insufficient of knowledge of the systems (Kontogeorgis and Gani, 2004). In general, the performance or accuracy of property prediction models is evaluated and shown in terms of statistical performance indicators. Some of the commonly used pointers include standard deviation (σ), average absolute error (AAE), average relative error (ARE) and coefficient of determination (R^2). As target properties of the product are estimated by using property prediction models, the predefined targeted property ranges are highly dependent on the accuracy of property prediction model. Hence, the optimal solution for a design problem might differ according to the predefined targeted property ranges. The effect of property prediction model accuracy in determining the optimal solution will be termed as property robustness in this chapter. Property robustness is the measure of accuracy of a target property to be predicted by property prediction model. In this chapter, optimal product is defined as the one which has the highest property superiority and property robustness. In addition, in order to design an optimal product, multiple product properties are needed to be considered and optimised simultaneously. For example, during the design of an effective solvent, the solvent should be designed with maximum separability or solubility. Besides, the designed solvent should possess flammability which is within the safety operating limit and toxicity which follows environmental regulation. Refrigerant design is another example where multiple product properties are important. In order to design an effective refrigerant, the volumetric heat capacity for the designed refrigerant should be high so that the amount of refrigerant required is reduced for the same refrigeration duty. Besides, the designed refrigerant should have a low viscosity to achieve low pumping power requirement. Since more than one target properties are involved together with the consideration of property superiority and property robustness in designing these products, the design problems have to be solved as multi-objective optimisation problem.

Multi-objective optimisation approach is a mathematical optimisation approach for problems which require simultaneous optimisation of more than one objective function. Accordingly to Kim and de Weck (2006), most of the multi-objective decision-making optimisation problems are solved by using the weighted sum method first developed by Fishburn (1967). The method converts multiple individual objectives into an aggregated scalar objective function by assigning a weighting factor or relative importance to each of the individual objective function. An optimised overall objective function can be determined by summing up all the contributions of each individual objective. While utilising the conventional multi-objective optimisation methods to solve a decision-making problem, the weighting factors for each individual goal are assumed to be deterministic/crisp (Deckro and Hebert, 1989). However, in the context of chemical product design, the relative importance of each target property to be optimised in a design problem is not always definable. Therefore, the contribution of each target property in designing an optimal product is generally uncertain. Moreover, in some situations, these objectives/target properties might be contradictory to each other in nature. These unclear and conflicting objectives are uncertain/fuzzy instead of deterministic/crisp (Deporter and Ellis, 1990). In addition, the fuzzy weighting factors are normally known intuitively by the product designer/decision maker based on available knowledge or personal preferences (Clark and Westerberg, 1983). Fuzzy set theory is developed to solve a decision-making problem under fuzzy environment (Zadeh, 1965). The developed theory systematically defines and quantifies vagueness or fuzziness. Fuzzy optimisation approach is later developed by Bellman and Zadeh (1970) to determine the preferred alternative of a decision-making problem by solving an objective function subjected to a set of constraints. Zimmermann (1976) then developed an approach based on fuzzy set theory for linear programming problems by solving the problems under fuzzy goals and constraints. Later, Zimmermann (1978) further extended the approach to address linear programming problems which involve multiple objectives by integrating multiple objectives into a single objective. An optimised solution can then be obtained by solving the overall objective based on the predefined target property ranges. This approach provides flexibility and efficacy for solving and improving multi-objective decision-making problems.

11.3 Two-Stage Optimisation Approach for Optimal Product Design from Biomass

This section presents a systematic methodology for simultaneous process and product design for an integrated biorefinery. In the presented methodology, a two-stage optimisation approach is utilised to identify the optimal product as well as the optimal conversion pathway for the product. In the first stage, signature-based molecular design technique is employed to design the optimal product that satisfies the target properties. In the second

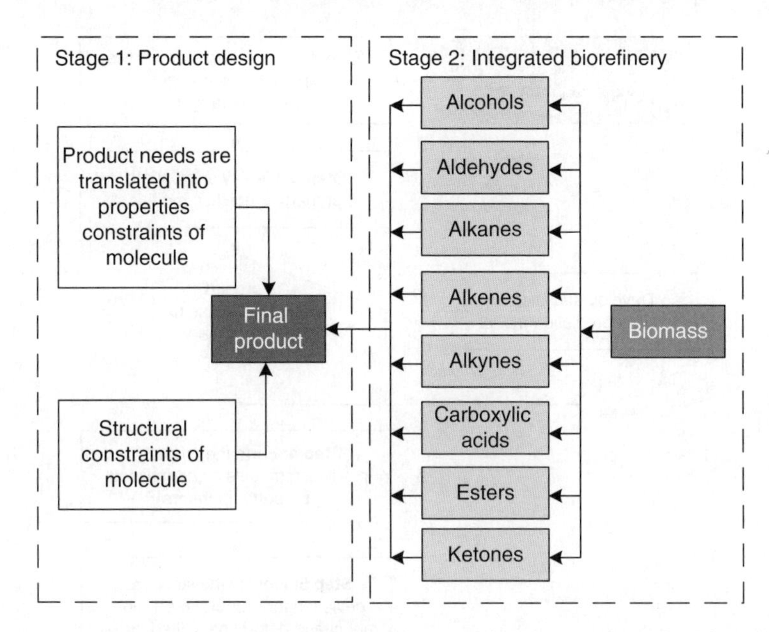

Figure 11.2 *Two-stage optimisation approach to produce optimal product from biomass*

stage, an integrated biorefinery is synthesised based on the components designed for the optimal product in the first stage. The general representation of the two-stage optimisation approach can be shown in Figure 11.2.

As shown in Figure 11.2, optimum product that fulfils target properties is first identified in the first stage. Based on the customer requirements, the product needs are translated into a set of property constraints which represents product specification. Structural constraints are applied to guarantee complete formation of the molecular structure of the product. The optimal product which satisfies property and structural constraints is identified by utilising signature-based molecular design technique developed by Chemmangattuvalappil et al. (2010). After the identification of optimal product, the optimal conversion pathway that converts biomass into the final product is identified in the second stage. Based on the available conversion pathway and technologies, a superstructure is constructed as a representation of integrated biorefinery. By using superstructural mathematical optimisation approach, optimal conversion pathway based on different design goals such as economic potential, production yield, environmental impact, etc. can be determined in this stage. By integrating product design with the synthesis of an integrated biorefinery, the proposed two-stage optimisation approach is able to determine the optimum conversion pathway that converts biomass into chemical that meets target properties. The following subsections further elaborate the proposed two-stage optimisation approach.

11.3.1 Stage 1: Product Design

In this stage, the optimal product is designed by utilising signature-based molecular design techniques. The steps involved in the optimal product design are represented by using a flowchart as shown in Figure 11.3.

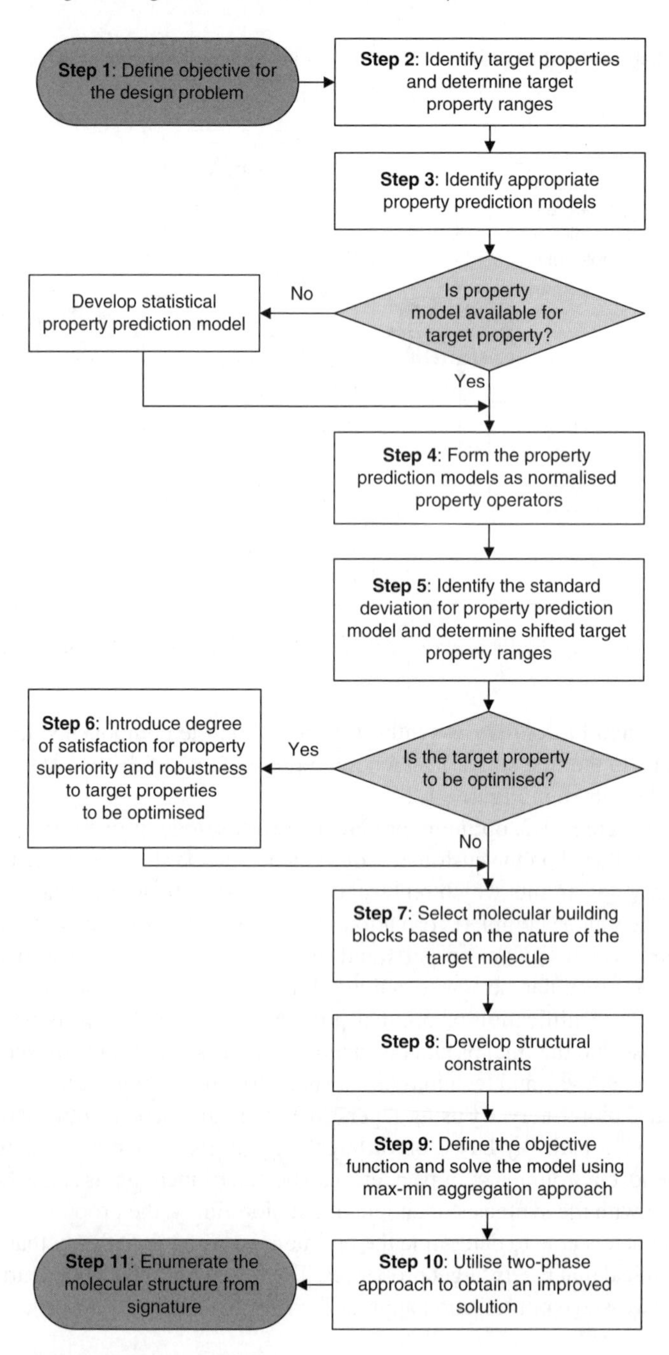

Figure 11.3 *Procedure for solving a multi-objective chemical product design problem*

Note that the procedure is designed specifically for product design problem where different classes of property prediction models are used and the molecular structure of the product is represented by using molecular signature descriptors. The detail of each step is discussed as follows.

11.3.1.1 Define objective for product design problem

The objective for the product design problem is determined by identifying the product needs. These product needs can be extracted from the operating conditions of an industrial process or from customer requirements. The product needs cover the physical properties which are responsible for a particular functionality of the product. Properties that make sure that the product fulfils the environmental and safety regulations can be considered as well. For example, in order to design an effective refrigerant, the performance of the refrigerant should be high, while the power requirement for the refrigerant is preferred to be low. Besides, it has to make sure that the refrigerant is not harmful to the environment and safe to be used. Hence, the objective of the design problem can be the optimisation of any target property or performance criterion.

11.3.1.2 Identify target properties and determine target property ranges

Once the product needs and the objective of the product design problem have been identified, the identified qualitative product needs are translated into measurable quantitative target properties. For example, during the design of refrigerant, the performance of the refrigerant can be expressed as volumetric heat capacity, which should be high so that the amount of refrigerant required is reduced for the same refrigeration duty. The power requirement of the refrigerant can be measured as viscosity, which is preferred to be low to achieve low pumping power requirement. Properties which can measure toxicity such as median lethal dose/concentration (LD_{50}/LC_{50}) can be used to ensure that the designed refrigerant is environmentally benign and safe to be used. These target properties are then expressed as property specification, which can be written as a set of property constraints bounded by upper and lower limit. For example, while designing a gasoline blend, the Reid vapour pressure is designed to fall within 45 and 60 kPa, the desired viscosity should fall within 0.30 and 0.60 cP, while the preferred density should be within 0.720 and 0.775 g/cm^3. The property specifications for a product design problem can be generalised and shown in Equation 11.5:

$$v_p^{\mathrm{L}} \le V_p \le v_p^{\mathrm{U}} \quad \forall p \in P \tag{11.5}$$

Here, p is the index for target property, V_p is the target property value, v_p^{L} is the lower bound, and v_p^{U} is the upper bound for product target property. By following Equation 11.5, the optimal solution is identified within the predefined target property ranges while solving a product design problem.

11.3.1.3 Identify appropriate property prediction models

After the identification of target property ranges, suitable property prediction models which estimate the target properties of the product are identified. Different classes of property prediction models such as property prediction models developed from GC method or TIs are utilised for the prediction of target properties. For target properties where property prediction models are unavailable, models which combined experimental data and available property prediction models can be developed to estimate the respective property.

11.3.1.4 Form property prediction models as normalised property operators

The identified property prediction models are then expressed and formed as normalised property operators. Normalised property operators are dimensionless property operators, which are required so that different target properties can be expressed and compared together on the same property platform (Shelley and El-Halwagi, 2000). According to Shelley and El-Halwagi (2000), property operators are functions of the original properties tailored to obey linear mixing rules. Hence, no matter whether the original property is linear or not, the property operator will follow simple linear mixing rules. Property specification in Equation 11.5 can be written as normalised property operators as shown in Equation 11.6:

$$\Omega_p^{\mathrm{L}} \leq \Omega_p \leq \Omega_p^{\mathrm{U}} \quad p = 1, 2, \ldots P \tag{11.6}$$

Here, Ω_p is the normalised property operator for property p, Ω_p^{L} is the lower limit for the normalised property operator, and Ω_p^{U} is the upper limit for the normalised property operator. As signature-based molecular design technique is employed in this developed methodology, normalised property operators are used to express molecules as linear combinations of atomic signatures.

11.3.1.5 Identify standard deviation, determine shifted target property ranges and introduce degree of satisfaction for property superiority and robustness for properties to be optimised

As steps 5 and 6 are closely related, both steps are discussed together under this subsection. To consider the effect of property model accuracy (property robustness), the effect of property model accuracy in the form of standard deviation of property prediction model for target property p, σ_p, is considered. σ_p is chosen as it is the measure of average variation between the measured and estimated values in regression analysis. An example to estimate a molecule with target property range for boiling point $80°\mathrm{C} < T_b < 100°\mathrm{C}$ by utilising a property prediction model with σ_p of 10°C is shown here. In order to take the allowance of property prediction model accuracy in terms of σ_p into consideration, the property target range is revised as $70°\mathrm{C} < T_b < 110°\mathrm{C}$. This allows the effect of property prediction model accuracy to be considered during the generation of the molecule. Hence, together with the optimality of product property, the effect of the accuracy of property prediction models can be taken into consideration during the generation of the optimal product.

After taking the allowance of property prediction model accuracy, target property ranges are shifted and divided into three different regions to improve the estimation of target properties with consideration of the property prediction model accuracy. The three regions after shifting and dividing the target property ranges are certain region (CR), lower uncertain region (LUR) and upper uncertain region (UUR). LUR is the region below the CR, while UUR is the region above it. Depending on the accuracy of property prediction model provided from the source literature of the property prediction model, the confidence level that the predicted target property will fall within the CR is higher compared to the confidence level that the predicted target property will fall within the URs. The lower and upper bounds for these regions can be obtained by adding and subtracting σ_p for the respective property prediction model from v_p^{L} and v_p^{U} of the target property. LUR is bounded by lower lower bound (v_p^{LL}) and lower upper bound (v_p^{LU}); CR is bounded by lower upper bound (v_p^{LU}) and upper lower bound (v_p^{UL}); UUR is bounded by upper lower bound (v_p^{UL}) and upper upper bound (v_p^{UU}). This is summarised in Table 11.1.

Hence, the lower and upper bounds of the product specification as shown in Equation 11.5 are shifted as shown in Equation 11.7:

$$v_p^{\mathrm{LL}} \leq V_p \leq v_p^{\mathrm{UU}} \quad \forall p \in P \tag{11.7}$$

For a chemical product design problem, other than target properties to be optimised, there are target properties which are used as property constraints to be fulfilled without optimising the target property. It is worth noting that the shifting of target property ranges applies to these target properties which are not optimised as well. By taking the accuracy of property prediction model into consideration for target properties which are not optimised, it is made sure that the developed approach is consistent towards every property prediction models utilised in the product design problem. Hence, the shifting of target property ranges as shown in Equation 11.7 applies to every target properties, while shifting of target property ranges into regions with different confidence level as shown in Table 11.1 applies only to target properties to be optimised.

To maximise the optimality of product property (property superiority), target properties to be optimised are first identified and expressed as target property range, bounded by upper and lower limit. Then, the comparison and trade-off between target properties to be optimised and potentially conflicting target properties is done by introducing a degree of satisfaction for property superiority, λ^{s}, to each of the properties. This can be achieved through writing λ^{s} as a linear membership function bounded by lower and upper bounds of the target property. The mathematical representation of the relationship of degree of satisfaction for property superiority is shown by Equations 11.8 and 11.9. Note that Equation 11.8 is used for property to be minimised, while Equation 11.9 is used for property to be maximised:

$$\lambda_p^{\mathrm{s}} = \begin{cases} 0 \text{ if } V_p \geq v_p^{\mathrm{U}} \\ \dfrac{v_p^{\mathrm{U}} - V_p}{v_p^{\mathrm{U}} - v_p^{\mathrm{L}}} \text{ if } v_p^{\mathrm{L}} \leq V_p \leq v_p^{\mathrm{U}} \quad \forall p \in P \\ 1 \text{ if } V_p \leq v_p^{\mathrm{L}} \end{cases} \tag{11.8}$$

Table 11.1 Lower and upper bounds for regions with different uncertainty

Region	Lower bound	Upper bound
Lower uncertain region	$v_p^{\mathrm{LL}} = v_p^{\mathrm{L}} - \sigma_p$	$v_p^{\mathrm{LU}} = v_p^{\mathrm{L}} + \sigma_p$
Upper uncertain region	$v_p^{\mathrm{UL}} = v_p^{\mathrm{U}} - \sigma_p$	$v_p^{\mathrm{UU}} = v_p^{\mathrm{U}} + \sigma_p$

Reproduced with permission from Ng et al. (2014), © 2014, Elsevier.

$$\lambda_p^{\mathrm{s}} = \begin{cases} 0 \text{ if } V_p \le v_p^{\mathrm{L}} \\ \dfrac{V_p - v_p^{\mathrm{L}}}{v_p^{\mathrm{U}} - v_p^{\mathrm{L}}} \text{ if } v_p^{\mathrm{L}} \le V_p \le v_p^{\mathrm{U}} \quad \forall p \in P \\ 1 \text{ if } V_p \ge v_p^{\mathrm{U}} \end{cases} \tag{11.9}$$

Here, λ_p^{s} is the degree of satisfaction for property superiority. As shown in Equation 11.8, λ_p^{s} is bounded within the interval of 0–1. The interval of 0–1 represents the level of satisfaction of the target property value V_p within the predefined target property range (v_p^{L} and v_p^{U}). The higher the λ_p^{s}, the better the product is in terms of property superiority. For property to be minimised, as lower value is preferred, when the property approaches the lower bound, the value of λ_p^{s} approaches 1; when the property approaches the upper bound, the value approaches 0. Vice versa applies to property to be maximised since higher value is desired. This can be explained graphically in Figure 11.4a and b.

In order to maximise property robustness, trade-off between target properties to be optimised is done by introducing a degree of satisfaction for property robustness, λ^{r}, to each of the target properties. As mentioned earlier, after making allowances for the accuracy of property prediction models, the confidence level that the predicted target property will fall within the CR is higher compared to the confidence level that the predicted target property will fall within LUR or UUR Thus, a value of 1 is given to λ^{r} when the target property falls within CR and 0 when it falls outside of the range of LUR or UUR. Within LUR and UUR, the nearer the property falls from the CR, the better it is in terms of property robustness. As CR, assigned with λ^{r} of 1, is bounded by two linear membership functions, which are λ^{r} of LUR and UUR, this forms an isosceles trapezoidal shape as shown in Figure 11.4c. Hence, this can be modelled as two-sided fuzzy optimisation problem by using trapezoidal fuzzy membership function (Zimmermann, 2001). This is described mathematically by the following equations:

$$\lambda_p^{\mathrm{r}} = \begin{cases} 0 \text{ if } V_p \ge v_p^{\mathrm{UU}} \\ \dfrac{v_p^{\mathrm{UU}} - V_p}{v_p^{\mathrm{UU}} - v_p^{\mathrm{UL}}} \text{ if } v_p^{\mathrm{UL}} \le V_p \le v_p^{\mathrm{UU}} \\ 1 \text{ if } v_p^{\mathrm{LU}} \le V_p \le v_p^{\mathrm{UL}} \quad \forall p \in P \\ \dfrac{V_p - v_p^{\mathrm{LL}}}{v_p^{\mathrm{LU}} - v_p^{\mathrm{LL}}} \text{ if } v_p^{\mathrm{LL}} \le V_p \le v_p^{\mathrm{LU}} \\ 0 \text{ if } V_p \le v_p^{\mathrm{LL}} \end{cases} \tag{11.10}$$

where λ_p^{r} is the degree of satisfaction for property robustness for property p. When V_p falls within LUR, λ_p^{r} approaches 0 as V_p approaches v_p^{LL}, and it approaches 1 when V_p approaches v_p^{LU}; when V_p falls within UUR, λ_p^{r} approaches 0 as V_p approaches v_p^{UU}, and it approaches 1 when V_p approaches v_p^{UL}; and when V_p falls within CR, λ_p^{r} remains as 1. Figure 11.4c shows the graphical representation and comparison of fuzzy linear functions for property superiority and property robustness.

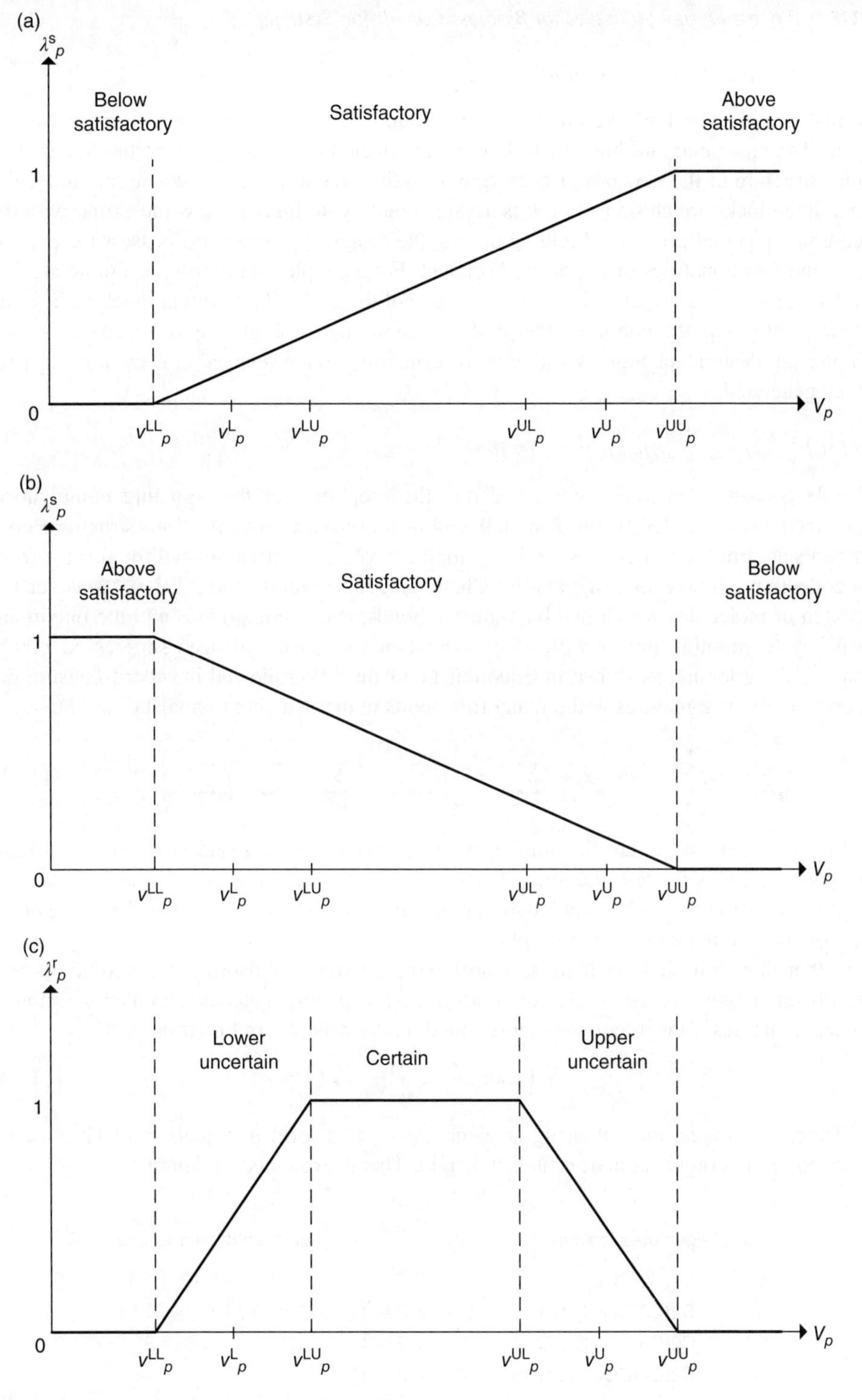

Figure 11.4 *Fuzzy membership functions for (a) property superiority of property to be maximised, (b) property superiority of property to be minimised and (c) property robustness. Reproduced with permission from Ng et al. (2014), © 2014, Elsevier*

11.3.1.6 Select molecular building blocks

Suitable molecular building blocks for the product design problem are determined in this step. The molecular building blocks have to be selected such that the properties and molecular structure of the new product are similar to the available product where the molecular building blocks are chosen from. It is assumed that by designing a new molecular with the chosen molecular groups as building blocks, the designed product will possess the properties and functionalities of the desired product. For example, in order to design an alcohol solvent, molecular group –OH is chosen as one of the molecular building blocks as it is the functional group of alcohol. As the product design methodology employs signature-based molecular design technique, signatures corresponding to the selected molecular groups are then generated.

11.3.1.7 Develop structural constraints

While property constraints guarantee that the properties of the signature combination obtained from the design problem fall within the product specification, structural constraints are imposed to make sure that complete molecular structures will be formed from a collection of molecular signatures (Chemmangattuvalappil et al., 2010). Consider the design of molecules which involve multiple bonds, cyclic compounds and the maximum valency (combining power with other atoms) of four in a hydrogen-suppressed graph, handshaking lemma as shown in Equation 11.11 must be followed in order to ensure the connectivity of signatures without any free bonds in the structure (Trinajstić, 1992):

$$\sum_{i=1}^{n_1} x_i + 2\sum_{n_1}^{n_2} x_i + 3\sum_{n_2}^{n_3} x_i + 4\sum_{n_3}^{n_4} x_i = 2\left[\left(\sum_{i=1}^{N} x_i + \frac{1}{2}\sum_{i=0}^{N_{Di}} x_i + \sum_{i=0}^{N_{Mi}} x_i + \sum_{i=1}^{N_{Ti}} x_i\right) - 1 + R\right] \quad (11.11)$$

where n_1, n_2, n_3 and n_4 are the number of signatures of valency one, two, three and four, respectively; N is the total number of signatures in the molecule; N_{Di}, N_{Mi} and N_{Ti} are the signatures with one double bond, two double bonds and one triple bond; and R is the number of circuits in the molecular graph.

Other than handshaking lemma, handshaking dilemma (Wilson, 1986) should also be fulfilled to make sure the number of bonds in each signature matches with the bonds in the other signatures. This dilemma is represented mathematically in Equation 11.12:

$$\sum \left(l_i \rightarrow l_j\right)_h = \sum \left(l_j \rightarrow l_i\right)_h \quad (11.12)$$

Here, $(l_i \rightarrow l_j)_h$ is one colouring sequence $l_i \rightarrow l_j$ at a level h. Equation 11.12 must be obeyed for all colour sequences at each height. This is explained in Figure 11.5.

List of signatures [colouring sequence]			Handshaking dilemma	
S1	C1(C2(CC))	[1 → 2]		
S2	C2(C2(CC)C1(C))	[2 → 2, 2 → 1]	$\Sigma(l_1 \rightarrow l_2)$	$= \Sigma(l_2 \rightarrow l_1)$
S3	C2(C3(CCC)C1(C))	[2 → 3, 2 → 1]	S1	= S2 + S3 + S4
S4	C2(C4(CCCC)C1(C))	[2 → 4, 2 → 1]		

Figure 11.5 *Explanation of handshaking dilemma*

In Figure 11.5, the edges of the signatures have the colours of 1 and 2. The reading of colouring sequence for signature S1 will be $1 \rightarrow 2$, $2 \rightarrow 2$ and $2 \rightarrow 1$ for signature S2, $2 \rightarrow 3$ and $2 \rightarrow 1$ for signature S3 and $2 \rightarrow 4$ and $2 \rightarrow 1$ for signature S4. Hence, by following Equation 11.12, the handshaking dilemma can be written. Each colour sequence (e.g. $1 \rightarrow 2$) has to be complemented with another colouring sequence in reverse order (e.g. $2 \rightarrow 1$) to ensure linkage and consistency of the signatures. The use of molecular signatures in molecular design and the connectivity rules of signatures are discussed in detail by Chemmangattuvalappil and Eden (Chemmangattuvalappil and Eden, 2013).

11.3.1.8 Generate feasible solutions by using max–min aggregation approach

To obtain the optimal solution in terms of property superiority and property robustness, the multi-objective chemical product design problem is solved by utilising fuzzy optimisation approaches in two stages. In the first stage, max–min aggregation approach is adapted to obtain the least satisfied degree of satisfaction. The objective of max–min aggregation approach is to make sure that every individual objective or fuzzy constraint will be satisfied partially to at least the degree λ. Therefore, each individual objective has an associated fuzzy function, and the optimum overall objective is obtained by maximising the least satisfied objective (Zimmermann, 1978). The objective function of max–min aggregation approach is shown in Equation 11.13, subjected to constraint of Equation 11.14:

$$\text{Maximise } \lambda \tag{11.13}$$

$$\lambda \leq \lambda_p \quad \forall p \in P \tag{11.14}$$

Note that the degree of satisfaction for target property p, λ_p, applies to both property superiority and property robustness. By solving Equations 11.13 and 11.14, a solution with maximised least satisfied degree of satisfaction is generated. This solution is the optimal product in terms of property superiority and property robustness. Additional feasible solutions can be generated by using integer cuts. Integer cuts work by adding additional constraints in the mathematical programming model to ensure the generated solution (in terms of combination of molecular signatures) will not appear again when the model is solved. This step may be continued until no feasible solution can be found.

Max–min aggregation approach aims to maximise the least satisfied degree of satisfaction so that the difference among all degrees of satisfaction would be reduced. However, this approach is unable to distinguish between solutions that possess similar value of least satisfied degree of satisfaction (Dubois et al., 1996). As other fuzzy goals might be overly relaxed in order to maximise the least satisfied goal, there is still room to search for better solutions in terms of overall degree of satisfaction. Hence, two-phase approach is adapted in this work to address this limitation (Guu and Wu, 1999).

11.3.1.9 Generate feasible solutions by using two-phase approach

After the application of max–min aggregation approach, two-phase approach is applied to solve the optimisation problem. The overall objective for two-phase approach is maximising the summation of all degrees of satisfaction. This means that all of the individual objectives are contributing to the objective function and optimised as whole. The objective function of the two-phase approach is shown in Equation 11.15. Moreover, to generate improved solutions and at the same time differentiate between solutions with identical least satisfied degree of satisfaction, it must be ensured that the solution obtained in the second stage will not be worse than the solution initially obtained in the first phase. In order to guarantee so, Equation 11.16 is introduced:

$$\text{Maximise} \sum_{p} \lambda_p^* \tag{11.15}$$

$$\lambda_p^* \geq \lambda_p \quad \forall p \in P \tag{11.16}$$

where λ_p^* is the new degree of satisfaction identified in second stage and λ_p is the degree of satisfaction identified in the first stage. As explained earlier, λ_p^* and λ_p apply to both property superiority and property robustness. From Equation 11.15, the individual degree of satisfaction obtained in the second stage must be equal or greater than the one obtained in the first stage in order to obtain a feasible solution. Thus, if the two-phase approach does not generate a feasible solution in the second stage, the solution obtained in the first stage by using max–min aggregation approach is the optimal solution for the multi-objective product design problem.

11.3.1.10 Enumerate molecular structure

With the signatures obtained from solving the design problem, molecular graph can now be generated based on the graph signature enumeration algorithm by Chemmangattuvalappil and Eden (2013). By using the graph enumeration algorithm, molecular structures are generated from the list of signatures, and the name of the new solvent are identified.

11.3.2 Stage 2: Integrated Biorefinery Design

In the first stage of the proposed two-stage optimisation approach, optimal product candidates that satisfy the product needs are identified. The second stage determines the optimal biomass conversion pathway to produce the identified optimal product by utilising superstructural mathematical optimisation approach. A superstructure which includes all the possible conversion pathways and technologies that process the biomass into the intermediate and convert the intermediates into the final products is constructed as the representation of an integrated biorefinery. Following this, all alternatives are mathematically modelled based on a generic systematic methodology shown in Figure 11.6.

Figure 11.6 illustrates a general superstructure of an integrated biorefinery with biomass feedstock b converted through pathways q to produce intermediates s and further processed via pathways q' to produce products s'. The mathematical model which relates the flow of biomass through different conversion pathways to produce the products is explained and discussed as follows.

As shown in Figure 11.6, biomass feedstock b can be split into their respective flow rate F_{bq}^{I} to biomass conversion pathway q with Equation 11.17:

$$B_b^{\mathrm{Bio}} = \sum_{q} F_{bq}^{\mathrm{I}} \quad \forall b \tag{11.17}$$

where B_b^{Bio} is the total flow rate of biomass feedstock b. Biomass feedstock with flow rate F_{bq}^{I} is converted to intermediate s based on the conversion rate of biomass conversion

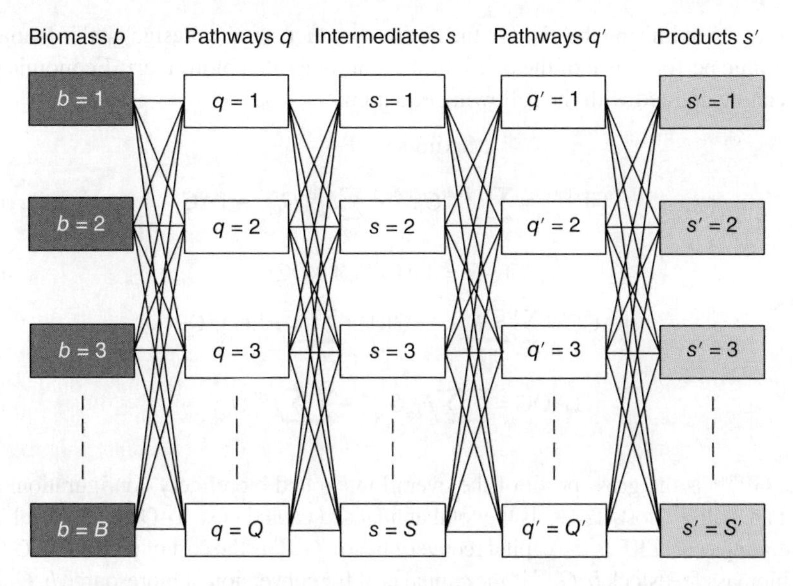

Figure 11.6 *Superstructure for an integrated biorefinery. Reproduced with permission from Andiappan et al. (2015), © 2015, John Wiley*

pathway q, R^{I}_{bqs}. This gives a total production rate of intermediate T^{Inter}_{s} as shown in the following equation:

$$T^{\mathrm{Inter}}_{s} = \sum_{q}\sum_{b}\left(F^{\mathrm{I}}_{bq}R^{\mathrm{I}}_{bqs}\right) \quad \forall s \tag{11.18}$$

Subsequently, the intermediate s is then further converted to product s' via upgrading pathway q'. The splitting of flow rate of intermediate T^{Inter}_{s} to all possible upgrading pathway q' with flow rate $F^{\mathrm{II}}_{sq'}$ can be calculated by Equation 11.19:

$$T^{\mathrm{Inter}}_{s} = \sum_{q'}F^{\mathrm{II}}_{sq'} \quad \forall s \tag{11.19}$$

The total production rate of product s', $T^{\mathrm{Prod}}_{s'}$, can be determined based on given conversion rates of upgrading pathway q', $R^{\mathrm{II}}_{sq's'}$ via Equation 11.20:

$$T^{\mathrm{Prod}}_{s'} = \sum_{q'}\sum_{s}\left(F^{\mathrm{II}}_{sq'}R^{\mathrm{II}}_{sq's'}\right) \quad \forall s' \tag{11.20}$$

By following Equations 11.17–11.20, the material balance of the biomass, intermediates and final products can be performed. Thus, an integrated biorefinery can be represented by using the developed superstructure.

The objective of this stage is to determine the optimal conversion pathways that convert biomass into the optimal product identified in the first stage of the methodology. The optimality of the conversion pathways can be aimed at maximising the yield of the desired product, as shown in the following equation:

$$\text{Maximise } T^{\mathrm{Prod}}_{s'} \tag{11.21}$$

Other than maximising the yield of the desired product, another design goal is to maximise the economic performance of the configuration of integrated biorefinery. Economic performance can be defined with the following equation:

$$\text{Maximise } \text{GP}^{\text{Total}}, \tag{11.22}$$

$$\text{GP}^{\text{Total}} = \sum_{s'} T_{s'}^{\text{Prod}} G_{s'}^{\text{Prod}} - \sum_{b} B_b^{\text{Bio}} G_b^{\text{Bio}} - \text{TAC} \tag{11.23}$$

$$\text{TAC} = \text{TACC} + \text{TAOC} \tag{11.24}$$

$$\text{TACC} = \sum_{q}\sum_{b} F_{bq}^{\text{I}} G_{bq}^{\text{Cap}} \text{CRF} + \sum_{q'}\sum_{s} F_{sq'}^{\text{II}} G_{sq'}^{\text{Cap}} \text{CRF} \tag{11.25}$$

$$\text{TAOC} = \sum_{q}\sum_{b} F_{bq}^{\text{I}} G_{bq}^{\text{Opr}} + \sum_{q'}\sum_{s} F_{sq'}^{\text{II}} G_{sq'}^{\text{Opr}} \tag{11.26}$$

Here, GP^{Total} is the gross profit of the overall integrated biorefinery configuration, TAC is the total annualised cost, TACC is the total annualised capital cost, TAOC is the total annualised operating cost, CRF is the capital recovery factor, $G_{s'}^{\text{Prod}}$ is the cost of product s', G_b^{Bio} is the cost of biomass feedstock b, G_{bq}^{Cap} is the capital cost for conversion of bioresource b, $G_{sq'}^{\text{Cap}}$ is the capital cost for conversion of intermediate s, G_{bq}^{Opr} is the operating cost for conversion of bioresource b, and $G_{sq'}^{\text{Opr}}$ is the operating cost for conversion of intermediate s. As such, the optimal conversion pathway that leads to the desired optimal product can be determined in this stage. For cases where the conversion pathway leads to the formation of products as a mixture of several components, separation processes are included. These separation processes are taken into account to refine and separate the final product from the other by-products based on the results obtained from the design of product in stage 1 of the methodology. In addition, please note that different objectives (e.g. economic performance, environmental impact, process safety, etc.) can be considered and included in the development of the superstructure.

11.4 Case Study

A product design problem of producing bio-based fuel from biomass is solved to illustrate the methodology. In the first stage, the bio-based fuel with optimal target properties is designed. Molecular signatures are used to represent different classes of property prediction models, while property superiority and property robustness are considered and optimised by using fuzzy optimisation approach. In the second stage, the optimal conversion pathways in terms of different production objectives that convert biomass into the designed bio-based fuel are identified. In order to demonstrate the efficacy of the methodology, the conversion pathways of an integrated biorefinery are synthesised for two scenarios: conversion pathways for maximum product yield and conversion pathways for maximum economic potential.

11.4.1 Design of Optimal Product

11.4.1.1 Step 1: Define design goal and objective

Bio-based fuel with multiple improved properties is designed from palm-based biomass. It is aware that biofuel is a mixture of different hydrocarbons. For the ease of illustration, the biofuel is targeted and designed as a single-component biofuel.

11.4.1.2 Steps 2 and 3: Identify target properties and appropriate property prediction models

The bio-based fuel is designed in terms of different product needs. The first is engine efficiency, which can be measured as octane rating. Octane rating is a measure of a fuel's ability to resist autoignition and knock in a spark-ignited engine conditions. Higher octane rating helps vehicle to run smoothly and keep the vehicles' fuel system clean for optimal performance. In this case study, octane rating is expressed as research octane number (RON). In addition to RON, engine efficiency can be measured as energy content in terms of higher heating value (HHV). As the energy of a fuel is determined by the heat content of the compounds, HHV is identified as the measurement of energy content of the fuel. In order to increase the engine efficiency, the HHV of the fuel should be high so that the energy content of the fuel is high. Furthermore, the bio-based fuel should be less toxic. The toxicity of the fuel is considered during the design stage to make sure the fuel is safe to be utilised. The toxicity of the fuel is measured as lethal concentration ($\log LC_{50}$) in this case study. Meanwhile, autoignition temperature (T_{ig}) and viscosity (η) of the fuel are the other target properties that are considered during the product design stage to ensure the consistency of the fuel flow as well as the safety and stability of the bio-based fuel. In this case study, RON, HHV and $\log LC_{50}$ are the target properties to be optimised while designing the bio-based fuel. Hence, this problem is formulated as a multi-objective optimisation problem to design the optimal bio-based fuel which fulfils the design goals.

Following the design procedure, after identifying the target properties for the product, property prediction models for each target properties are identified. In order to illustrate the ability of the methodology to utilise different classes of property prediction models in a design problem, property prediction models based on GC methods and CI are chosen to estimate the target properties. For RON and T_{ig} of the bio-based fuel, a reliable GC is available (Albahri, 2003a, 2003b):

$$f(X) = a + b\left(\sum_i N_i C_i\right) - c\left(\sum_i N_i C_i\right)^2 + d\left(\sum_i N_i C_i\right)^3 + e\left(\sum_i N_i C_i\right)^4 + f/\left(\sum_i N_i C_i\right) \quad (11.27)$$

Here, *a*, *b*, *c*, *d*, *e*, *f* are the correlation constants. Since the values of the constant *c*, *d*, *e*, *f* are relatively insignificant, only the first two terms of Equation 11.27 will be considered for this case study. For the prediction of HHV, a GC model developed by Yunus (2014) as shown in Equation 11.28 is utilised:

$$\text{HHV} = \text{HHV}_0 + \sum_i N_i C_i + w\sum_j M_j D_j + z\sum_k O_k E_k \quad (11.28)$$

where HHV_0 is the adjustable parameter for the prediction of HHV. For η, reliable GC models developed by Conte, Martinho, Matos and Gani (2008) are utilised:

$$\ln \eta = \sum_i N_i C_i + w\sum_j M_j D_j + z\sum_k O_k E_k \quad (11.29)$$

For the prediction of $\log LC_{50}$, a valence CI of order zero (Jurić et al., 1992) as shown in Equation 11.30 is utilised:

$$\log \text{LC}_{50} = 4.115 - 0.762\left({}^0\chi^v\right) \quad (11.30)$$

where ${}^0\chi$ is the zero-order connectivity index.

11.4.1.3 Step 4: Transform property prediction models into normalised property operators

With the identification of property prediction models, the next step is to transform the property prediction models into their respective normalised property operators. Property prediction models as shown in Equations 11.31–11.35 are written as normalised property operators as shown below:

$$\Omega_{\mathrm{RON}} = \frac{\mathrm{RON} - 103.6}{0.231} \tag{11.31}$$

$$\Omega_{\mathrm{HHV}} = \mathrm{HHV} - 146.826 \tag{11.32}$$

$$\Omega_{\log \mathrm{LC}_{50}} \frac{4.115 - \log\ \mathrm{LC}_{50}}{0.762} \tag{11.33}$$

$$\Omega_{T_{\mathrm{ig}}} = \frac{T_{\mathrm{ig}} - 780.42}{26.78} \tag{11.34}$$

$$\Omega_{\eta} = \ln \eta \tag{11.35}$$

11.4.1.4 Steps 5 and 6: Identify standard deviation and shift target property ranges

RON, HHV and log LC_{50} are target properties to be optimised, while T_{ig} and η are property constraints to be fulfilled. The target property ranges, standard deviation of the property prediction models and shifted target property ranges for each of the target property are identified and shown in Table 11.2.

Once the shifted target property ranges for all target properties are identified, the next step is to formulate the design problem as a multi-objective optimisation problem. In order to achieve so, property superiority and property robustness of target properties to be optimised are written as fuzzy linear functions. A degree of satisfaction is then assigned to each of them, as shown in Table 11.3.

From Table 11.3, it can be seen that there is no fuzzy linear function written for T_{ig} and η as they are not optimised during the product design process, but as constraints to be fulfilled.

Table 11.2 *Target property ranges for design problem*

Property	σ_p	Target range		Shifted target range			
		v_p^{L}	v_p^{U}	v_P^{LL}	v_p^{LU}	v_p^{UL}	v_p^{UU}
RON	5.00	95	108.00	90.00	100.00	103.00	113.00
HHV (kJ/mol)	26.96	5000.00	6600.00	4973.04	5026.96	6573.04	6626.96
log LC_{50}	0.193	1.000	1.900	0.807	1.193	1.707	2.093
T_{ig} (K)	28.00	600.00	800.00	572.00	628.00	772.00	828.00
η (cP)	0.89	1.00	3.00	0.11	1.89	2.11	3.89

Table 11.3 *Fuzzy membership functions of properties of interest*

Property	Interest	Fuzzy membership functions		
		Property superiority, λ_p^s	Property robustness, λ_p^r	
			Lower uncertain	Upper uncertain
RON	Maximise	$\frac{\Omega_{RON} - 90.00}{113.00 - 90.00}$	$\frac{\Omega_{RON} - 90.00}{100.00 - 90.00}$	$\frac{113.00 - \Omega_{RON}}{113.00 - 103.00}$
HHV	Maximise	$\frac{\Omega_{HHV} - 4973.04}{6626.96 - 4973.04}$	$\frac{\Omega_{HHV} - 4973.04}{5026.96 - 4973.04}$	$\frac{6626.96 - \Omega_{HHV}}{6626.96 - 6573.04}$
$\log LC_{50}$	Maximise	$\frac{\Omega_{\log LC_{50}} - 0.807}{2.093 - 0.807}$	$\frac{\Omega_{\log LC_{50}} - 0.807}{1.193 - 0.807}$	$\frac{2.093 - \Omega_{\log LC_{50}}}{2.093 - 1.707}$
T_{ig}	Constraint	—	—	—
η	Constraint	—	—	—

11.4.1.5 Step 7: Select appropriate molecular building blocks

The next step is to select the suitable molecular building blocks for the design problem. As the objective of this design problem is to design a bio-based fuel, the target molecule category is identified as alkanes. Therefore, only carbon (C) and hydrogen (H) atom are considered. As molecular signature descriptors are utilised in solving the chemical product design problem, only signatures with single bond are considered in this design problem to design the bio-based fuel. Signatures of height one are required since property prediction models of zeroth-order CI are utilised. The generated signatures can be classified into first-order groups of carbon with zero (C–), one (CH–), two (CH_2–) and three (CH_3–) hydrogen atoms. For signature C–, as it is bonded with zero hydrogen atoms, it can be connected to four other matching signatures. Same concept applies for other signatures as well, where signature CH– can be connected to three other matching signatures, signature CH_2– can be connected to two other matching signatures, and signature CH_3– can be connected to one matching signature. The generated signatures for the design problem are shown in Table 11.4.

11.4.1.6 Steps 8 and 9: Develop structural constraints and solve the model using max–min aggregation approach

The bio-based fuel design problem can now be formulated as a mixed-integer linear programming (MILP) model. The objective function in Equation 11.36 and constraint in Equation 11.37 are applied, while the design problem is solved by using max–min aggregation approach, where the objective is to maximise the least satisfied degree of satisfaction among all property superiority and robustness.

Objective functions

$$\text{Maximise } \lambda \tag{11.36}$$

Table 11.4 *List of signatures*

Number	Signature
1	C(C)
2	C(CC)
3	C(CCC)
4	C(CCCC)

Table 11.5 *Possible design of bio-based fuel*

Solution	Name	Property				
		RON	HHV (kJ/mol)	Log LC_{50}	T_{ig} (K)	η (cP)
A	2,2,3,3-Tetramethylbutane	105.91	5461.33	1.448	767.98	0.58
B	2,3,3,4-Tetramethylpentane	103.07	6097.81	1.199	707.12	0.49
C	2,2,4,4-Tetramethylpentane	103.96	6114.17	1.179	729.84	0.72
D	4,4-Dimethylheptane	96.41	6126.12	1.323	599.21	0.64
E	3,5-Dimethylheptane	95.53	6109.77	1.343	576.39	0.43

subject to

$$\lambda \leq \lambda_p \quad \forall p \in P \tag{11.37}$$

The model is solved via fuzzy optimisation-based inverse design techniques together with Equations 11.27 and 11.28, fuzzy linear functions as shown in Table 11.3 and structural constraints as shown in Chemmangattuvalappil et al. (2010). The solution is obtained in terms of molecular signatures. Integer cuts have been applied to generate different feasible alternatives for the design of alkyl substituent. The best five solutions are obtained in terms of signatures and summarised in Table 11.5. Table 11.6 shows the list of solutions arranged according to their least satisfied degree of satisfaction.

The solutions are named as solution A to solution E. It is seen from Table 11.5 that all of the solution properties fall between the boundaries that represent product needs (see Table 11.2). Note that these values are the optimised target properties subjected to the properties and structural constraints. From Table 11.6, it is noted that the least satisfied degree of satisfaction is not always from the same target property, and it is not restricted to only property superiority or property robustness of a property. For example, the least satisfied fuzzy goal in solution A is the property superiority of HHV, λ^{s}_{HHV} with the value of 0.30, while the least satisfied fuzzy goal in solution E is the property superiority of RON, λ^{s}_{RON} with the value of 0.24. This indicates that the methodology identifies the priorities of each goal to be optimised without the presence of decision maker. As long as the degree of satisfaction for the least satisfied goal is maximised, the generated solution is a candidate for bio-based fuel.

Furthermore, as shown in Table 11.6, the least satisfied goal for solution A and solution B has the same value of 0.30. Although the least satisfied goal for solution A is property

Table 11.6 *Comparison of λ_p between different designs of bio-based fuel*

Solution	RON		HHV		Log LC_{50}	
	λ_p^s	λ_p^r	λ_p^s	λ_p^r	λ_p^s	λ_p^r
A	0.69	0.71	0.30	1.00	0.50	1.00
B	0.57	0.99	0.68	1.00	0.30	1.00
C	0.61	0.90	0.69	1.00	0.29	0.96
D	0.28	0.64	0.70	1.00	0.40	1.00
E	0.24	0.55	0.69	1.00	0.42	1.00

superiority of HHV λ_{RON}^s while the least satisfied goal for solution B is property superiority of log LC_{50}, $\lambda_{\log LC_{50}}^s$, it is difficult to identify the better solution as they have similar value for least satisfied fuzzy goal. As mentioned earlier, the major drawback of max–min aggregation approach is its lack of discriminatory power to distinguish between solutions which have different levels of satisfaction other than the least satisfied goal. While the least satisfied goal is maximised, other goals might be overly relaxed or curtained, thus leaving room for improvement to search for better solutions. Due to this limitation, max–min aggregation approach does not guarantee to yield a Pareto-optimal solution (Jiménez and Bilbao, 2009). In order to discriminate these solutions to refine the order of solutions and at the same time to ensure Pareto-optimal solution, a two-phase approach is utilised.

11.4.1.7 Step 10: Solve the model by using two-phase approach

The optimisation model is solved again by using two-phase approach with the aim of maximising the summation of all degrees of satisfaction. The objective function is shown by Equation 11.38. As the least satisfied goal identified by using max–min aggregation approach is 0.59, the degree of satisfaction for the second stage should at least be equal or higher than this value to ensure that the solution obtain in the second stage is not worse than that obtained in the first stage. Hence, Equation 11.39 is introduced.

Objective functions

$$\text{Maximise } \sum_p \lambda_p^* \tag{11.38}$$

subject to

$$0.30 \le \lambda_p^* \le 1 \tag{11.39}$$

The MILP model is solved by using the same computational software and hardware specification as mentioned previously. The top five solutions generated by using two-phase approach are arranged accordingly to their summation of degrees of satisfaction as shown in Table 11.7.

From Table 11.7, although the best five solutions remain the same as solution A to solution E, it can be seen that the ranking of solutions changes significantly compared to that obtained by using max–min aggregation approach. As two-phase approach identifies the best solution which cannot be worse than the solution identified earlier by using max–min

Table 11.7 *Comparison of λ_p between different designs of bio-based fuel*

Solution	RON		HHV		Log LC_{50}		$\sum\lambda_p$
	λ_p^s	λ_p^r	λ_p^s	λ_p^r	λ_p^s	λ_p^r	
B	0.57	0.99	0.68	1.00	0.30	1.00	4.55
C	0.61	0.90	0.69	1.00	0.29	0.96	4.45
A	0.69	0.71	0.30	1.00	0.50	1.00	4.19
D	0.28	0.64	0.70	1.00	0.40	1.00	4.02
E	0.24	0.55	0.69	1.00	0.42	1.00	3.90

aggregation, the least satisfied degree of satisfaction among all the best five solutions is still the property robustness of RON, λ_{RON}^r, of solution E, which has the value of 0.24. From Table 11.7, although both solution A and solution B have the same value for least satisfied fuzzy goal, solution A is now the third best, while solution B is now the best solution. This is because the objective for two-phase approach is the maximisation of the summation of degrees of satisfaction of all fuzzy goals. Thus, solutions with similar value of least satisfied goal are distinguished.

For this case study, the best product is solution B, with λ of 0.30 and $\Sigma\lambda_p$ of 4.55, while the other solution with λ of 0.30, solution A being now ranked third with its $\Sigma\lambda_p$ of 4.19 as shown in Table 11.7. Furthermore, it can be observed from Table 11.7 that by using two-phase approach, the least satisfied degree of satisfaction is still the highest for the best product (solution B). This proves that even though the two-phase approach is able to discriminate the solutions with similar least satisfied fuzzy goal, the approach does not compromise the degree of satisfaction of that goal. Hence, utilisation of two-phase approach after max–min aggregation approach ensures the generation of optimal results without worsening any other goal. It is noted that in order to utilise two-phase approach, the problem must first be solved by using max–min aggregation approach to obtain the least satisfied degree of satisfaction. The least satisfied degree of satisfaction is then used as a constraint as shown in Equation 11.39 to make sure that the two-phase approach seeks for improved solution without worsening any of the degrees of satisfaction.

11.4.1.8 Step 11: Enumerate molecular structure from signature A

With the generated solutions, molecular graphs can be generated based on the graph signature enumeration algorithm by Chemmangattuvalappil and Eden (2013). The generated molecular structures of the bio-based fuel are shown in Table 11.8.

11.4.2 Selection of Optimal Conversion Pathway

Once the optimal product is designed, the optimal conversion pathways that convert biomass into the bio-based fuel are identified in the second stage of the methodology. In this case study, palm-based biomass known as empty fruit bunches (EFB) is chosen as feedstock of the integrated biorefinery. The lignocellulosic composition of the EFB is shown in Table 11.9.

From the first stage of the methodology, it is known that the end product of the integrated biorefinery is alkane with carbon number of 9. A list of possible conversion pathways that

Table 11.8 Molecular structures for the possible designs of bio-based fuel

Solution	Name	Molecular structure
A	2,2,3, 3-Tetramethylbutane	CH_3 CH_3 $CH_3 - C - C - CH_3$ CH_3 CH_3
B	2,3,3, 4-Tetramethylpentane	CH_3 CH_3 CH_3 $CH_3 - CH - C - CH - CH_3$ CH_3
C	2,2,4, 4-Tetramethylpentane	CH_3 CH_3 $CH_3 - C - CH_2 - C - CH_3$ CH_3 CH_3
D	4,4-Dimethylheptane	CH_3 $CH_3 - CH_2 - CH_2 - C - CH_2 - CH_2 - CH_3$ CH_3
E	3,5-Dimethylheptane	CH_3 CH_3 $CH_3 - CH_2 - CH - CH_2 - CH - CH_2 - CH_3$

Table 11.9 Lignocellulosic composition of EFB

Components	Composition (% of dry matter)
Lignin	39
Cellulose	22
Hemicellulose	29

produce alkanes from biomass is shown in Table 11.10. These pathways can be categorised into reactions from biochemical and thermochemical platforms. For illustration purpose, the end products—alkanes of the integrated biorefinery—are represented as straight-chain products without considering the formation of isomers. For example, the optimal bio-based fuel 2,3,3,4-tetramethylpentane is represented as alkane with carbon number C_9 in this case

Table 11.10 *List of conversion pathways and yield*

Pathway	Process	Product	Conversion (%)	Selectivity (%)
1	Ammonia explosion	Sugars, lignin	98.0	—
2	Steam explosion	Sugars, lignin	49.2	—
3	Organosolv separation	Lignin	79.0[a]	—
4	Organosolv separation	Sugars	97.0[a]	—
5	Autohydrolysis	HMF	90.9	—
6	Dehydration of sugars	Furfural	40.9	—
7	Yeast fermentation	Ethanol	61.9	—
8	Bacterial fermentation	Ethanol	41.0	—
9	Hydrogenation of furfural	THFA	98.2	—
10	Hydrogenation of THFA 1	Pentanediol	99.0	95.0
		Pentanol		4.0
11	Hydrogenation of THFA 2	Pentanediol	60.0	51.0
		Pentanol		22.0
12	Pyrolysis	Syngas	94.0	—
13	Gasification	Syngas	90.0	—
14	Anaerobic digestion	Methane	40.0	—
15	Water gas shift reaction	Syngas	100.0	—
16	Fischer–Tropsch 1	Hydrocarbon C_2–C_4	40.0	16.0
		Hydrocarbon C_5–C_9		27.0
		Hydrocarbon C_{10}		26.0
17	Fischer–Tropsch 2	Hydrocarbon C_2–C_4	75.0	23.0
		Hydrocarbon C_5–C_9		19.0
		Hydrocarbon C_{10}		9.7
18	Conversion of syngas 1	Methanol	25.1	2.6
		Ethanol		61.4
19	Conversion of syngas 2	Methanol	24.6	3.9
		Ethanol		56.1
20	Hydrogenation of CO	Methanol	28.8	20.7
		Ethanol		23.8
		Propanol		14.1
		Butanol		7.5
21	Monsanto process	Ethanoic acid	99.0	—
22	Dehydration of alcohols 1	Hydrocarbon C_2	67.0	—
23	Dehydration of alcohols 2	Hydrocarbon C_3	59.0	28.8
		Hydrocarbon C_4		37.3
24	Dehydration of alcohols 3	Hydrocarbon C_5	64.0	15.2
		Hydrocarbon C_6		5.5
		Hydrocarbon C_7		5.6
		Hydrocarbon C_8		4.6
25	Decarboxylation of acids	Hydrocarbon C_2	62.0	21.3
26	Fractional distillation of alkanes	Hydrocarbon C_9	99.0[a]	—
		Hydrocarbon C_2–C_8, C_{10}	99.0[a]	—

[a] Separation efficiency.

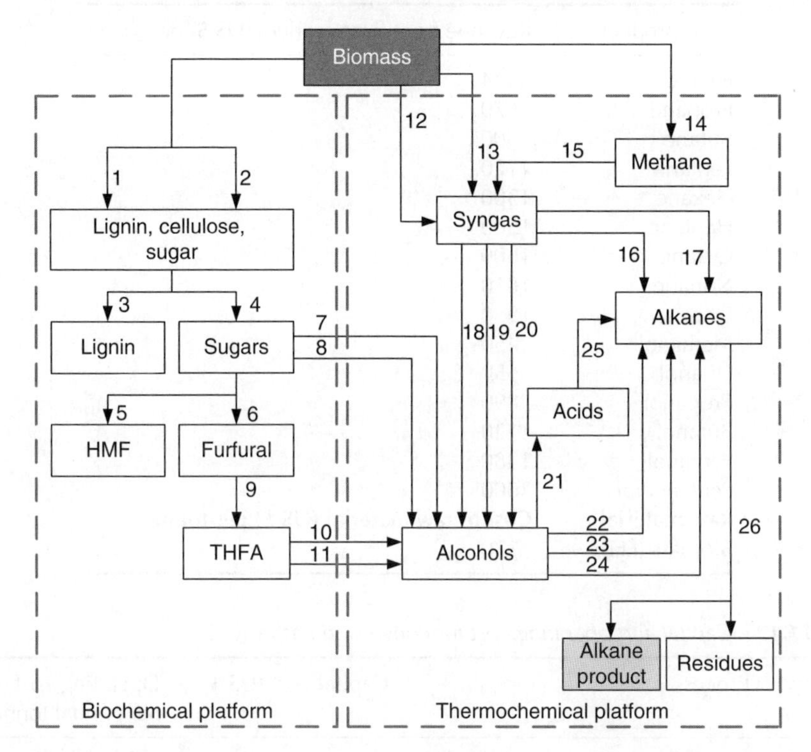

Figure 11.7 *Production of additives made from alkane and alcohol from lignocellulosic biomass*

study. Figure 11.7 presents a superstructure developed based on the conversion pathways in Table 11.10. It is noted that the developed superstructure can be revised to include more conversion pathways and technologies in synthesising an integrated biorefinery.

In this case study, two scenarios of different production objectives are considered in synthesising the integrated biorefinery:

1. Design for maximum product yield
2. Design for maximum economic potential

Table 11.11 shows the market price of the products and biomass feedstock, while Table 11.12 contains the capital and operating costs for each conversion pathway. Note that the prices of the products, feedstock and conversion pathways can be revised according to the market prices to produce an up-to-date economic analysis.

In this case study, other than the revenue generated by producing the bio-based fuel, the revenue obtained from the generation of by-products along with the product is included in the overall economic potential of the integrated biorefinery as well. With a feed of 50,000 tonnes per year of EFB, superstructural optimisation model is formulated and solved for the following production objectives.

11.4.2.1 Scenario 1: Design for maximum product yield

In this scenario, an integrated biorefinery is synthesised by solving the optimisation model using the optimisation objective in Equation 11.40. Note that the optimum bio-based

Table 11.11 Price of products and raw materials

Final product	Revenue from final product (US $) per tonne
Ethane	424
Propane	670
Butane	900
Pentane	1120
Hexane	1300
Heptane	1500
Octane	1600
Nonane	1810
Decane	1950
Methanol	450
Ethanol	650
Propanol	890
Butanol	1120
Pentanol	1480
Pentanediol	3000
Raw material	**Cost of raw material (US $) per tonne**
Biomass (EFB)	170

Table 11.12 Capital and operating cost for conversion pathways

Pathway	Process	Capital cost (US $)	Operating cost (US $) per annual tonne
1	Ammonia explosion	7.47×10^6	11.30
2	Steam explosion	5.29×10^6	7.97
3	Organosolv separation	1.55×10^7	23.30
4	Organosolv separation	1.55×10^7	23.30
5	Autohydrolysis	2.41×10^7	36.40
6	Dehydration of sugars	1.05×10^7	15.80
7	Yeast fermentation	1.54×10^7	22.00
8	Bacterial fermentation	1.20×10^7	18.00
9	Hydrogenation of furfural	1.15×10^7	17.30
10	Hydrogenation of THFA 1	1.65×10^7	24.90
11	Hydrogenation of THFA 2	1.73×10^7	26.00
12	Pyrolysis	2.39×10^7	36.00
13	Gasification	3.29×10^7	55.00
14	Anaerobic digestion	9.98×10^6	15.00
15	Water gas shift reaction	5.57×10^6	8.66
16	Fischer–Tropsch 1	7.36×10^7	111.00
17	Fischer–Tropsch 2	6.92×10^7	104.00
18	Conversion of syngas 1	1.47×10^7	22.10
19	Conversion of syngas 2	1.56×10^7	23.60
20	Hydrogenation of CO	1.53×10^7	23.00
21	Monsanto process	1.55×10^7	23.30
22	Dehydration of alcohols 1	1.54×10^7	23.20
23	Dehydration of alcohols 2	1.43×10^7	21.50
24	Dehydration of alcohols 3	1.31×10^7	19.70
25	Decarboxylation of acids	1.75×10^7	26.30
26	Fractional distillation of alkanes	6.52×10^7	98.20

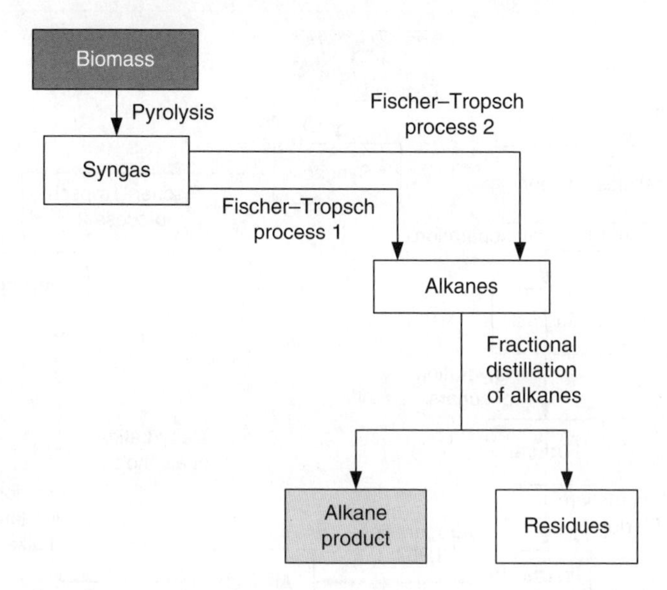

Figure 11.8 *Flow diagram of synthesised integrated biorefinery (maximum product yield)*

fuel 2,3,3,4-tetramethylpentane is represented as alkane of carbon number C_9 (Alkane_{C_9}) in the case study:

$$\text{Maximise } T^{\text{Prod}}_{\text{Alkane}_{C_9}} \tag{11.40}$$

Based on the obtained result, the maximum yield for Alkane_{C_9} is 5645.74 t/y. Alkanes with different carbon number are produced as by-products together with Alkane_{C_9}. The GP^{Total} for the scenario is found to be US $23.63 million (per annum). The conversion pathways selected for the scenario are illustrated in the synthesised integrated biorefinery as shown in Figure 11.8.

From Figure 11.8, it can be seen that Alkane_{C_9} is produced from biomass in the conversion pathway sequence of pyrolysis and Fischer–Tropsch processes 1 and 2 followed by fractional distillation of alkanes, which are all thermochemical pathways. It is worth pointing out that specific separation processes that suit the identified product can be chosen and included in the integrated biorefinery to refine and separate the final product from by-products. Hence, separation processes for alkanes are chosen based on the results of the product design identified in stage 1 of the methodology. The performance of the separation processes is then taken into consideration in identifying the product yield and economic potential of the overall conversion pathway.

11.4.2.2 Scenario 2: Design for maximum economic potential

In this scenario, an integrated biorefinery configuration with maximum economic potential is determined. Similar constraints in scenario 2 are applied in solving the optimisation objective as shown in Equation 11.41:

$$\text{Maximise } GP^{\text{Total}} \tag{11.41}$$

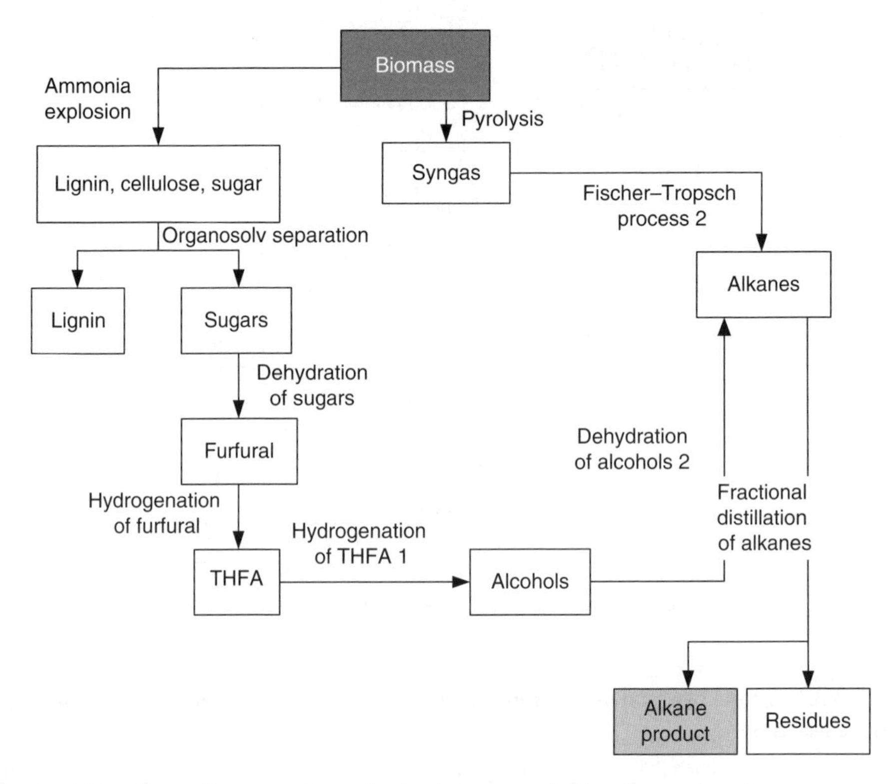

Figure 11.9 *Flow diagram of synthesised integrated biorefinery (maximum economic potential)*

Based on the generated optimisation result, the maximum GP^{Total} for the scenario is identified to be US \$24.07 million (per annum) with the annual production for $Alkane_{C_9}$ of 2643.42 tonnes. As the objective of this scenario is to synthesise an integrated biorefinery with maximum economic potential, alcohols are produced and sold as by-products together with the main product $Alkane_{C_9}$. The conversion pathways chosen for the scenario are presented in the synthesised integrated biorefinery as shown in Figure 11.9.

From Figure 11.9, it can be seen that alcohols are produced from biomass in the conversion pathway sequence of ammonia explosion, organosolv separation, dehydration of sugars, hydrogenation of furfural and hydrogenation of THFA 1. $Alkane_{C_9}$ is produced from fractional distillation of alkanes, which are produced from pyrolysis of biomass, followed by Fischer–Tropsch process 2 together with dehydration of alcohols 2. The selected conversion pathways consist of both biochemical and thermochemical pathways. The comparison of the results generated for scenario 1 and 2 is summarised in Table 11.13.

It can be seen from Table 11.13 that as the objective of scenario 1 is to identify the conversion pathways which produce maximum yield of $Alkane_{C_9}$, the production rate of $Alkane_{C_9}$ in scenario 1 is higher compared to scenario 2. In scenario 2, significant amount of alcohols are generated together with $Alkane_{C_9}$. Although the production rate of $Alkane_{C_9}$

Table 11.13 *Comparison of results for scenario 1 and 2*

Scenario	1	2
GP^{Total} (U.S $/y)	23.63×10^6	24.07×10^6
$Alkane_{C_9}$ $Alkane_{C_9}$ production rate (t/y)	5645.74	2643.42
Alkane by-product production rate (t/y)	41354.26	22885.15
Alcohol production rate (t/y)	0.00	6908.48

in scenario 2 is lower compared to scenario 1, the GP^{Total} generated is higher. This is because the revenue obtained from the generation of by-products along with the main product $Alkane_{C_9}$ is included in the overall economic potential of the integrated biorefinery.

11.5 Conclusions

This chapter discusses the integration of product design with synthesis of integrated biorefinery by introducing a systematic two-stage optimisation approach. In the first stage, product design is done by CAMD technique. Signature-based molecular design techniques are utilised such that different classes of property prediction models such as GC models and TIs can be applied together to estimate the molecular structure from a set of target properties. In the second stage, the optimal conversion pathway is determined via superstructural mathematical optimisation approach. Note that the optimum conversion pathway based on different optimisation objective (e.g. highest product yield, highest economic performance, lowest environmental impact, etc.) can be determined by utilising the optimisation approach.

11.6 Future Opportunities

Future effort can be focused on improving the completeness of the methodology. Product stability analysis and verification can be investigated to improve the accuracy of the product design methodology. Furthermore, process design considerations such as safety, environmental and health consideration can be taken into consideration while developing a comprehensive product and process design. Moreover, in order to enhance the methodology, a detailed business model could be incorporated into the methodology. This is possible by investigating the effects and influence of corporate, business and engineering stakeholders in the overall product and process design.

Nomenclature

CI	connectivity index
GC	group contribution

QSPR/QSAR	quantitative structure–property/activity relationships
TI	topological index
χ	connectivity index
G	molecular sub-graph
h	height of signature
N_i	number of occurrence of first-order group of type *i*
M_j	number of occurrence of second-order group of type *j*
O_k	number of occurrence of third-order group of type *k*
C_i	contribution of the first-order group of type *i*
D_j	contribution of the second-order group of type *j*
E_k	contribution of the third-order group of type *k*
Ω_p	normalised property operator for target property *p*
v_p^{LL}	lower lower limit for target property *p*
v_p^{L}	lower limit for target property *p*
v_p^{LU}	lower upper limit for target property *p*
v_p^{UL}	upper lower limit for target property *p*
v_p^{U}	upper limit for target property *p*
v_p^{UU}	upper upper limit for target property *p*
V_p	value for target property *p*
λ_p	degree of satisfaction for target property *p*
λ_p^{r}	degree of satisfaction for property robustness for target property *p*
λ_p^{s}	degree of satisfaction for property superiority for target property *p*
λ_p^{*}	degree of satisfaction for target property *p* in two-phase approach
σ_p	standard deviation for property prediction model for target property *p*
B_b^{Bio}	flow rate of biomass feedstock *b* in t/y
F_{bq}^{I}	flow rate of bioresource *b* to pathway *q* in t/y
$F_{sq'}^{\text{II}}$	flow rate of intermediate *s* to pathway *q′* in t/y
R_{bqs}^{I}	conversion of bioresource *b* to intermediate *s*
$R_{sq's'}^{\text{II}}$	conversion of intermediate *s* to product *s′*
T_s^{Inter}	total production rate of intermediate *s* in t/y
$T_{s'}^{\text{Prod}}$	total production rate of product *s′* in t/y
GP^{Total}	total gross profit in US $/y
G_b^{Bio}	cost of biomass feedstock *b*
$G_{s'}^{\text{Prod}}$	cost of product *s′*
G_{bq}^{Cap}	capital cost for conversion of bioresource *b* through pathway *q*
$G_{sq'}^{\text{Cap}}$	capital cost for conversion of intermediate *s* through pathway *q′*
G_{bq}^{Opr}	operating cost for conversion of bioresource *b* through pathway *q*
$G_{sq'}^{\text{Opr}}$	operating cost for conversion of intermediate *s* through pathway *q′*
TAC	total annualised cost
TACC	total annualised capital cost
TAOC	total annualised operating cost
CRF	capital recovery factor

Appendix

!Stage 1: Design of optimal product;

```
!Abbreviations
sdR     =       Standard deviation
lR      =       Lower bound
uR      =       Upper bound
uu      =       Upper upper bound
ul      =       Upper lower bound
lu      =       Lower upper bound
ll      =       Lower lower bound
dss     =       Degree of satisfaction for property superiority
dsr     =       Degree of satisfaction for property robustness
R       =       Research octane number
A       =       Auto ignition temperature
HHV     =       Higher heating value
DV      =       Dynamic viscosity
LC      =       Toxicity;

!Objective function;
max=lamda;

!Constraints;
lamda   <       dssR;
lamda   <       dsrR;
lamda   <       dssHHV;
lamda   <       dsrHHV;
lamda   <       dssLC;
lamda   <       dsrLC;

!Property constraints;
!Research octane number (R);
R       =       103.6+0.231*((ron1*(x1))+(ron2*(x2))+(ron3*(x3))+(ron4*(x4)));
!Contribution of groups (ron);
ron1    =       −2.315;
ron2    =       −8.448;
ron3    =       −0.176;
ron4    =       11.94;
@free(ron1);@free(ron2);@free(ron3);

!Target property ranges;
sdR     =       5;
lR      =       95;
uR      =       108;
uuR     =       uR+sdR;
ulR     =       uR−sdR;
```

```
luR     =    lR+sdR;
llR     =    lR–sdR;
R       >    llR;
R       <    uuR;

!Degree of satisfaction;
dssR    =    ((R–llR)/(uuR–llR));
dsrR    <    ((uuR–R)/(uuR–ulR));
dsrR    <    ((R–llR)/(luR–llR));
dsrR    >    0;
dsrR    <    1;

!Auto ignition temperature (A);
A       =    780.42+26.78*((ait1*(x1))+(ait2*(x2))+(ait3*(x3))+(ait4*(x4)));

!Contribution of groups (ait);
ait1    =    –0.8516;
ait2    =    –1.4207;
ait3    =    0.0249;
ait4    =    2.3226;
@free(ait1);@free(ait2);

!Target property ranges;
sdA     =    28;
lA      =    600;
uA      =    800;
uuA     =    uA+sdA;
ulA     =    uA–sdA;
luA     =    lA+sdA;
llA     =    lA–sdA;
A       >    llA;
A       <    uuA;

!Higher heating value (HHV);
HHV     =    146.826+((hhvc1*(x1))+(hhvc2*(x2))+(hhvc3*(x3))+(hhvc4*(x4)));

!Contribution of groups (hhvc);
hhvc1   =    710.6822;
hhvc2   =    652.8408;
hhvc3   =    580.8447;
hhvc4   =    525.2059;

!Target property ranges;
sdHHV   =    26.96;
lHHV    =    5000;
uHHV    =    6600;
uuHHV   =    uHHV+sdHHV;
ulHHV   =    uHHV–sdHHV;
luHHV   =    lHHV+sdHHV;
llHHV   =    lHHV–sdHHV;
```

```
HHV     >       llHHV;
HHV     <       uuHHV;

!Degree of satisfaction;
dssHHV=         ((HHV−llHHV)/(uuHHV−llHHV));
dsrHHV<         ((uuHHV−HHV)/(uuHHV−ulHHV));
dsrHHV<         ((HHV−llHHV)/(luHHV−llHHV));
dsrHHV>0;
dsrHHV<1;

!Dynamic viscosity (DV);
DV      =       ((dvc1*(x1))+(dvc2*(x2))+(dvc3*(x3))+(dvc4*(x4)));

!Contribution of groups (dvc);
dvc1    =       −1.0278;
dvc2    =       0.2125;
dvc3    =       1.318;
dvc4    =       2.8147;
@free(dvc1);@free(DV);@free(llDV);
!Target property ranges;
sdDV    =       0.89;
lDV     =       0;
uDV     =       1.0986;
uuDV    =       1.3584;
ulDV    =       0.7467;
luDV    =       0.6366;
llDV    =       −2.2073;
DV      >       llDV;
DV      <       uuDV;

!logLC50 (LC);
LC      =       4.115−0.762*(lc1*x1+lc2*x2+lc3*x3+lc4*x4);

!TI values for each root of atomic signature;
lc1     =       0.5;
lc2     =       0.35355;
lc3     =       0.28868;
lc4     =       0.25;

!Target property ranges;
sdLC    =       0.193;
lLC     =       1;
uLC     =       1.9;
uuLC    =       uLC+sdLC;
ulLC    =       uLC−sdLC;
luLC    =       lLC+sdLC;
llLC    =       lLC−sdLC;
LC      >       llLC;
LC      <       uuLC;
```

```
!Degree of satisfaction;
dssLC  =     ((LC−llLC)/(uuLC−llLC));
dsrLC  =     ((uuLC−LC)/(uuLC−ulLC));
dsrLC  =     ((LC−llLC)/(luLC−llLC));
dsrLC  >     0;
dsrLC  <     1;

!Structural constraints;
(x1)+2*(x2)+3*(x3)+4*(x4)     =     2*(x1+x2+x3+x4−1);

!Integers constraints;
@GIN(x1);@GIN(x2);@GIN(x3);@GIN(x4);

!Positive constraints;
x1>0;x2>0;x3>0;x4>0;
end

!Stage2: Design of optimal conversion pathways;
!Abbreviations
X      = Composition,
A      = Conversion/Yield/Separation efficiency,
AGC    = Annualised capital cost,
ACO    = Annualised operating cost,
R      = Conversion
F      = Flow rate

!Objective function;
Max    =     Profit;

!Biomass Feedstock Flowrate Input (tonne/y);
B      =     50000;

! Biomass Composition Input (Please Input Data);
XL     =     0.29;
XC     =     0.39;
XHC    =     0.22;

! Conversion (or yield if there is no selectivity in the process);
R1     =     0.98;
R2     =     0.492;
R3     =     0.79; !Separation efficiency;
R4     =     0.97; !Separation efficiency;
R5     =     0.909;
R6     =     0.409;
R7     =     0.619;
R8     =     0.41;
```

```
R9      =       0.982;
R10     =       0.99;
R11     =       0.60;
R12     =       0.94;
R13     =       0.90;
R14     =       0.40;
R15     =       1.00;
R16     =       0.40;
R17     =       0.75;
R18     =       0.251;
R19     =       0.246;
R20     =       0.288;
R21     =       0.99;
R22     =       0.67;
R23     =       0.59;
R24     =       0.64;
R25     =       0.62;
R26     =       0.99;

!Annualised capital cost (per annual tonne);
AGCF1 =         19.64;
AGCF2 =         13.90;
AGCF3 =         40.68;
AGCF4 =         40.68;
AGCF5 =         63.46;
AGCF6 =         27.62;
AGCF7 =         40.62;
AGCF8 =         31.43;
AGCF9 =         30.22;
AGCF10 =        43.52;
AGCF11 =        45.45;
AGCF12 =        62.86;
AGCF13 =        86.43;
AGCF14 =        26.23;
AGCF15 =        15.11;
AGCF16 =        193.41;
AGCF17 =        181.93;
AGCF18 =        38.56;
AGCF19 =        41.10;
AGCF20 =        40.19;
AGCF21 =        40.68;
AGCF22 =        40.50;
AGCF23 =        37.47;
AGCF24 =        34.45;
AGCF25 =        45.94;
AGCF26 =        169.48;
```

```
!Operating cost (per annual tonne);
AGOF1 =      11.30;
AGOF2 =      7.97;
AGOF3 =      23.30;
AGOF4 =      23.30;
AGOF5 =      36.40;
AGOF6 =      15.80;
AGOF7 =      22.00;
AGOF8 =      18.00;
AGOF9 =      17.30;
AGOF10 =     24.90;
AGOF11 =     26.00;
AGOF12 =     36.00;
AGOF13 =     55.00;
AGOF14 =     15.00;
AGOF15 =     8.66;
AGOF16 =     111.00;
AGOF17 =     104.00;
AGOF18 =     22.10;
AGOF19 =     23.60;
AGOF20 =     23.00;
AGOF21 =     23.30;
AGOF22 =     23.20;
AGOF23 =     21.50;
AGOF24 =     19.70;
AGOF25 =     26.30;
AGOF26 =     98.20;

!Price (USD) of ton of feedstock or Products;
Gbiomass       = 170;
Gethane        = 424;
Gpropane       = 670;
Gbutane        = 900;
Gpentane       = 1120;
Ghexane        = 1300;
Gheptane       = 1500;
Goctane        = 1600;
Gnonane        = 1810;
Gdecane        = 1950;
Gmethanol      = 450;
Gethanol       = 650;
Gpropanol      = 890;
Gbutanol       = 1120;
Gpentanol      = 1480;
Gpentanediol   = 3000;
```

```
!Flowrates (Into each layer);
Tlcs1    =    (XC+XHC)*R1*F1+(XC+XHC)*R2*F2;
Tlcs2    =    XL*F1+XL*F2;
Tlcs     =    Tlcs1+Tlcs2;
Tlignin  =    R3*F3;
Tsugar   =    R4*F4;
Thmf     =    R5*F5;
Tf       =    R6*F6;
Tthfa    =    R9*F9;
Ts       =    R12*F12+R13*F13+R15*F15;
Tm       =    R14*F14;
Talc1    =    R7*F7+R8*F8;
Talc2    =    R10*F10+R11*F11;
Talc3    =    R18*F18+R19*F19+R20*F20;
Talc     =    Talc1+Talc2+Talc3;
Talk1    =    R16*F16+R17*F17;
Talk2    =    R22*F22+R23*F23+R24*F24;
Talk3    =    R25*F25;
Talk     =    Talk1+Talk2+Talk3;
Talk     =    F26;
Tac      =    R21*F21;

!Flowrates (Out from each layer);
B        =    F1+F2+F12+F13+F14;
F1       <=   B;
F2       <=   B;
F12      <=   B;
F13      <=   B;
F14      <=   B;
Tlcs2    =    F3;
F3       <=   Tlcs2;
Tlcs1    =    F4;
F4       <=   Tlcs1;
Tsugar   =    F5+F6+F7+F8;
F5       <=   Tsugar;
F6       <=   Tsugar;
F7       <=   Tsugar;
F8       <=   Tsugar;
Tf       =    F9;
F9       <=   Tf;
Tthfa    =    F10+F11;
F10      <=   Tthfa;
F11      <=   Tthfa;
Tm       =    F15;
F15      <=   Tm;
```

```
Ts      =    F16+F17+F18+F19+F20;
F16     <=   Ts;
F17     <=   Ts;
F18     <=   Ts;
F19     <=   Ts;
F20     <=   Ts;
Talc    =    F21+F22+F23+F24;
F21     <=   Talc;
F22     <=   Talc;
F23     <=   Talc;
F24     <=   Talc;
Tac     =    F25;
F25     <=   Tac;

!Production rates of alkanes;
ethane  =    R26*(R16*F16*0.16+R17*F17*0.23+R22*F22*0.103
             +R25*F25*0.213);
propane =    R26*(R16*F16*0.16+R17*F17*0.23+R23*F23*0.288);
butane  =    R26*(R16*F16*0.16+R17*F17*0.23+R23*F23*0.373);
pentane =    R26*(R16*F16*0.27+R17*F17*0.19+R24*F24*0.152);
hexane  =    R26*(R16*F16*0.27+R17*F17*0.19+R24*F24*0.055);
heptane =    R26*(R16*F16*0.27+R17*F17*0.19+R24*F24*0.056);
octane  =    R26*(R16*F16*0.27+R17*F17*0.19+R24*F24*0.042);
nonane  =    R26*(R16*F16*0.27+R17*F17*0.19);
decane  =    R26*(R16*F16*0.26+R17*F17*0.097);
Talk    >=   ethane+propane+butane+pentane+hexane+heptane+octane+nonane
             +decane;

!Production rates of alcohols;
methanol       =    R18*F18*0.026+R19*F19*0.039+R20*F20*0.207;
ethanol        =    R7*F7+R8*F8+R18*F18*0.614+R19*F19*0.561
                    +R20*F20*0.238;
propanol       =    R20*F20*0.141;
butanol        =    R20*F20*0.075;
pentanol       =    R10*F10*0.04+R11*F11*0.22;
pentanediol    =    R10*F10*0.95+R11*F11*0.51;
Talc           >=   methanol+ethanol+propanol+butanol+pentanol
                    +pentanediol;

!Revenue of alkanes;
EPethane       =    R16*F16*0.16*Gethane+R17*F17*0.23*Gethane+
                    R22*F22*0.103*Gethane+R25*F25*0.213*Gethane;
EPpropane      =    R16*F16*0.16*Gpropane+R17*F17*0.23*Gpropane
                    +R23*F23*0.288*Gpropane;
EPbutane       =    R16*F16*0.16*Gbutane+R17*F17*0.23*Gbutane
                    +R23*F23*0.373*Gbutane;
```

```
EPpentane       =  R16*F16*0.27*Gpentane+R17*F17*0.19*Gpentane+
                   R24*F24*0.152*Gpentane;
EPhexane        =  R16*F16*0.27*Ghexane+R17*F17*0.19*Ghexane
                   +R24*F24*0.055*Ghexane;
EPheptane       =  R16*F16*0.27*Gheptane+R17*F17*0.19*Gheptane
                   +R24*F24*0.056*Gheptane;
EPoctane        =  R16*F16*0.27*Goctane+R17*F17*0.19*Goctane
                   +R24*F24*0.042*Goctane;
EPnonane        =  R16*F16*0.27*Gnonane+R17*F17*0.19*Gnonane;
EPdecane        =  R16*F16*0.26*Gdecane+R17*F17*0.097*Gdecane;
EPalk           =  EPethane+EPpropane+EPbutane+EPpentane+EPhexane
                   +EPheptane+EPoctane+EPnonane+EPdecane;

!Revenue of alcohols;
EPmethanol      =  R18*F18*0.026*Gmethanol+R19*F19*0.039*Gmethanol
                   +R20*F20*0.207*Gmethanol;
EPethanol       =  R7*F7*Gethanol+R8*F8*Gethanol+R18*F18*0.614*Gethanol
                   +R19*F19*0.561*Gethanol+R20*F20*0.238*Gethanol;
EPpropanol      =  R20*F20*0.141*Gpropanol;
EPbutanol       =  R20*F20*0.075*Gbutanol;
EPpentanol      =  R10*F10*0.04*Gpentanol+R11*F11*0.22*Gpentanol;
EPpentanediol   =  R10*F10*0.95*Gpentanediol+R11*F11*0.51*Gpentanediol;
EPalc           =  EPmethanol+EPethanol+EPpropanol+EPbutanol+EPpentanol
                   +EPpentanediol;

!Total revenue;
Revenue         =  EPalc+EPalk;

!Cost for biomass;
CBiomass        =  B*Gbiomass;

!Total capital cost;
TACC  =  F1*AGCF1+F2*AGCF2+F3*AGCF3+F4*AGCF4+F5*AGCF5
         +F6*AGCF6+F7*AGCF7+F8*AGCF8+F9*AGCF9+F10*AGCF10
         +F11*AGCF11+F12*AGCF12+F13*AGCF13+F14*AGCF14
         +F15*AGCF15+F16*AGCF16+F17*AGCF17+F18*AGCF18
         +F19*AGCF19+F20*AGCF20+F21*AGCF21+F22*AGCF22
         +F23*AGCF23+F24*AGCF24+F25*AGCF25+F26*AGCF26;

!Total operating cost;
TAOC  =  F1*AGOF1+F2*AGOF2+F3*AGOF3+F4*AGOF4+F5*AGOF5
         +F6*AGOF6+F7*AGOF7+F8*AGOF8+F9*AGOF9+F10*AGOF10
         +F11*AGOF11+F12*AGOF12+F13*AGOF13+F14*AGOF14
         +F15*AGOF15+F16*AGOF16+F17*AGOF17+F18*AGOF18
         +F19*AGOF19+F20*AGOF20+F21*AGOF21+F22*AGOF22
         +F23*AGOF23+F24*AGOF24+F25*AGOF25+F26*AGOF26;
```

```
!Profit;
Profit    =        Revenue–CBiomass–TACC–TAOC;
@free(profit);
<!--<![CDATA[--><![CDATA

end
```

References

Achenie, L.E.K., Gani, R., Venkatasubramanian, V., 2003. *Computer Aided Molecular Design: Theory and Practice*. Elsevier, Amsterdam.

Albahri, T.A., 2003a. Structural Group Contribution Method for Predicting the Octane Number of Pure Hydrocarbon Liquids. *Ind. Eng. Chem. Res.* **42**, 657–662.

Albahri, T.A., 2003b. Flammability characteristics of pure hydrocarbons. *Chem. Eng. Sci.* **58**, 3629–3641.

Ambrose, D., 1978. *Correlation and Estimation of Vapour–Liquid Critical Properties. I: Critical Temperatures of Organic Compounds, Volume 1*. National Physical Laboratory, Middlesex.

Andiappan, V., Ko, A.S.Y., Ng, L.Y., Ng, R.T.L., Chemmangattuvalappil, N.G., Ng, D.K.S., 2015. Synthesis of sustainable integrated biorefinery via reaction pathway synthesis: economic, incremental enviromental burden and energy assessment with multiobjective optimization. *AIChE J.* **61**, 132–146.

Bao, B., Ng, D.K.S., El-Halwagi, M.M., Tay, D.H.S., 2009. Synthesis of Technology Pathways for an Integrated Biorefinery, in: 2009 AIChE Annual Meeting. Nashville, TN, USA.

Bellman, R.E., Zadeh, L.A., 1970. Decision-making in a fuzzy environment. *Manage. Sci.* **17**, 141–164.

Camarda, K.V., Maranas, C.D., 1999. Optimization in polymer design using connectivity indices. *Ind. Eng. Chem. Res.* **38**, 1884–1892.

Chemmangattuvalappil, N.G., Eden, M.R., 2013. A novel methodology for property-based molecular design using multiple topological indices. *Ind. Eng. Chem. Res.* **52**, 7090–7103.

Chemmangattuvalappil, N.G., Solvason, C.C., Bommareddy, S., Eden, M.R., 2010. Reverse problem formulation approach to molecular design using property operators based on signature descriptors. *Comput. Chem. Eng.* **34**, 2062–2071.

Churi, N., Achenie, L.E.K., 1996. Novel mathematical programming model for computer aided molecular design. *Ind. Eng. Chem. Res.* **35**, 3788–3794.

Clark, P.A., Westerberg, A.W., 1983. Optimization for design problems having more than one objective. *Comput. Chem. Eng.* **7**, 259–278.

Constantinou, L., Gani, R., 1994. New group contribution method for estimating properties of pure compounds. *AIChE J.* **40**, 1697–1710.

Conte, E., Martinho, A., Matos, H.A., Gani, R., 2008. Combined group-contribution and atom connectivity index-based methods for estimation of surface tension and viscosity. *Ind. Eng. Chem. Res.* **47**, 7940–7954.

Cussler, E.L., Moggridge, G.D., 2001. *Chemical Product Design*. Cambridge University Press, New York.

Deckro, R.F., Hebert, J.E., 1989. Resource constrained project crashing. *Omega* **17**, 69–79.

Deporter, E.L., Ellis, K.P., 1990. Optimization of project networks with goal programming and fuzzy linear programming. *Comput. Ind. Eng.* **19**, 500–504.

Dubois, D., Fargier, H., Prade, H., 1996. Refinements of the maximin approach to decision-making in a fuzzy environment. *Fuzzy Sets Syst.* **81**, 103–122.

El-halwagi, A.M., Rosas, C., Ponce-Ortega, J.M., Jiménez-Gutiérrez, A., Mannan, M.S., El-halwagi, M.M., 2013. Multiobjective optimization of biorefineries with economic and safety objectives. *AIChE J.* **59**, 2427–2434.

Estrada, E., 1995. Edge adjacency relationships in molecular graphs containing heteroatoms: a new topological index related to molar volume. *J. Chem. Inf. Comput. Sci.* **35**, 701–707.

Faulon, J.-L., Churchwell, C.J., Visco, D.P., 2003. The signature molecular descriptor. 2. Enumerating molecules from their extended valence sequences. *J. Chem. Inf. Comput. Sci.* **43**, 721–734.

Fernando, S., Adhikari, S., Chandrapal, C., Murali, N., 2006. Biorefineries: current status, challenges, and future direction. *Energy Fuels* **20**, 1727–1737.

Fishburn, P.C., 1967. Additive utilities with incomplete product sets: application to priorities and assignments. *Oper. Res.* **15**, 537–542.

Gani, R., Nielsen, B., Fredenslund, A., 1991. A group contribution approach to computer-aided molecular design. *AIChE J.* **37**, 1318–1332.

Gani, R., O'Connell, J.P., 2001. Properties and CAPE: from present uses to future challenges. *Comput. Chem. Eng.* **25**, 3–14.

Gani, R., Pistikopoulos, E.N., 2002. Property modelling and simulation for product and process design. *Fluid Phase Equilib.* **194–197**, 43–59.

Gebreslassie, B.H., Slivinsky, M., Wang, B., You, F., 2013. Life cycle optimization for sustainable design and operations of hydrocarbon biorefinery via fast pyrolysis, hydrotreating and hydrocracking. *Comput. Chem. Eng.* **50**, 71–91.

Gong, J., You, F., 2014. Optimal design and synthesis of algal biorefinery processes for biological carbon sequestration and utilization with zero direct greenhouse gas emissions: MINLP model and global optimization algorithm. *Ind. Eng. Chem. Res.* **53**, 1563–1579.

Guu, S., Wu, Y., 1999. Two-phase approach for solving the fuzzy linear programming problems. *Fuzzy Sets Syst.* **107**, 191–195.

Halasz, L., Povoden, G., Narodoslawsky, M., 2005. Sustainable processes synthesis for renewable resources. *Resour. Conserv. Recycl.* **44**, 293–307.

Harper, P.M., Gani, R., 2000. A multi-step and multi-level approach for computer aided molecular design. *Comput. Chem. Eng.* **24**, 677–683.

Harper, P.M., Gani, R., Kolar, P., Ishikawa, T., 1999. Computer-aided molecular design with combined molecular modeling and group contribution. *Fluid Phase Equilib.* **158–160**, 337–347.

Jiménez, M., Bilbao, A., 2009. Pareto-optimal solutions in fuzzy multi-objective linear programming. *Fuzzy Sets Syst.* **160**, 2714–2721.

Jurić, A., Gagro, M., Nikolić, S., Trinajstić, N., 1992. Molecular topological index: an application in the QSAR study of toxicity of alcohols. *J. Math. Chem.* **11**, 179–186.

Karunanithi, A.T., Achenie, L.E.K., Gani, R., 2005. A new decomposition-based computer-aided molecular/mixture design methodology for the design of optimal solvents and solvent mixtures. *Ind. Eng. Chem. Res.* **44**, 4785–4797.

Kier, L.B., 1985. A shape index from molecular graphs. *Quant. Struct. Relationships* **4**, 109–116.

Kier, L.B., Hall, L.H., 1986. *Molecular Connectivity in Structure–Activity Analysis*. Research Studies Press, Herefordshire.

Kim, I.Y., de Weck, O.L., 2006. Adaptive weighted sum method for multiobjective optimization: a new method for Pareto front generation. *Struct. Multidiscip. Optim.* **31**, 105–116.

Kokossis, A.C., Yang, A., 2010. On the use of systems technologies and a systematic approach for the synthesis and the design of future biorefineries. *Comput. Chem. Eng.* **34**, 1397–1405.

Kontogeorgis, G.M., Gani, R., 2004. Chapter 1: Introduction to computer aided property estimation. *Comput. Aided Chem. Eng.* **19**, 3–26.

Maranas, C.D., 1997. Optimal molecular design under property prediction uncertainty. *AIChE J.* **43**, 1250–1264.

Marrero, J., Gani, R., 2001. Group-contribution based estimation of pure component properties. *Fluid Phase Equilib.* **183–184**, 183–208.

Martinez-Hernandez, E., Sadhukhan, J., Campbell, G.M., 2013. Integration of bioethanol as an in-process material in biorefineries using mass pinch analysis. *Appl. Energy* **104**, 517–526.

Murillo-Alvarado, P.E., Ponce-Ortega, J.M., Serna-González, M., Castro-Montoya, A.J., El-Halwagi, M.M., 2013. Optimization of pathways for biorefineries involving the selection of feedstocks, products, and processing steps. *Ind. Eng. Chem. Res.* **52**, 5177–5190.

Ng, D.K.S., Pham, V., El-Halwagi, M.M., Jiménez-Gutiérrez, A., Spriggs, H.D., 2009. A Hierarchical Approach to the Synthesis and Analysis of Integrated Biorefineries, in: Design for Energy and the Environment. Proceedings of Seventh International Conference on Foundations of Computer-Aided Process Design. CRC Press, Boca Raton, FL.

Ng, R.T.L., Hassim, M.H., Ng, D.K.S., 2013. Process synthesis and optimization of a sustainable integrated biorefinery via fuzzy optimization. *AIChE J.* **59**, 4212–4227.
Ng, R.T.L., Ng, D.K.S., 2013. Systematic approach for synthesis of integrated palm oil processing complex. Part 1: Single owner. *Ind. Eng. Chem. Res.* **52**, 10206–10220.
Ng, L.Y., Chemmangattuvalappil, N., Ng, D.K.S., 2014. A multiobjective optimization-based approach for optimal chemical product design. *Ind. Eng. Chem. Res.* **53**, 17429–7444.
Odele, O., Macchietto, S., 1993. Computer aided molecular design: a novel method for optimal solvent selection. *Fluid Phase Equilib.* **82**, 47–54.
Pham, V., El-halwagi, M., 2012. Process synthesis and optimization of biorefinery configurations. *AIChE J.* **58**, 1212–1221.
Ponce-Ortega, J.M., Pham, V., El-Halwagi, M.M., El-Baz, A.A., 2012. A disjunctive programming formulation for the optimal design of biorefinery configurations. *Ind. Eng. Chem. Res.* **51**, 3381–3400.
Randić, M., 1975. Characterization of molecular branching. *J. Am. Chem. Soc.* **97**, 6609–6615.
Randić, M., Mihalić, Z., Nikolić, S., Trinajstić, N., 1994. Graphical bond orders: novel structural descriptors. *J. Chem. Inf. Comput. Sci.* **34**, 403–409.
Sahinidis, N.V., Tawarmalani, M., Yu, M., 2003. Design of alternative refrigerants via global optimization. *AIChE J.* **49**, 1761–1775.
Sammons, N., Eden, M., Yuan, W., Cullinan, H., Aksoy, B., 2007. A flexible framework for optimal biorefinery product allocation. *Environ. Prog.* **26**, 349–354.
Sammons, N.E., Yuan, W., Eden, M.R., Aksoy, B., Cullinan, H.T., 2008. Optimal biorefinery product allocation by combining process and economic modeling. *Chem. Eng. Res. Des.* **86**, 800–808.
Samudra, A., Sahinidis, N.V., 2013. Design of heat-transfer media components for retail food refrigeration. *Ind. Eng. Chem. Res.* **52**, 8518–8526.
Santibañez-Aguilar, J.E., González-Campos, J.B., Ponce-Ortega, J.M., Serna-González, M., El-Halwagi, M.M., 2011. Optimal planning of a biomass conversion system considering economic and environmental aspects. *Ind. Eng. Chem. Res.* **50**, 8558–8570.
Shelley, M.D., El-Halwagi, M.M., 2000. Component-less design of recovery and allocation systems: a functionality-based clustering approach. *Comput. Chem. Eng.* **24**, 2081–2091.
Siddhaye, S., Camarda, K., Southard, M., Topp, E., 2004. Pharmaceutical product design using combinatorial optimization. *Comput. Chem. Eng.* **28**, 425–434.
Svensson, E., Harvey, S., 2011. Pinch analysis of a partly integrated pulp and paper mill, in: Moshfegh, B. (Ed.), *World Renewable Energy Congress*. Linköping Electronic Conference Proceedings, Linköping, Sweden, pp. 1521–1528.
Tay, D.H.S., Kheireddine, H., Ng, D.K.S., El-Halwagi, M.M., 2010. Synthesis of an integrated biorefinery via the C–H–O ternary diagram. *Chem. Eng. Trans.* **21**, 1411–1416.
Trinajstić, N., 1992. *Chemical Graph Theory*. CRC Press, Boca Raton, FL.
Venkatasubramanian, V., Chan, K., Caruthers, J.M., 1994. Computer-aided molecular design using genetic algorithms. *Comput. Chem. Eng.* **18**, 833–844.
Visco, D.P., Pophale, R.S., Rintoul, M.D., Faulon, J.-L., 2002. Developing a methodology for an inverse quantitative structure–activity relationship using the signature molecular descriptor. *J. Mol. Graph. Model.* **20**, 429–438.
Wang, B., Gebreslassie, B.H., You, F., 2013. Sustainable design and synthesis of hydrocarbon biorefinery via gasification pathway: integrated life cycle assessment and technoeconomic analysis with multiobjective superstructure optimization. *Comput. Chem. Eng.* **52**, 55–76.
Wiener, H., 1947. Structural determination of paraffin boiling points. *J. Am. Chem. Soc.* **69**, 17–20.
Wilson, R.J., 1986. *Introduction to Graph Theory*, 4th ed. Pearson Education Limited, Harlow.
Yunus, N.A., 2014. *Systematic Methodology for Design of Tailor-Made Blended Products: Fuels and Other Blended Products*. Technical University of Denmark, Kongens Lyngby.
Zadeh, L.A., 1965. Fuzzy sets. *Inf. Control* **8**, 338–353.
Zimmermann, H.-J., 1976. Description and optimization of fuzzy systems. *Int. J. Gen. Syst.* **2**, 209–215.
Zimmermann, H.-J., 1978. Fuzzy programming and linear programming with several objective functions. *Fuzzy Sets Syst.* **1**, 45–55.
Zimmermann, H.-J., 2001. *Fuzzy Set Theory – and Its Applications*. Springer Science & Business Media, Norwell, MA.
Zondervan, E., Nawaz, M., de Haan, A.B., Woodley, J.M., Gani, R., 2011. Optimal design of a multiproduct biorefinery system. *Comput. Chem. Eng.* **35**, 1752–1766.

12

The Role of Process Integration in Reviewing and Comparing Biorefinery Processing Routes: The Case of Xylitol

Aikaterini D. Mountraki, Konstantinos R. Koutsospyros, and Antonis C. Kokossis

School of Chemical Engineering, National Technical University of Athens, Athens, Greece

12.1 Introduction

The effective utilization of biomass is essential for setting up sustainable biorefineries. There are numerous options for producing fuels and chemicals from biomass, and the best ones are yet to be selected. In the conventional petrochemical industry, both the supply chain and the process design have been optimized through thorough research, but for the bio-based processes, most of the research is still underway. There is a necessity to screen the different types of biomass feedstock, reaction yields, technologies, and target products along with environmental and economic objectives. To address these challenges, a systematic approach has been developed to screen biomass processing options and to identify the optimal production paths (Kokossis & Yang, 2010; Kokossis et al., 2010). Each production path starts from a specific type of biomass feedstock, followed by the choice of process treatment and ending up with a final product. Considering the great number of options, screening analysis can classify each route according to the criteria that are set (economical, environmental, social). Energy consumption clearly affects all three criteria, and therefore, heat integration emerges with an apparent importance in the process design. Thus, a

Process Design Strategies for Biomass Conversion Systems, First Edition. Edited by Denny K. S. Ng, Raymond R. Tan, Dominic C. Y. Foo, and Mahmoud M. El-Halwagi.
© 2016 John Wiley & Sons, Ltd. Published 2016 by John Wiley & Sons, Ltd.

question arises whether the heat integration can affect the results of this screening process. In order to study the effect of energy integration on the selection rating of a process among the rest of the competition, the case of xylitol production has been selected. Xylitol seems to cope well with the economic criteria set by the screening analysis, but in order to make the most efficient design, a more in-depth economical analysis is required that will take into account several other factors.

12.2 Motivating Example

The growing demand for sugar-free and low-calorie food products has attracted the attention of xylitol production from hemicelluloses (Pal et al., 2013; Fatehi et al., 2014; Lima et al., 2014). Xylitol is a substance of great commercial value as a food additive, with a lot of health benefits (Wisniak et al., 1974; Parajó et al., 1998a; Aranda-Barradas et al., 2010). Its share in the global market is expected to reach 242,000 metric tons valued at just above US$1 billion by 2020 (Research and Markets, 2014), making it an important coproduct in biorefinery. Furthermore, xylitol valorizes the C5 sugar syrup of hydrolyzed biomass, and as a result, it is not competitive to the production of bioethanol from C6 sugars (Cheng et al., 2014). In the case studied, the production path for xylitol product is based on a real-life biorefinery (Mountraki et al., 2011) which valorizes the biomass feedstock into three main products (cellulose, hemicellulose, and lignin) through the organosolv technology developed by CIMV. The CIMV Process™ has its origins in the pulp and paper industry and is applicable to a wide range of feedstock including cereal straws, sugarcane bagasse, sweet sorghum bagasse, and hardwood. Its major advantage, which is considered a breakthrough, is that it allows separation without degradation of celluloses, hemicelluloses, and lignins, while silica imposes no obstacles and could actually add value to the overall process (Delmas, 2008). The fully developed biorefinery offers paths to an impressive list of chemicals, biofuels, food and feed ingredients, fibers, and energy products that contribute to GHG emission reduction (Figure 12.1).

In order to evaluate the most remunerative paths for incorporation to the biorefinery, we need to examine the energy requirements of the selected pretreatment process (CIMV Process) as well as the energy requirements of each candidate process. Then, we also need to examine their integration opportunities with the rest of the biorefinery. The grand composite curve (GCC) of CIMV Process (Figure 12.2) reveals a threshold problem since a single utility is required—in this case hot utilities. This fact implies that the processes which will serve as heat sink (those who mainly have heating requirements) will have an advantage over the processes that will serve as heat source (those who mainly have cooling requirements).

12.3 The Three-Layer Approach

The design of a biorefinery requires the selection of the feedstock type and the valorization processes while keeping the cost low. All these options introduce the need for a systematic approach for the design of the future biorefinery against "haphazard" or "shortcut" choices. The role of a systems approach is to screen chemistries and new processes and to select products and integration scenarios in order to reach a conclusion

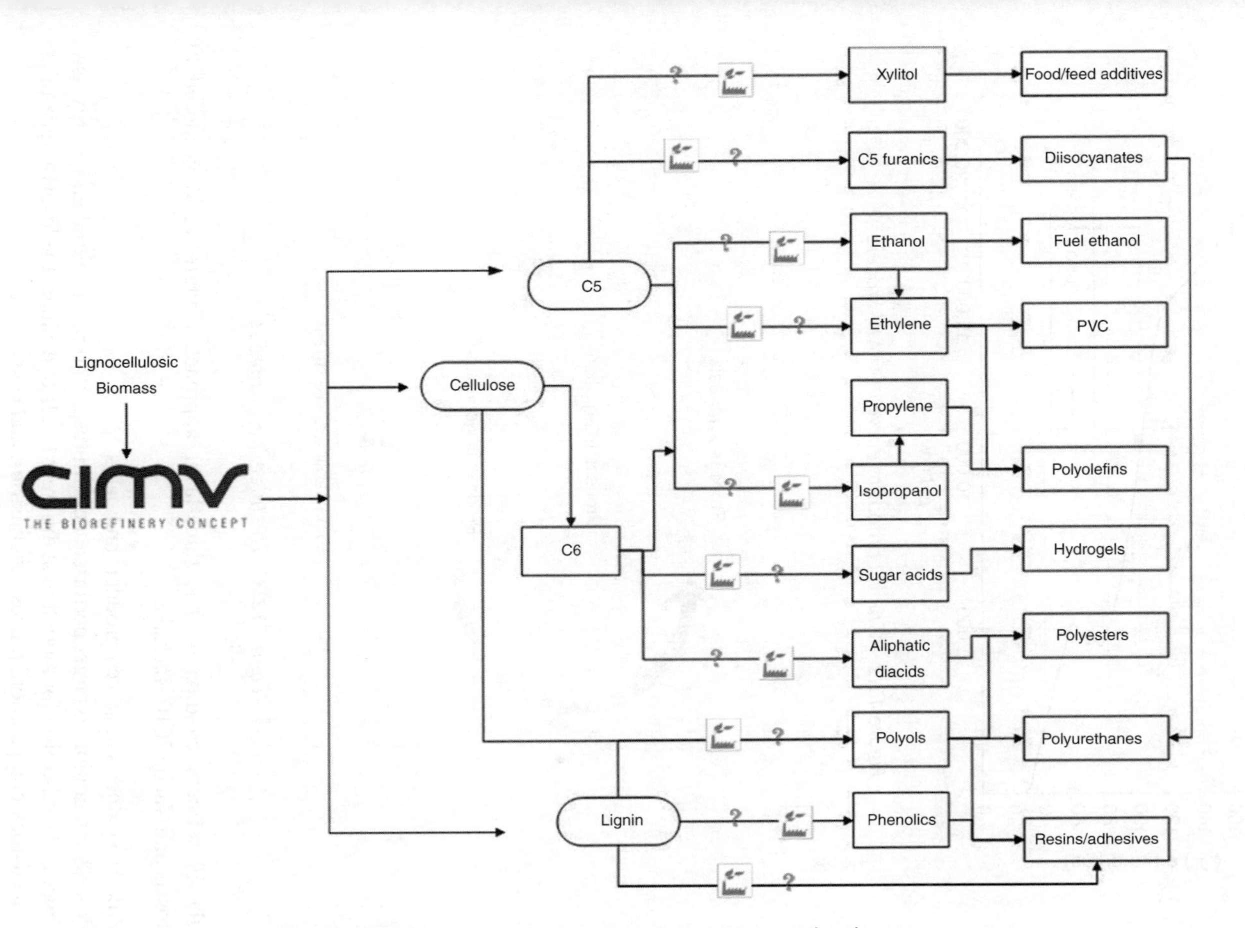

Figure 12.1 *Organosolv value chain (CIMV technology)*

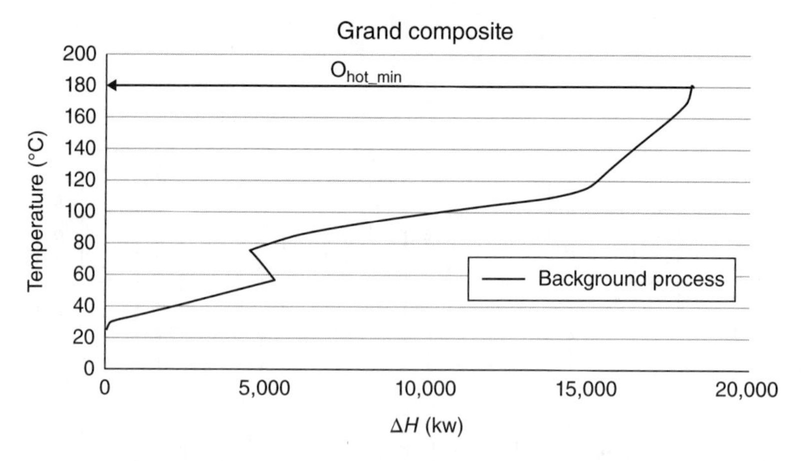

Figure 12.2 Grand composite curve of CIMV Process^{TM}

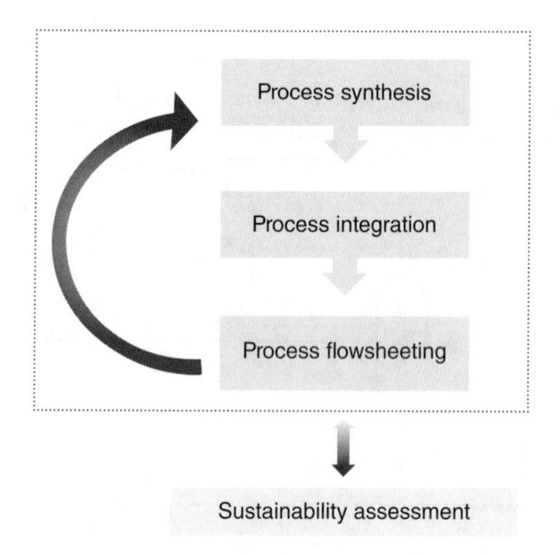

Figure 12.3 The three-layer approach

on the biorefinery candidates. The three-layer approach (Figure 12.3) proposed by Kokossis and Yand (2010) includes:

1. Process synthesis to screen products and paths
2. Process integration to target performance with respect to raw material and energy use
3. Process flowsheeting for modeling at different levels to evaluate the level of uncertainty in feedstock composition, process chemistries, and costs

First, we have to define the problem by developing a process flowsheet. Then, the flowsheets developed can be used to produce scenarios for different feedstock, different products, and detailed costing for CAPEX and OPEX at different scales. The bottom layer can provide every piece of information required by the first layer. Process synthesis is applied to search for the industrial chemistries and product portfolios, while process integration aims to the minimization of energy and raw materials requirements. The analysis of the second layer may propose improvements to the process flowsheeting, which in return triggers a new turnover. Inner loops feed information back to the experimental groups, pointing to the products worthy of study and further targeting conversions critical to achieve, while an outer loop assesses LCA and the overall sustainability.

Tsakalova and Kokossis (2013) studied the general problem of selecting paths and products from multiple choices and options for the synthesis problem of the aforementioned biorefinery. Their synthesis approach has been applied to select products and paths for:

1. The complete set of products at nominal prices and maximum demands
2. Repeated versions of case (1) under constraints on demands, prices, and processing parameters

The synthesis problem assesses the value of these chemistries from a holistic point of view, ranking and valorizing the chemical products. An impressive observation that derives from their results is that high market price of the product does guarantee a high ranking in the final listing, while the process of xylitol seems to be profitable in all of their case studies.

12.4 Production Paths to Xylitol

In our effort to evaluate the impact of energy integration on the rating of a process among the competition, we compare different production paths to xylitol before and after the heat integration analysis. To develop process flowsheets for xylitol production, we have cooperated with experimental teams. The production of xylitol in industry follows two different process paths: (i) the catalytic process and (ii) the biotechnological process. Biotechnological production of xylitol is a batch or semibatch process. Yeasts of *Candida* are widely preferred because of their high selectivity and their high yields (Parajó et al., 1998a). The common operating conditions in the bioreactor are 30°C and 1 atm under microaerobic conditions (0.4–1 vvm) with slightly acidic or neutral pH (pH 5–7). Total xylose consumption can reach up to 95%, while the residence time in the bioreactor ranges from 35 to over 100 h (Nigam & Singh, 1995; Parajó et al., 1998a). Catalytic hydrogenation of xylose to xylitol takes place at high pressure (40–70 bars) and high temperature (80–140°C). The reactor used is a batch, and the residence time is approximately 2–3 h (Mikkola et al., 1999). The most widely used catalyst is Raney nickel. However, neither of these processes is viable at a biorefinery investment unless integrated, as they show great energy demands for heating and cooling, mainly for evaporation and crystallization. Thus, important decisions should be made for the energy exploitation of the steam produced. The water input requirements are also high, and therefore, all the recycling opportunities should be examined. The evaluation and integration with the upstream and parallel processes in the plant should be considered as another necessity, since the plethora of opportunities for steam recycling that may arise could provide better energy exploitation solutions in comparison with individual integration. Furthermore, in-depth economic evaluation should be held for each scenario in order to make better decisions, including the most feasible processes in the biorefinery. This analysis should be made for every possible process, assessing the scope for individual integration as well as integration with upstream and parallel processes.

Both catalytic and biotechnological processes consist of three major sections. They include:

1. Pretreatment section
2. Reaction–condensation section
3. Crystallization section

The overall mass balances for the two processes are presented in Figures 12.4 and 12.5.

The thermodynamic model used is the nonrandom two liquid (NRTL), which can be used to describe vapor–liquid and liquid–liquid equilibrium of strongly nonideal solutions. The NRTL model can handle any combination of polar and nonpolar compounds, up to very strong nonideality. In addition, many parameters for xylitol pure component were not available in the databanks of Aspen Plus and had to be acquired from the literature and from regression of experimental data (Table 12.1).

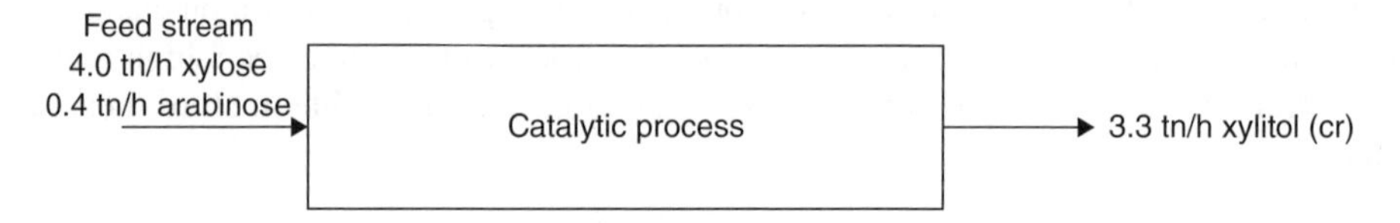

Figure 12.4 *Conceptual flows for the catalytic production*

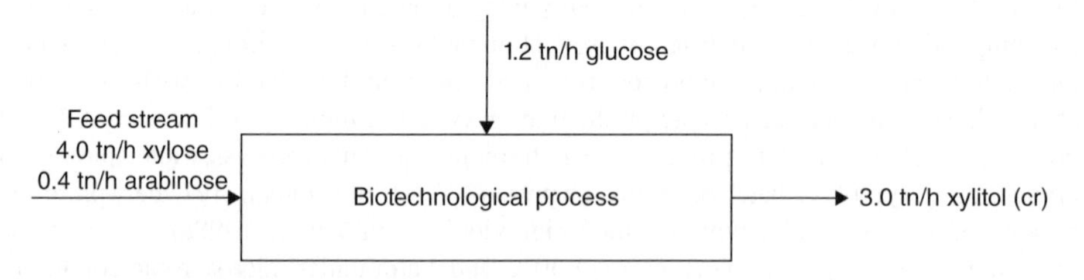

Figure 12.5 *Conceptual flows for the biotechnological production*

Table 12.1 *Parameters imported by user for xylitol component*

Parameters	Component
MW	152,150
DHFORM (kJ/kmol)	−1,219,300,000
DGFORM (kcal/mol)	−189,200
DHVLB (kJ/kmol)	37,400,000
DHSFRM (kJ/kmol)	161,000,000
VC (cc/mol)	498,722
OMEGA	1,216
ZC	0,467
FREEZEPT (C)	92,880
TB (C)	366,419
TC (C)	595,850

12.4.1 Catalytic Process

Figure 12.6 presents the process flow diagram of the catalytic production of xylitol. Three main sections are (i) pretreatment, (ii) reaction–condensation, and (iii) crystallization. The pretreatment section deals with the pH adjustment and the removal of the impurities contained in the feed stream, soluble and insoluble solids, and remnants of the hydrolysis of biomass. But the main goal is the removal of the alkalis, whose presence catalyzes the conversion of xylose to xylonic acid (Cannizzaro mechanism). The first step of the downstream processes (unit P1) involves a series of treatment by ionic resins, activated carbon column, and chromatographic separations. Then, the pH is adjusted (unit R1) in a range of 5–6 through the neutralization of acids (mainly acetic acid) by 1 M solution of calcium hydroxide ($Ca(OH)_2$):

$$2C_2H_4O_2 + 2Ca(OH)_2 \leftrightarrow 2H_2O + Ca(CH_3COO)_2 \tag{12.1}$$

The salt calcium propionate formed has also to be removed (unit P2). Before entering the reaction section, the feed is mixed with part of the gas hydrogen feed (at 87°C) and preheated in a column or a tank (unit E3) at 95°C and 1 atm for 15 min. This ensures better mixing and more secure preheating of the mixed feed, really close to the reaction temperature. Thereby, higher yields and better control are achieved. The rest of the hydrogen gas is preheated at 95°C and furnished directly in the reactor (unit R2). The reactor operates at 100°C and 50 bar, and the xylose concentration in the feed stream must be 40–60%, so that the highest yields and selectivity are achieved. Higher temperatures favor the by-product reactions and the degradation of xylose to furfural. The main reaction that takes place is the hydrogenation of xylose to xylitol (12.2) by 95%. The rest of the xylose is turned into xylulose (12.3), which is further converted to arabinitol (12.4) by (at a rate of) 90% (Mikkola et al., 1999, 2000; Mikkola & Salmi, 2001):

$$C_5H_{10}O_5 + H_2 \leftrightarrow C_5H_{12}O_5 \tag{12.2}$$

$$C_5H_{10}O_5 \leftrightarrow C_5H_{10}O_5 \tag{12.3}$$

$$C_5H_{10}O_5 + H_2 \leftrightarrow C_5H_{12}O_5 \tag{12.4}$$

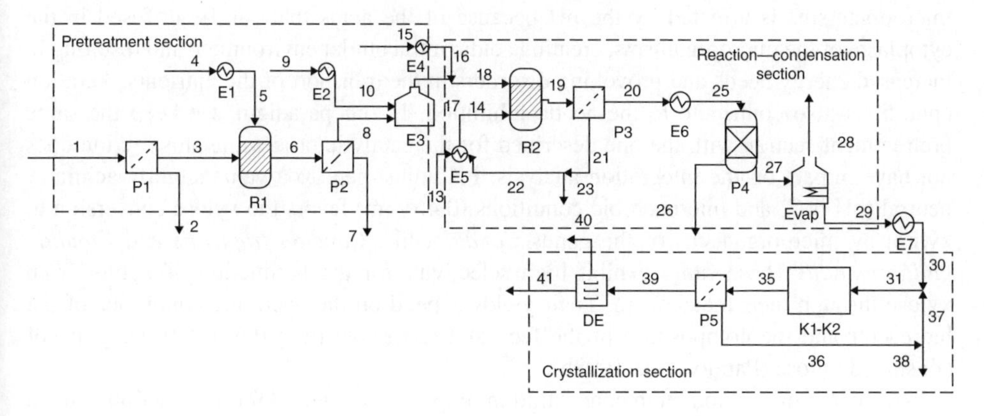

Figure 12.6 *Process flow diagram of the catalytic process*

The catalyst is introduced in a mass ratio of 5% of xylose. After the end of the reaction, 95% of the catalyst used is recycled, reactivated through alcoholic washing (mainly ethanol), and resupplied after the stabilization of the operating conditions to the reactor. The catalyst Raney nickel must not come into contact with ambient air and must remain in an inert environment because it is strongly oxidized with a high risk of fire.

The product stream is cooled down to 80°C and purified through ion-exchange membrane treatment and activated carbon column at pH 6–7 for 60 min, so that all substances that may contaminate the final product can be removed, in order to qualify for food additive. This purification treatment is simultaneously a discoloration treatment, because these substances color the final xylitol crystal and alter its characteristics. Discoloration is followed by an evaporation step before the final crystallization of xylitol. The desired concentration of xylitol after the evaporator ranges from 650 to 900 g/l and strongly affects the yields of crystallization (Misra et al., 2011). During the whole process, the temperature should not exceed the 105°C in order to avoid the degradation of xylitol and xylose. For this reason, the evaporator operates at vacuum pressure.

In the crystallization section, xylitol is crystallized in two steps. First, it enters the crystallizer where under continuous gentle stirring it is cooled down to 8°C. Due to the temperature decrease, the solubility of xylitol in the main solvent is decreased and finally separated from the liquid phase in the form of crystals. Commercial xylitol crystals may also be added (1 g per liter of feed) in order to act as crystallization nuclei. Then, after 24 h, the temperature is further decreased at −20°C for another 72 h. This two-step crystallization process may yield up to 70%, assisted by the crystallization nuclei and the recycling of noncrystallized product (Guex et al., 1981; Sampaio et al., 2006; Misra et al., 2011). For our case, 0.81 kg xylitol crystals/kg xylose is achieved. Data on the solubility of xylitol can be found in literature (Wang et al., 2007, 2013; Martínez et al., 2009). After the crystallization, the xylitol crystals are purified by centrifugation or sedimentation from the mother liquor. The amount of remaining impurities is significantly decreased in the mother liquor by fractionation on parallel ion-exchange columns. The purified crystallized product is finally dried.

12.4.2 Biotechnological Process

In the biotechnological process, the feed stream also needs to be pretreated. The activity of microorganisms is affected by the pH because of the acids that can be diffused in the cytoplasm of the microorganisms, creating acidic intracellular environment and resulting in increased energy needs and provoking disorders in the transport of the nutrients. Aeration conditions also contribute to the acetic inhibition. In our paradigm, we keep the same pretreatment section with the one described for the catalytic process, as this section does not have impact on the integration analysis. PH adjustment to a value slightly acidic or neutral (pH 6–7) and microaerobic conditions (0.46 vvm) favors the xylose conversion to xylitol by microorganisms of the genus *Candida*, like *Candida tropicalis* and *Candida guilliermondii*. These fungi exhibit high selectivity for the production of xylitol from xylose through their metabolism. Their yields depend on the operating conditions of the bioreactor and the composition of the feed and can range from 0.6 to 0.9 per gram of consumed xylose (Parajó et al., 1998b).

In the catalytic method, high concentration of xylose is required for the reaction. On the contrary, the biotechnological method works with highly diluted solutions. The optimal

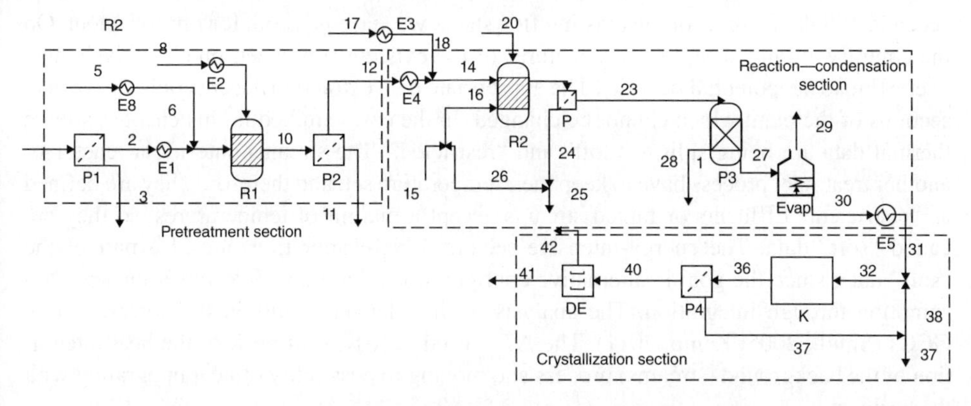

Figure 12.7 *Process flow diagram of the biotechnological process*

xylose concentration depends on the organism used and in our case on the fungi of the genus *Candida* and lies from 50 to 100 g/l (Faria et al., 2002; Branco et al., 2007). In addition, the presence of sugars other than xylose can boost the activity of the microorganism. For this purpose, glucose is added in a mass ratio of 10% of xylose (Faria et al., 2002; Branco et al., 2007). Other nutrients should also be furnished, such as nitrogen source (e.g., NH_3, NH_4OH, NH_4SO_4) and phosphorus (e.g., potassium phosphate). The bioreactor (unit R2) operates at 1 atm and 30°C. These mild conditions limit the production of by-products, such as furfurals, which are toxic for the microorganism (Silva et al., 1996; Jiang & Wu, 1998; Parajó et al., 1998a; Cheng et al., 2014). The reactions used for the description of xylitol production (95% consumption of xylose) (12.5) and the metabolic activity of microorganisms for cell growth and respiration (glucose is assumed to be 100% consumed) (12.6) are given below:

$$100C_5H_{10}O_5 + 8.75O_5 + 7NH_3 + 37H_2 \leftrightarrow 35CH_{1.8}O_{0.5}N_{0.2} + 35CO_2 + 86C_5H_{12}O_5 \quad (12.5)$$

$$C_6H_{12}O_6 + 3O_2 + 0.6NH_3 + 3H_2 \leftrightarrow 3CH_{1.8}O_{0.5}N_{0.2} + 3CO_2 + 4.5H_2O \quad (12.6)$$

The xylitol produced is discolored, crystallized, and dried in the same equipment and conditions as the one described in the catalytic process. The productivity achieved for the biotechnological process is 0.73 kg xylitol crystals per 1 kg of xylose. Figure 12.7 presents the process flow diagram for the biotechnological production of xylitol.

12.5 Scope for Process and Energy Integration

The opportunities offered to integrate each of the aforementioned xylitol processes stand quite different as there are important dissimilarities and the process streams are at different conditions. The proper exploitation of the energy and the resources in each case can result in autonomy and therefore achieve feasibility for both methods, making them a great option for a biorefinery investment plan. The most critical data for pinch analysis are those of the thermal stream heating and cooling information along with the utility requirements. For this reason, extreme caution is required when extracting the data for heat integration. By accepting all the features of the existing flowsheet, we leave no room for improvement. On the contrary, by not accepting any features of the existing flowsheet, pinch analysis may overestimate the potential benefits. The appropriate data extraction accepts only the critical sections of the plant which cannot be changed. In the cases studied in this chapter, stream thermal data are being split to "soft" and "restricted." The streams entering a reactor or another treatment process have to keep their temperature set, and therefore, they are defined as "restricted." Effluents or mixed streams accept a margin of temperatures, so they are called "soft" data. The energy-intensive sectors (e.g., evaporation) are also part of the "soft" data, since the goal is alternative configurations that may yield lower energy consumption through integration. The analysis in this chapter results in the corresponding GCCs (Smith, 2005; Kemp, 2011). The ΔT_{min} used is 10 K, starting with the heat integration of the background (streams) process and moving to possibility of heat integrating with the basic units.

12.5.1 Catalytic Process

For the catalytic process, we first examine the integration of the background processes and then the scenario of steam recycling. The hot and cold streams of the process are given in Table 12.2, and the GCC is shown in Figure 12.8.

The GCC of the catalytic process indicates that the minimum energy required for hot utilities is 1280 kW and for cold utilities 30 kW. The pinch temperature for the overall process is 293 K. Unlike hot utilities, the requirements for cold utilities remain low. Energy integration could save up to 18% of the total heating requirements. This amount of heat is not sufficient, considering the greater needs for hot utilities. The dryer (DE; Figure 12.9) can be further optimized through integration, but its operating conditions should respect the temperature constraints in regard to the sugar degradation issue (operating temperature at <378 K). Stream 28 (Table 12.3) has a mass flow of 3625 kg/h and a pressure of 0.6 atm. It is cooled from 362 to 288 K and condensed to water. Investigating the possibility of heating the cold streams via stream 28 in order to reduce the hot utility required, we observe that the minimum requirements for heating have now changed to 1193 and to 2559 kW for cooling, while the pinch temperature is raised to 358 K (Figure 12.10).

Table 12.2 *Stream properties for the heat integration of the chemical process*

Stream	Type	T_{inlet} (K)	T_{target} (K)	CP (kW/K)
4	Cold	288	360	2.6
11	Cold	349	368	22.1
9	Cold	288	360	0.1
13–22	Cold	288	368	0.0
17	Cold	362	373	83.7
15	Cold	288	368	0.1
20	Hot	373	353	6.7
29	Hot	382	281	1.8

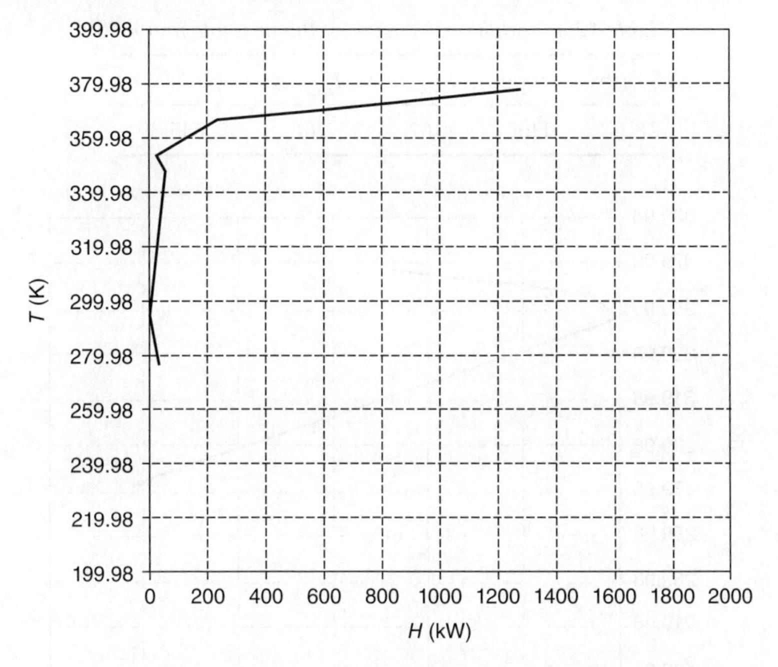

Figure 12.8 *Grand composite curve of the catalytic process*

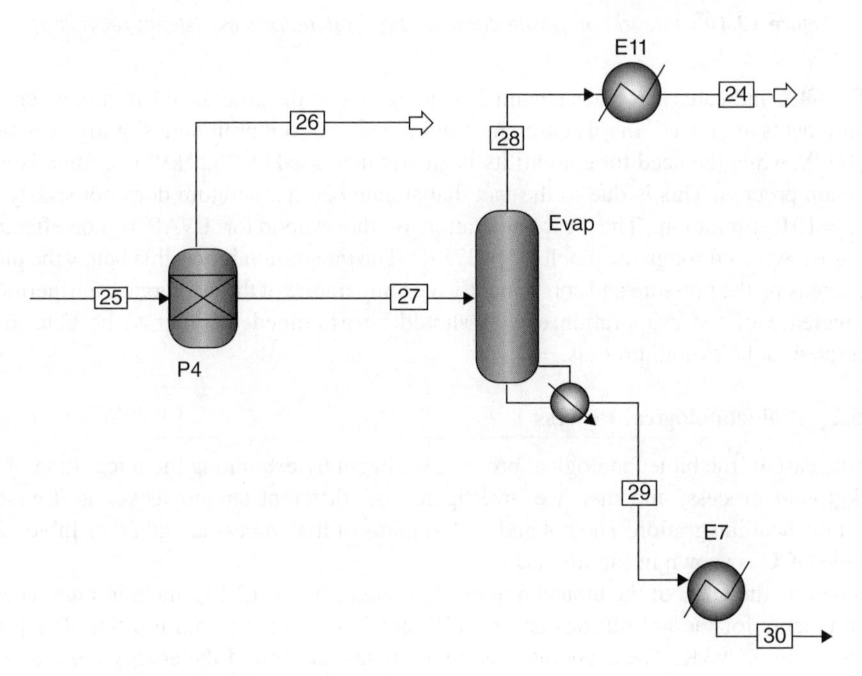

Figure 12.9 *Cooling simulation of stream 28 in Aspen Plus 7.1*

Table 12.3 Additional stream for the heat integration

Stream	Type	T_{inlet} (K)	T_{target} (K)	CP (kW/K)
28	Hot	362	288	35.3

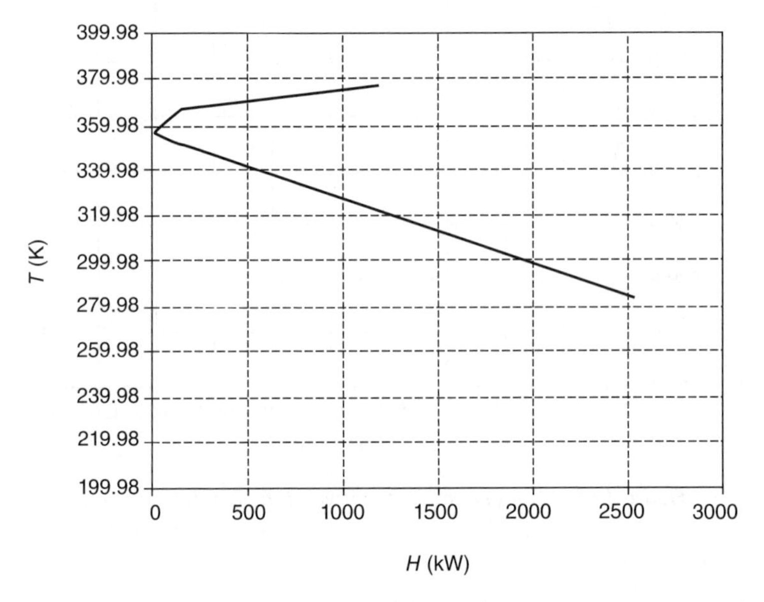

Figure 12.10 *Grand composite curve of the catalytic process (steam recycling)*

The GCC indicates that when stream 28 is integrated in the process, the minimum energy requirements are raised. In this case, the requirements for hot utility are slightly decreased by 87 kW, while the need for cold utility is greatly increased by 2529 kW in comparison to the main process. This is due to the fact that streams 28 temperature does not satisfy the $\Delta T_{min} = 10$ K limitation. The heat integration of the evaporator EVAP is not effective, because it works through the pinch (356–373 K). The integration is possible below the pinch by decreasing the pressure and consequently the temperature of the evaporator. Furthermore, the increase of the evaporation effects should be examined in order to be able to be integrated to the overall process.

12.5.2 Biotechnological Process

For the case of the biotechnological process, we begin by examining the integration of the background process, and then we investigate how different stream recycling scenarios affect the heat integration. The hot and cold streams of the process are given in Table 12.4, and the GCC is shown in Figure 12.11.

Based on the GCC of the biotechnological process (Figure 12.11), the minimum energy requirements for the hot utilities are 228 kW and 26 kW for the cold utilities. The pinch temperature is 293 K. The evaporator uses the biggest portion of the energy requirements (27.6 MW), and when integrated, the energy conservation achieved is 21 kW or about 24%.

Table 12.4 Hot and cold streams of the biotechnological process

Stream	Type	T_{inlet} (K)	T_{target} (K)	CP (kW/K)
5	Cold	288	298	0.1
8	Cold	288	298	1.7
17	Cold	288	303	40.5
12	Cold	298	303	5.0
15	Cold	288	303	0.0
2	Hot	360	298	4.8
30	Hot	377	288	1.6

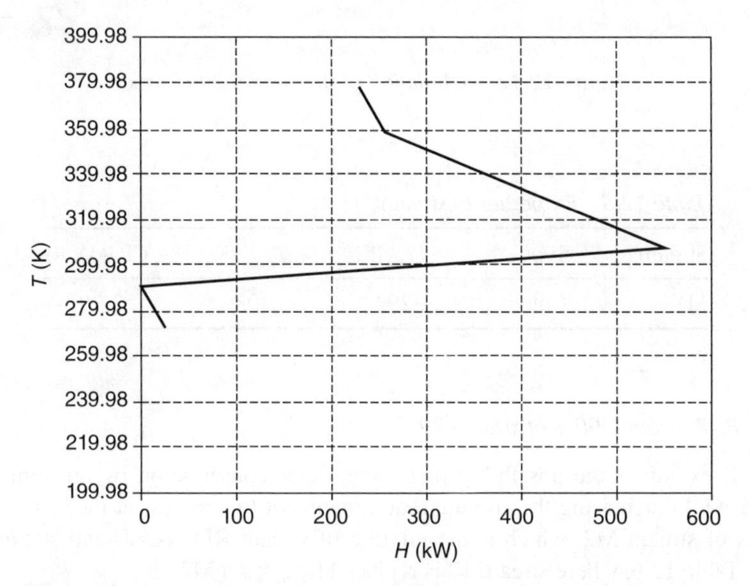

Figure 12.11 Grand composite curve of the biotechnological process

Scenario A: Recycling 0.8% of stream 29

The evaporator condensate (stream 29) can be recycled in the water input section (Figure 12.12). The first scenario examined is the recycling of only 0.8 % of the stream 29. This stream is also used for preheating the freshwater stream (stream 17). The properties of stream M1, which is the mixture of stream RECYCLE and stream 17, are shown in Table 12.5, where stream 17 is replayed by stream M1.

The recycling of stream 29, in this scenario, conserves 314 kg/h of freshwater. Hot utilities are also conserved since the water inflow is heated at 294 K by mixing it directly with the recycling stream. The new GCC (Figure 12.13) shows that the minimum requirements are now 25 kW for heating and 50 kW for cooling, while the pinch temperature is slightly raised to 298 K.

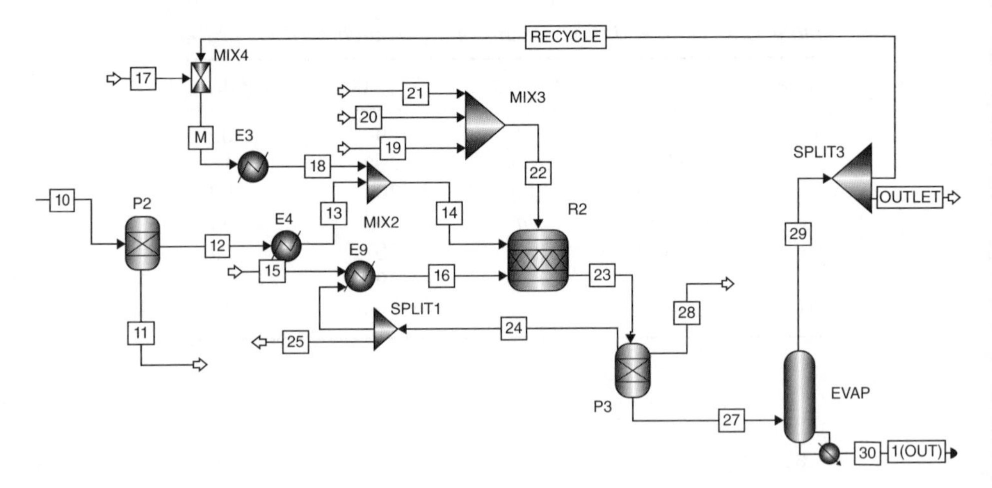

Figure 12.12 Simulation of the stream 29 recycle

Table 12.5 Properties of stream M1

Stream	Type	T_{inlet} (K)	T_{target} (K)	CP (kW/K)
M1	Cold	294	303	37.9

Scenario B: Recycling 90% of stream 29

Scenario B examines the possibility of greater water conservation by recycling 90% of stream 29 while increasing the cooling requirements of the process at the same time. The properties of stream M2, which is the mixture of stream RECYCLE and stream 17, are shown in Table 12.6, where stream 17 is replaced by stream M2.

The new GCC (Figure 12.14) indicates a threshold problem, since there are no heating requirements, while the cooling requirement is 25,085 kW. The pinch temperature is raised to 373 K.

12.5.3 Summarizing Results

Table 12.7 summarizes the results of heat integration for every scenario studied. The catalytic process has higher requirements for hot and cold utility than the biotechnological one. This is due to the fact that the catalytic process has higher stream temperatures. On the other hand, the great energy requirements for evaporation in the biotechnological process make the catalytic method more attractive, although the heat integration of the background process gives worse results.

The scenario of the biotechnological process with 0.8% recycle of the evaporation condensates achieves the highest energy recovery for both hot and cold utilities. The

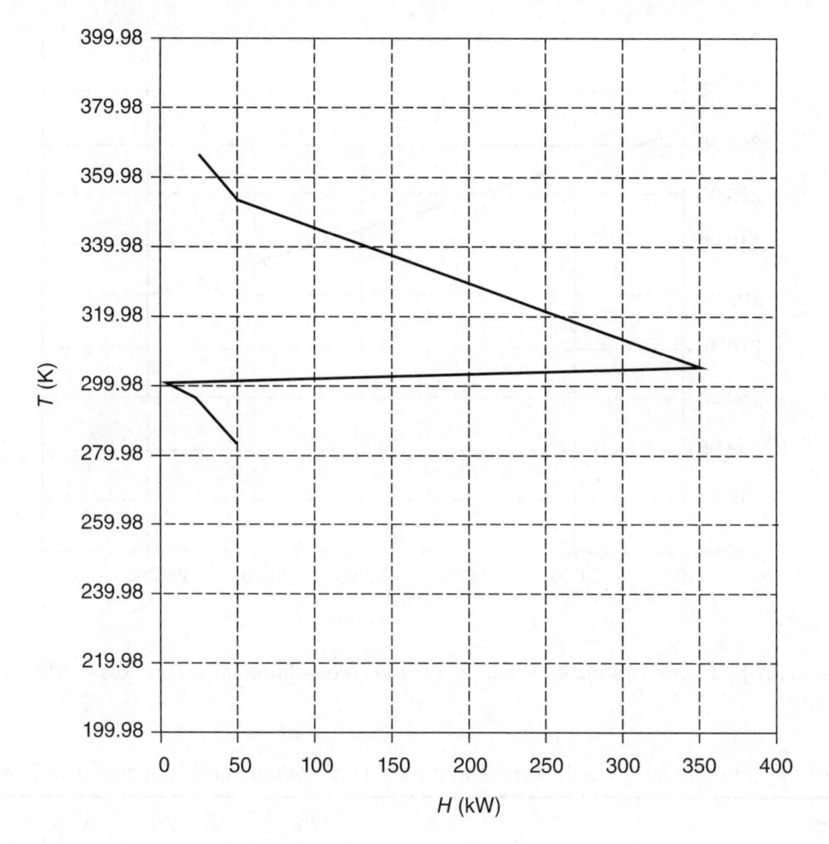

Figure 12.13 *Grand composite curve of the biotechnological process—0.8% steam recycling*

Table 12.6 *Properties of stream M2*

Stream	Type	T_{inlet} (K)	T_{target} (K)	CP (kW/K)
M2	Hot	367	303	383.8

biotechnological process scenario with 90% recycle of the evaporation condensates is a threshold problem, indicating that the condensate recycle is an important parameter for the integration. However, the heat energy required for evaporation is 27.6 MW. This amount of energy is far more important than the energy conservation achieved through the integration. On the other hand, the corresponding evaporation energy requirement in the catalytic process is 2.4 MW. The best energy recovery scenario in catalytic process is that of the main process, showing that there are no better alternatives. Integrating the crystallization process did not result in any energy saving for the cooling requirements since the target temperatures are below 293 K.

Having concluded the part of process flowsheeting and the process integration analysis, we need to examine the process synthesis part. Process synthesis will be used to evaluate

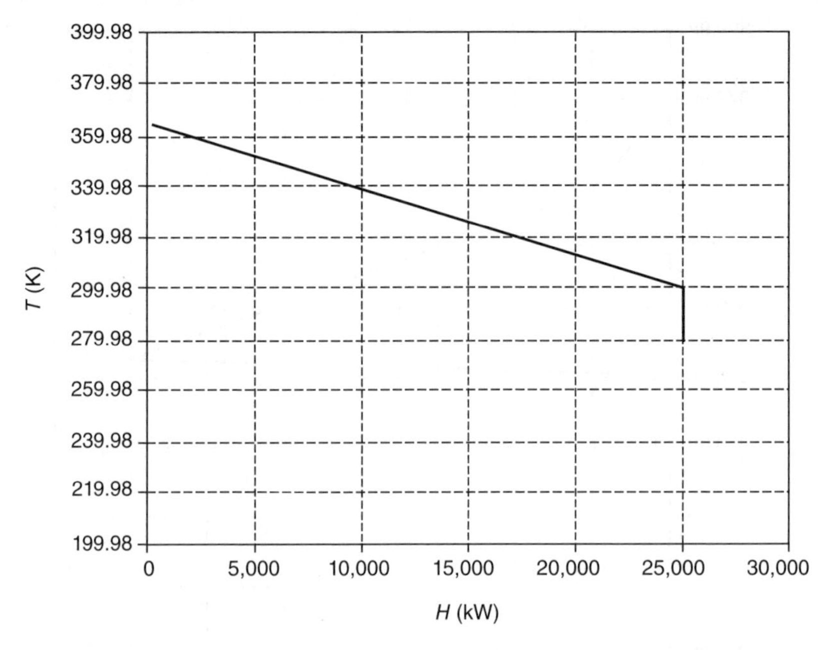

Figure 12.14 *Grand composite curve of the biotechnological process—90% steam recycling*

Table 12.7 *Summarizing results regarding the heat integration in every scenario and case*

Process	T_{pinch} (K)	Qc_{min} (kW)	Qh_{min} (kW)	E (kW)	C%	T%
Catalytic	293	30	1280	283	90.3	18
Steam recycling	357	2,559	1193	379	13.0	24
Biotechnological	293	27	228	423	93.9	65
Steam recycling (0.8%)	298	50	25	401	88.9	94
Steam recycling (90%)	372	25,085	0	43	0.2	100

C%, percentage % of conservation of the cooling needs of the streams; *E*, energy conservation; *T*%, percentage % of conservation of the heating needs of the streams.

the integration potential of these processes within the overall concept of biorefinery. Using the transshipment method, excess heat from a process will be used to cover the heat shortage in another. The excess heat in the catalytic process is 166 kW, while in the biotechnological it is 149 kW. If the xylitol process is expected to be integrated with the other processes of the biorefinery plant, it must be registered as a heat source process. The utilization of this heat demands the usage of a heat pump for the production of hot streams that will be able to satisfy heating needs over the pinch temperature. Nevertheless, catalytic method needs an extreme temperature increase, resulting in low pump performance. Through the site-to-site energy integration, proposals for process modification may emerge. New process design can contribute to the viability of the biorefinery for the valorization and selection of chemical paths integrated with the rest of biorefinery (Stefanakis et al.,

2013; Koufolioulios et al., 2014). The different scenarios studied in this chapter for catalytic and biotechnological path offer different integration profiles (GCCs) that one way or another may be more efficient to integrate with the core process (CIMV Process). Higher energy savings can be obtained by integrating plants together rather than independently.

12.6 Conclusion

Comparing the two different paths for the xylitol production, we observe that the productivity of the catalytic process is 0.81 kg crystal xylitol/kg xylose, while the corresponding value for the biotechnological path is 0.73 kg crystal xylitol/kg xylose. Even though productivity is an important factor for the economic viability of the process, the energy cost and safety factors may shift the choice. The main disadvantage of the catalytic process is the increased safety measures required due to the necessity for storage, transportation, and usage of hydrogen. Moreover, the high pressures and temperatures in the main reactor, in combination with the great flammability of the catalyst used, are another safety issue that has to be taken into account. The disadvantages of the biotechnological process are the high requirements for water for the dilution of the feed stream. Heat integration of the background process indicates that satisfactory energy recovery can be achieved. Better results occur for the biotechnological method, while the most efficient scenario includes recycling, condensation, and reuse of a small part (0.8%) of the steam produced in the evaporator, with the energy conservation reaching 89% of the cooling and 94% of the heating needs of the streams. Apart from the benefits of individual integrating, more possibilities of energy conservation arise, if the designed processes are combined with other processes. The excess energy from one process can be used to cover the energy shortage in another. Higher energy savings can be obtained by integrating plants together rather than independently.

Acknowledgment

Financial support from the Consortium of Marie Curie project RENESENG (FP7-607415) is gratefully acknowledged. The authors would also like to thank Bouchra Benjelloun Mlayah, Marilyn Wiebe and all the people working for the BIOCORE Project (FP7-241566), for their collaboration and excellent communication.

References

Aranda-Barradas, J.S., Garibay-Orijel, C., Badillo-Corona J.A., and Salgado-Manjarrez E. (2010) A stoichiometric analysis of biological xylitol production. *Biochemical Engineering Journal*, **50**, 1–9.

Branco, R.F., Santos, J.C., Murakami, L.Y., et al. (2007) Xylitol production in a bubble column bioreactor: Influence of the aeration rate and immobilized system concentration. *Process Biochemistry*, **42** (2), 258–262.

Cheng, K., Wu, J., Lin, J., and Zhang, J. (2014) Aerobic and sequential anaerobic fermentation to produce xylitol and ethanol using non-detoxified acid pretreated corncob. *Biotechnology for Biofuels*, **7** (1), 166.

Delmas, M. (2008) Vegetal refining and agrochemistry. *Chemical Engineering & Technology*, **31**(5), 792–797.

Faria, L.F.F., Pereira Jr, N., and Nobrega, R. (2002) Xylitol production from d-xylose in a membrane bioreactor. *Desalination*, **149** (1), 231–236.

Fatehi, P., Catalan, L., and Cave, G. (2014) Simulation analysis of producing xylitol from hemicelluloses of pre-hydrolysis liquor. *Chemical Engineering Research and Design*, **92**, 1563–1570.

Guex, W., Klaeui, H., Pauling, H., and Voirol, F. (1981) Reusable heat devices containing xylitol as the heat-storage material. U.S. Patent No. 4,295,517. Washington, DC: U.S. Patent and Trademark Office, October 20, 1981.

Jiang, Z. and Wu, P. (1998) Pseudo-steady state method on study of xylose hydrogenation in a trickle-bed reactor. *Catalysis Today*, **44** (1), 351–356.

Kemp, I.C. (2011) *Pinch Analysis and Process Integration: A User Guide on Process Integration for the Efficient Use of Energy*. Butterworth-Heinemann, Oxford.

Kokossis, A. and Yang, A. (2010) On the use of systems technologies and a systematic approach for the synthesis and the design of future biorefineries. *Computers & Chemical Engineering*, **34**, 1397–1405.

Kokossis, A., Yang, A., Tsakalova, M., and Lin, T.C. (2010) A systems platform for the optimal synthesis of biomass based manufacturing systems. *Computer Aided Chemical Engineering*, **28**, 1105–1110.

Koufolioulios, D., Nikolakopoulos, A., Pyrgakis, K., and Kokossis, A. (2014) A Mathematical Decomposition for the Synthesis and the Application of Total Site Analysis on Multi-product Biorefineries, Proceedings of the Eighth International Conference on Foundations of Computer-Aided Process Design – FOCAPD 2014, July 13–17, 2014, Cle Elum, WA, USA, Elsevier B.V., pp. 549–554.

Lima, T., José, I., Ribeiro, G., and Valderez, M. (2014) Biotechnological production of xylitol from lignocellulosic wastes: A review. *Process Biochemistry*, **49**, 1779–1789.

Martínez, E.A., Canilha, L., Alves, L.A., de Almeida, J.B., and Giulietti, M. (2009) Estudio de la solubilidad de la xilosa, glucosa y xilitol. *Ingeniería química*, **470**, 96–99.

Mikkola, J.P. and Salmi, T. (2001) Three-phase catalytic hydrogenation of xylose to xylitol—Prolonging the catalyst activity by means of on-line ultrasonic treatment. *Catalysis Today*, **64** (3), 271–277.

Mikkola, J.P., Sjöholm, R., Salmi, T., and Mäki-Arvela, P. (1999) Xylose hydrogenation: Kinetic and NMR studies of the reaction mechanisms. *Catalysis Today*, **48** (1), 73–81.

Mikkola, J.P., Vainio, H., Salmi, T., Sjöholm, R., Ollonqvist, T., and Väyrynen, J. (2000) Deactivation kinetics of Mo-supported Raney Ni catalyst in the hydrogenation of xylose to xylitol. *Applied Catalysis A: General*, **196** (1), 143–155.

Misra, S., Gupta, P., Raghuwanshi, S., Dutt, K., and Saxena, R.K. (2011) Comparative study on different strategies involved for xylitol purification from culture media fermented by *Candida tropicalis*. *Separation and Purification Technology*, **78** (3), 266–273.

Mountraki, A., Nikolakopoulos, A., Benjelloun, B., and Kokossis, A. (2011) BIOCORE – A systems integration paradigm in the real-life development of a lignocellulosic biorefinery. 21st European Symposium on Computer Aided Process Engineering – ESCAPE 21, May 29–June 1, 2011, Chalkidiki, Greece, Elsevier, pp. 1381–1385.

Nigam, P. and Singh, D. (1995) Processes of fermentative production of xylitol – A sugar substitute. *Process Biochemistry*, **30** (2), 117–124.

Pal, S., Choudhary, V., Kumar, A., Biswas, D., Mondal, A.K., and Sahoo, D.K. (2013) Studies on xylitol production by metabolic pathway engineered *Debaryomyces hansenii*. *Bioresource Technology*, **147**, 449–455.

Parajó, J.C., Domínguez, H., and Domínguez, J.M. (1998) Biotechnological production of xylitol. Part 1: Interest of xylitol and fundamentals of its biosynthesis. *Bioresource Technology*, **65** (3), 191–201.

Parajó, J.C., Dominguez, H., and Domínguez, J. (1998) Biotechnological production of xylitol. Part 3: Operation in culture media made from lignocellulose hydrolysates. *Bioresource Technology*, **66** (1), 25–40.

Research and Markets (2014) *Xylitol – A Global Market Overview*, report ID: 2846975. Available at: http://www.researchandmarkets.com/reports/2846975/xylitol-a-global-market-overview (accessed on July 17, 2015).

Sampaio, F.C., Passos, F.M.L., Passos, F.J.V., De Faveri, D., Perego, P., and Converti, A. (2006) Xylitol crystallization from culture media fermented by yeasts. *Chemical Engineering and Processing: Process Intensification*, **45** (12), 1041–1046.

Silva, S.S., Roberto, I.C., Felipe, M.G., and Mancilha, I.M. (1996) Batch fermentation of xylose for xylitol production in stirred tank bioreactor. *Process Biochemistry*, **31** (6), 549–553.

Smith, R. (2005) *Chemical Process Design and Integration*, second edition. John Wiley & Sons Ltd, Chichester.

Stefanakis, M., Pyrgakis, K., Mountraki, A., and Kokossis, A. (2013) The Total Site Approach as a synthesis tool for the selection of valorization paths in lignocellulosic biorefineries, ESCAPE 23, June 9–12, 2013, Lappeenranta, Finland.

Tsakalova, M. and Kokossis, A. (2013) On the systematic synthesis screening and integration of real-life biorefineries. In Symposium on Biorefinery for Food, Fuel, and Materials, April 7–10, 2013, Wagenigen, the Netherlands. Ton van Boxtel and Marieke Bruins, Wagenigen.

Wang, S., Li, Q.S., Li, Z., and Su, M.G. (2007) Solubility of xylitol in ethanol, acetone, N,N-dimethylformamide, 1-butanol, 1-pentanol, toluene, 2-propanol, and water. *Journal of Chemical & Engineering Data*, **52**(1), 186–188.

Wang, Z., Wang, Q., Liu, X., Fang, W., Li, Y., and Xiao, H. (2013) Measurement and correlation of solubility of xylitol in binary water + ethanol solvent mixtures between 278.00 K and 323.00 K. *Korean Journal of Chemical Engineering*, **30**(4), 931–936.

Wisniak, J., Hershkowitz, M., Leibowitz, R., and Stein, S. (1974) Hydrogenation of xylose to xylitol. *Industrial & Engineering Chemistry Product Research and Development*, **13** (1), 75–79.

13

Determination of Optimum Condition for the Production of Rice Husk-Derived Bio-oil by Slow Pyrolysis Process

Suzana Yusup, Chung Loong Yiin, Chiang Jinn Tan, and Bawadi Abdullah

Department of Chemical Engineering, Biomass Processing Laboratory, Center of Biofuel and Biochemical, Green Technology (MOR), Universiti Teknologi PETRONAS, Tronoh, Malaysia

13.1 Introduction

The depletion of fossil fuels along with the rising demand of energy from fossil fuels has become a threat to countries which are still relying on petroleum to generate electricity and transportation fuel. Recently, studies on bio-oil have captured international interest in developing it as an alternative energy option. Every year, a large number of biomass residues are left behind after a massive number of agricultural crops have been harvested. Disposing biomass residues by open burning could contribute to global warming and environmental pollution; hence, utilizing biomass residues as raw materials to generate sustainable and renewable energy for electricity and transportation fuels offers a better alternative. Undoubtedly, sustainable and renewable energy options generated from agricultural and biomass residues have high potential to become a substitute to petroleum fuels.

Biomass can generally be defined as any hydrocarbon material which mainly consists of carbon, hydrogen, oxygen, and nitrogen, with presence of sulfur in lesser proportions. Biomass is an abundant carbon-neutral renewable resource for the production of bioenergy

Process Design Strategies for Biomass Conversion Systems, First Edition. Edited by Denny K. S. Ng, Raymond R. Tan, Dominic C. Y. Foo, and Mahmoud M. El-Halwagi.
© 2016 John Wiley & Sons, Ltd. Published 2016 by John Wiley & Sons, Ltd.

and biomaterials (Ragauskas et al., 2006). There are several methods to convert biomass into energy. Biochemical and thermochemical conversion methods can be applied to biomass to make use of its energy potential for a sustainable renewable energy option (Yaman, 2004). Biochemical conversion methods involve converting biomass into alcohols or oxygenated products by biological activity. Besides that, thermochemical conversion methods such as pyrolysis, liquefaction, gasification, and supercritical fluid extraction methods can be applied by converting biomass into energy.

Among all the thermochemical conversion methods, pyrolysis has attracted the most attention (Domínguez & Menéndez, 2006). Pyrolysis is one of the thermochemical processes that convert the solid biomass into liquid (bio-oil), gas, and char in the absence of oxygen (Salleh et al., 2011). Biomass pyrolysis converts 80–95% of the feed material to gases and bio-oil. The pyrolysis of biomass is a complex process due to differences in the chemical composition, particle structure within the biomass material, and operating condition. Pyrolysis process needs to be optimized to maximize the production of liquids (tar and bio-oil) (Natarajan & Ganapathy, 2009). Bio-oil which is also known as pyrolysis oil or bio-fuel oil was produced by pyrolysis process without any additional oxygen. Bio-oil is a complex organic mixture with high oxygen content, containing more or less solid carbon particles, and its color is usually dark brown or black, having a strong pungent smell (Xiu et al., 2012). The pyrolysis bio-oil is a complex mixture of oxygenated aliphatic and aromatic compounds (Meier & Faix, 1999).

Pyrolysis bio-oil consists of hundreds of specific organic groups which include organic acids, esters, alcohols, ketones, aldehydes, phenols, aromatic hydrocarbons, nitrogen compounds, furans, and furodrosugars (Abu Bakar & James, 2012). Bio-oil has a high potential to be the substitute for fossil fuels as it can produce heat, electricity, and chemicals. However, raw bio-oil that has not been upgraded has several undesirable characteristics such as having high water and oxygen content, high viscosity, and low calorific value and being corrosive, unstable when under storage and heating conditions, and immiscible with petroleum fuels (Iliopoulou et al., 2012). The main reason for the instability of bio-oil is the presence of oxygenated compounds in the bio-oil. They tend to further react during storage or at high-temperature conditions (Gopakumar et al., 2011). As a result of the instability, bio-oil aging occurs, causing the viscosity of the bio-oil to increase. This instability can actually be minimized or decreased through either the stabilization process after pyrolysis through upgrading methods such as emulsification with diesel oil (Jiang & Ellis, 2010) or addition of antioxidants and solvents (Wanasundara & Shahidi, 2005; Udomsap et al., 2011). The characteristics of bio-oil such as density, pH, viscosity, and heating value should be monitored. The resulting characteristics are considered as desirable for bio-oil if they increase its potential to replace fossil fuels. For instance, the bio-oil should have comparable properties with petroleum fuel such as low moisture content, viscosity, and acidity as well as high heating value as shown in Table 13.1.

Among various biomass species, rice husk is a potential source of energy and a value-added by-product of the rice milling industry (Sharma & Rao, 1999). In recent years, utilization of agriculture waste and residues (for instance, rice husk) for energy generation has gained a lot of attention. Rice is one of the major food crops in the world. The production of rice generates a great amount of waste in the world, namely, rice husk. Rice husk is the outer covering of the paddy and accounts for 20–25% of its weight (Jenkins, 1989) which stood at 672 million tons in the year 2010 (GeoHive, 2011). Rice husk is produced in the first step of the milling process

Table 13.1 Properties of petroleum fuel

Properties	Value
Elemental composition (wt%)	85.00
Carbon	11.00
Hydrogen	1.00
Oxygen	0.30
Nitrogen	
Moisture content (wt%)	0.10
Heating value (MJ/kg)	40.00
Viscosity at 40°C (cP)	18.00

when the husk is removed from the grain in the husking stage of the rice mill. Rice husk usually ends up being burned in open air, therefore causing environmental pollution and wastes of various biomass resources (Ezzat et al., 2012).

Generally, rice husk is a cellulosic material, consisting of about 20–35% cellulose, 15–30% hemicelluloses, 5–10% lignin, and some materials. Rice husk is a renewable source of energy, with a high calorific value of about 4260 kcal/kg on dry basis, and is characterized by low bulk density and high ash content (18–22% by weight). The large amount of ash generated during combustion has to be continuously removed for a smooth operation of the system. Rice husk may be treated by pyrolysis methods to obtain fuels as discussed in previous papers (Islam & Ani, 2000; Williams & Nugranad, 2000). Biomass pyrolysis essentially converts 80–95% of the feed material to gases and bio-oil. The pyrolysis process converts biomass into high-energy-content biofuels, in the absence of oxygen/air, leading to the formation of solid (charcoal), liquid (tar and other organics), and gaseous products.

Pyrolysis can be divided into two types: fast pyrolysis and slow pyrolysis. These reactions are different mainly in terms of heating rates and maximum reaction temperatures (Brown et al., 2011). Heating rates for slow pyrolysis are typically below 100 K/min, whereas fast pyrolysis can achieve heating rates exceeding 1000 K/min. Reaction temperatures are about 300°C and 500°C for slow and fast pyrolysis, respectively. Slow pyrolysis requires several minutes or even hours, while fast pyrolysis is completed within 2 s. This difference in time results in dramatic differences in product distributions.

In this study, an attempt was given to produce bio-oil through slow pyrolysis of rice husk (RH) at different heating rates in order to determine the optimum reaction condition that will give maximum liquid yield. The characteristics of bio-oil produced at different heating rates are then analyzed. Apart from that, the properties of bio-oil that was produced at optimum operating condition are compared with those in literature.

13.2 Experimental Study

13.2.1 Biomass Preparation and Characterization

Raw rice husks were taken from local suppliers. The rice husk is then ground to smaller particle sizes using FRITSCH Cutting Mill. After the grinding process, the rice husk is reduced into the desired feedstock size within the range of 0.25–0.50 mm. The ground rice husk is sieved using CISA BA 500 N sieve shaker to obtain the desired feedstock size of 0.25–0.50 mm. Sieved rice husk is then dried in the oven at 100°C to remove the moisture content. Ultimate analysis of the RH is carried out using LECO 932 CHNS Analyzer and the higher heating value (HHV) is measured using IKA C5000 bomb calorimeter.

13.2.2 Experimental Procedure

The rice husk sample is placed in a borosilicate tube. Approximately 0.6 g of glass wool is placed above the rice husk sample to avoid the rice husk from being disseminated from the tube during pyrolysis reaction inside the furnace. The experiment is set up and nitrogen gas is purged for 5 min at a rate of 500 ml/min before the experiment. Based on the literature review, the pyrolysis temperature is set at 500°C and the nitrogen gas flow rate is at 100 ml/min in order to have the optimum production of bio-oil (Sukiran et al., 2009). The temperature of the heater and thermocouple is recorded every 2 min. The time at which the first drop of bio-oil is formed is recorded. When the temperature of the heater and thermocouple reaches a stable condition, the experiment is stopped. The reactor is left to cool down. The condenser is weighed to calculate the liquid yield. The bio-oil is collected for characterization analysis. A total of 16 runs were carried out at four different heating rates with four runs at 5°C/min, 10°C/min, 15°C/min, and 20°C/min, respectively. The liquid yield is calculated using Equation 13.1:

$$\text{Yield}_{\text{bio-oil}} = \frac{m_{\text{bio-oil}}}{m_{\text{biomass}}} \times 100\% \tag{13.1}$$

Meanwhile, the overall conversion of biomass into gas and oil products can be determined by using Equation 13.2:

$$\text{Conversion}\,\% = \frac{\text{decrease in weight of reaction mixture}}{m_{\text{biomass}}} \times 100\% \tag{13.2}$$

Apart from that, the gas yield and residue yield can be calculated using Equation 13.3 and Equation 13.4, respectively:

$$\text{Yield}_{\text{gas}} = \text{conversion} - \text{yield}_{\text{bio-oil}} \tag{13.3}$$

$$\text{Yield}_{\text{residue}} = 100\% - \text{conversion} \tag{13.4}$$

13.2.3 Equipment

The reactor used in this study is a semibatch reactor as shown in Figure 13.1.

In every run, about 15 g of dried RH is placed into the borosilicate glass tube and heated in a vertical furnace. Before the reaction starts, nitrogen gas is allowed to flow through the tube for a few minutes to drive out all the oxygen. The desired temperature of the reaction, heating rate of 20°C/min, and the flow rate of nitrogen gas are set. The pyrolysis vapor is carried out by the nitrogen gas and passes through an ice bath condenser, where the condensable vapor will condense to form bio-oil.

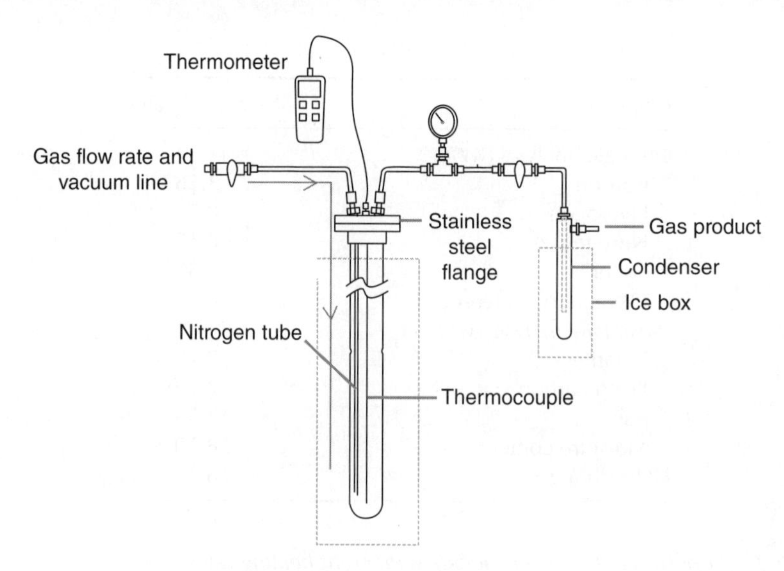

Figure 13.1 *Semibatch reactor*

13.2.4 Characterization of Bio-oil

The bio-oil produced at different heating rates is characterized. The physical properties of bio-oil such as density, pH, viscosity, and heating value are measured using Anton Paar DMA 4500 M density meter, EUTECH pH 510 pH/mV/°C meter, Brookfield CAP 2000+ viscometer, and IKA C5000 bomb calorimeter, respectively. Ultimate analysis using LECO CHNS Analyzer 932 is carried out to determine the elemental content of bio-oil. Besides that, field emission scanning electron microscope (FESEM) is used to study the morphology of raw RH and residue after pyrolysis. The chemical composition of the bio-oil produced is tested using Agilent GC/MS 5975 C.

13.3 Results and Discussion

13.3.1 Characterization of RH

The properties of raw RH sample are summarized in Table 13.2.

13.3.2 Characterization of Bio-oil

Table 13.3 shows the properties of the bio-oil produced at different heating rates. Normally, bio-oil is highly acidic and the pH of bio-oil is reported between the range of 2 and 3. The most acidic pH value obtained is 2.68 at a heating rate of 20°C/min. In this work, the pH value decreased with the increase in heating rates, as shown in Table 13.4. Hence, the bio-oil produced is more acidic at higher heating rates. Most of the acidity may arise from the presence of acetic acid, but other carboxylic acids, phenols, and other acidic compounds will also have a significant contribution. The viscosity of bio-oil produced increased with an increment of the heating rates. The viscosity values of bio-oil with the heating

Table 13.2 *Properties of RH (wt%)*

Properties	Measured value
Ultimate analysis (wt%)	41.34
Carbon	3.56
Hydrogen	0.37
Nitrogen	0.15
Sulfur	54.59
Oxygen (by difference)	
Proximate analysis (wt%)	
Volatiles	77.13
Fixed carbon	2.12
Ash	12.25
Moisture content	8.50
HHV (MJ/kg)	16.72

Table 13.3 *Properties of bio-oil produced at different heating rates*

Properties	Heating rate (°C/min)			
	5	10	15	20
pH	2.83	2.76	2.71	2.68
Viscosity at 40°C (cP)	42.35	44.56	45.21	47.83
Density at 20°C (kg/m^3)	1060.14	1064.85	1073.80	1082.69
Ultimate analysis (wt%)				
Carbon	20.45	22.13	23.87	26.76
Hydrogen	8.53	8.98	9.03	9.22
Nitrogen	0.18	0.22	0.25	0.31
Sulfur	0.27	0.35	0.41	0.44
Oxygen	70.75	68.32	66.44	63.27
HHV (MJ/kg)	9.89	11.26	12.13	13.69

Table 13.4 *Comparison of bio-oil properties with literatures*

Characterization parameters	This work	Abu Bakar and James (2012)	Zheng (2007)	Guo et al. (2011)
pH	2.68	3	2.8	3.36
Viscosity (cP)	47.83	1.79	152.32	82.43
Density (kg/m^3) at 20°C	1082.69	1065	1190	1210
Ultimate analysis (wt%)				
Carbon	26.76	23.38	41.7	35.63
Hydrogen	9.22	10.39	7.7	7.00
Nitrogen	0.31	0.51	0.3	—
Sulfur	0.44	0.09	0.2	—
Oxygen	63.27	65.63	50.3	57.37
HHV (MJ/kg)	13.69	13.61	17.42	13.36

rates of 5°C/min, 10°C/min, 15°C/min, and 20°C/min are 42.35 cP, 44.56 cP, 45.21 cP, and 47.83 cP, respectively. On the other hand, the measured density of bio-oil is increasing as the heating rates are elevated.

The highest density value of 1082.69°kg/m³ measured at 20°C is obtained at a heating rate of 20°C/min. Higher density value will attribute to an increase in the water content which does not favor the production of bio-oil. Apart from that, it can be seen from Table 13.2 that the carbon, hydrogen, nitrogen, and sulfur contents are increasing as the heating rates are increased. The nitrogen and sulfur contents were still considered low in all of the bio-oils, showing their potential as a clean fuel when used for combustion purposes. The elemental carbon, hydrogen, nitrogen, sulfur, and oxygen contents of bio-oil basically have low carbon content and high oxygen content. High oxygen content is not favored because it will lead to higher instability of bio-oil.

The heating value of bio-oil gives information related to the energy content of bio-oil, which has the potential of being upgraded to transportation fuels. The HHV of bio-oil can be estimated using the following formula with the unit of MJ/kg (Parikh et al., 2005):

$$\text{HHV} = 0.3491\text{C} + 1.1783\text{H} + 0.1005\text{S} - 0.1034\text{O} - 0.0151\text{N} - 0.0211\text{A}$$

where A represents ash; C, carbon; H, hydrogen; N, nitrogen; O, oxygen; and S, sulfur.

The result shows that the estimated HHV for the bio-oil produced is slightly lower compared to the estimated HHV of raw rice husk. The HHV rose with increase in the heating rates. The highest HHV of 13.69 MJ/kg is obtained at slow pyrolysis condition of heating rate at 20°C/min.

As shown in Table 13.4, the pH value and density of bio-oil in this work are comparable with other literatures. In addition, the bio-oil produced has moderate value of viscosity which favored the range of bio-oil application. Based on the ultimate analysis, the oxygen content measured in this work is almost the same in other findings which resulted in proportionate HHV. Hence, the slow pyrolysis bio-oil produced from this work has a high potential to substitute conventional fossil fuel produced from fast pyrolysis process.

13.3.3 Parametric Analysis

Table 13.5 shows the result of conversion, gas yield, and residue yield calculated for 16 runs.

To study the trend of the results, the average values of liquid yield, gas yield, residue yield, and conversion are calculated and plotted as shown in Figure 13.2.

Figure 13.2 shows that the liquid yield and conversion increased with elevation of heating rates. As the heating rate increases, the gas yield is slightly increased. However, the gas yield decreased at a heating rate of 20°C/min due to higher liquid yield and conversion produced. Besides, the residue yield decreases as the heating rates increase. The rate of mass or heat transfer in the complex matrix of biomass at low heating rates allows intraparticle cracking. Extended exposure times resulting from low heating rates favor secondary reactions such as cracking, repolymerization, and recondensation of radical components leading to the formation of char with limited oil and gaseous products (Gerçel, 2004). Thus, maximum liquid yield of 35.38 wt% and highest conversion value of 63.82 wt% are obtained at a heating rate of 20°C/min.

Table 13.5 Conversion, gas yield, and residue yield

Run	Heating rate (°C/min)	Decrease in weight of reaction mixture (g)	Conversion (wt%)	Gas yield (wt%)	Residue yield (wt%)
1	5	9.07	60.43	32.78	39.57
2	5	8.84	58.93	31.80	41.07
3	5	8.85	59.00	28.73	41.00
4	5	8.91	59.36	29.31	40.64
5	10	9.21	61.28	30.07	38.72
6	10	9.14	60.81	30.67	39.19
7	10	9.45	62.92	31.96	37.08
8	10	9.15	60.88	30.14	39.12
9	15	9.26	61.65	28.63	38.35
10	15	9.52	63.34	34.07	36.66
11	15	9.33	62.20	29.67	37.80
12	15	9.43	62.74	31.60	37.26
13	20	9.54	63.43	27.39	36.57
14	20	9.49	63.18	28.16	36.82
15	20	9.66	64.36	29.45	35.64
16	20	9.66	64.31	28.76	35.69

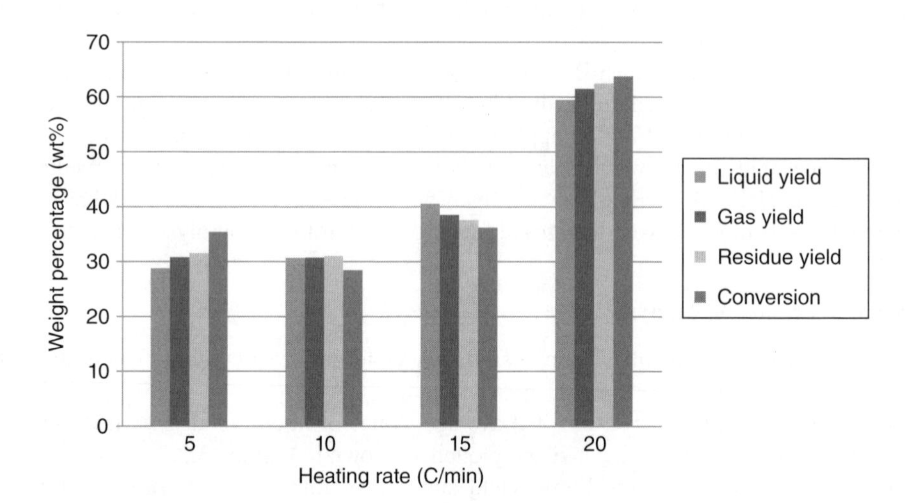

Figure 13.2 Average values of liquid, gas, residue yield, and conversion

13.3.4 Field Emission Scanning Electron Microscope

To gain insight into a particle's structure, the morphologies of raw rice husk and the residue left after slow pyrolysis at a heating rate of 20°C/min are examined by FESEM. FESEM micrographs of raw rice husk are presented in Figure 13.3.

FESEM micrographs of residue after slow pyrolysis in nitrogen atmosphere at a heating rate of 20°C/min are presented in Figure 13.4.

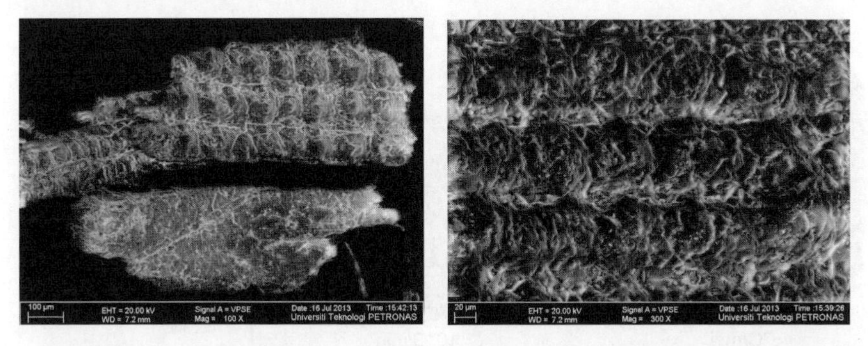

Figure 13.3 *Field emission scanning electron microscope of raw rice husk*

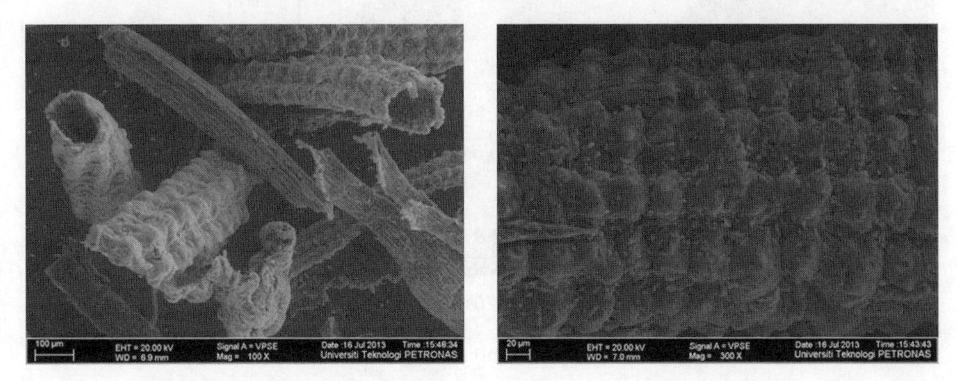

Figure 13.4 *Field emission scanning electron microscope of residue*

As shown in Figure 13.3, the structure of raw rice husk was well organized and originally in globular shape. The transverse section is dense, with no existence of pores. However, after slow pyrolysis in nitrogen atmosphere at a heating rate of 20°C/min as shown in Figure 13.4, the surface texture of rice husk residue changed. The globules shrunk and were densified due to the release of volatile products (Vlaev et al., 2003). Evaporation of volatile materials creates a large number of button-like structures or bumps interspaced with small pores formed on the particles with rough surface. The pores appear as channels from where the cellulose material was preferentially removed during pyrolysis (Jenkins, 1989).

13.3.5 Chemical Composition (GC–MS) Analysis

The chemical compound of bio-oil product was analyzed by using gas chromatography–mass spectroscopy (GC–MS). The bio-oil produced at different heating rates is analyzed for its chemical composition. The components of the bio-oil can be classified into aldehydes, acids, alcohols, ketones, phenols, furans, and sugars (Diebold, 1999). In this work, the area % of the GC–MS chromatogram was used to indicate the amount of the various chemical compounds in the bio-oil. Figure 13.5 summarizes the components contained in the bio-oil produced at different heating rates which are extracted from GC–MS chromatograms with the area % readings.

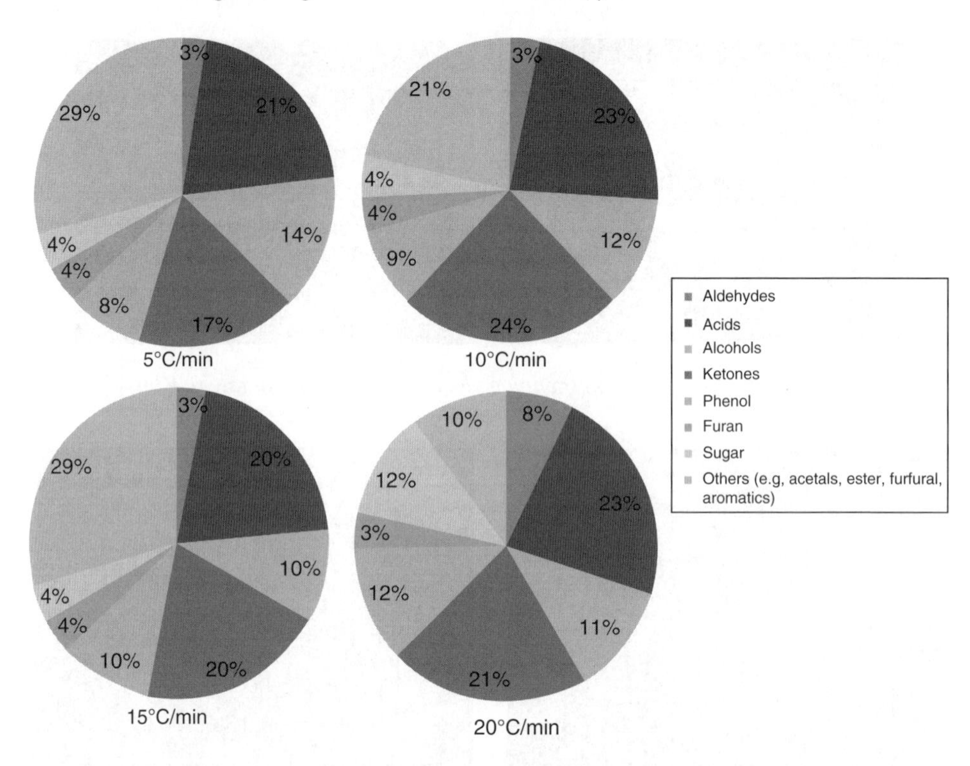

Figure 13.5 Chemical composition of bio-oil at different heating rates

As shown in Figure 13.5, the bio-oil produced at a heating rate of 20°C/min has more aldehydes, acids, ketones, phenol compounds, and more compounds of heavy molecular weight which cause high-viscosity reading. The higher number of phenol and acid compounds contributes to the higher pH number of bio-oil that was produced at a heating rate of 20°C/min.

13.4 Conclusion

In this study, bio-oil is produced from slow pyrolysis of RH. The optimum reaction condition result in maximum liquid yield is determined at a heating rate of 20°C/min, a temperature of 500°C, and a nitrogen flow rate of 100 ml/min. The maximum average liquid yield of bio-oil obtained is 35.38 wt% with conversion value of 63.82 wt%. The physical properties of bio-oil produced at the optimum condition are as follows: pH value of 2.68; viscosity value of 47.83 cP; elemental composition of 26.76 wt% carbon, 9.22 wt% hydrogen, 0.31 wt% nitrogen, 0.44 wt% sulfur, and 63.27 wt% oxygen; density value of 1082.69 kg/m^3 at 20°C; and HHV of 13.69 MJ/kg. The aforementioned desirable properties of bio-oil produced from slow pyrolysis in this show that bio-oil has a high potential as a substitute for conventional fossil fuel.

Acknowledgement

The authors would like to express their greatest gratitude to the Chemical Engineering Department of Universiti Teknologi PETRONAS and Long Term Research Grant Scheme (LRGS) for the financial and technical support.

References

Abu Bakar, M.S. and James, O.T. (2012) Catalytic pyrolysis of rice husk for bio-oil production. *Journal of Analytical and Applied Pyrolysis*, vol **103**, pp. 362–368.

Brown, T.R., Wright, M.M. and Brown, R.C. (2011) Estimating profitability of two biochar production scenarios: slow pyrolysis vs fast pyrolysis. *Biofuels, Bioproducts and Biorefining*, vol **5**, pp. 54–68.

Diebold, J.P. (1999) A Review of the Chemical and Physical Mechanisms of the Storage Stability of Fast Pyrolysis Bio-oils. Retrieved from website: http://home.rmi.net/diebolic (accessed 26 February 2013).

Domínguez, A., Menéndez, J.A., Inguanzo, M. et al. (2006) Production of bio-fuels by high temperature pyrolysis of sewage sludge using conventional and microwave heating. *Bioresource Technology*, vol **97** (10), pp. 265–271.

Ezzat, R., Shahebrahimi, S., Mostafa, F., et al. (2012) Optimization of synthesis and characterization of nanosilica produced from rice husk (a common waste material). *International Nano Letters*, vol **2**, (1), pp. 1–8.

GeoHive. (2011) World: Rice Production in Metric Tonnes. Retrieved on February 2013 from website: http://www.geohive.com/charts/ag_rice.aspx (accessed 3 July 2015).

Gerçel, H.F. (2004) Production and characterization of pyrolysis liquids from sunflower presses bagasse. *Bioresource Technology*, vol **85**, pp. 113–117.

Gopakumar, S.T., Adhikari, S., Gupta, R.B., et al. (2011) Production of hydrocarbon fuels from biomass using catalytic pyrolysis under helium and hydrogen environments. *Bioresource Technology*, vol **102**, pp. 6742–6749.

Guo, X.J., Wang, S.R., Wang, Q., et al. (2011) Properties of bio-oil from fast pyrolysis of rice husk. *Chinese Journal of Chemical Engineering*, vol **19** (1), pp. 116–121.

Iliopoulou, E.F., Stefanidis, S.D., Kalogiannis, K.G., et al. (2012) Catalytic upgrading of biomass pyrolysis vapors using transition metal-modified ZSM-5 zeolite. *Applied Catalysis B: Environmental*, vol **127**, pp. 281–290.

Islam, M.N. and Ani, F.N. (2000) Techno-economics of rice husk pyrolysis, conversion with catalytic treatment to produce liquid fuel. *Bioresource Technology*, vol **73**, pp. 67–75.

Jenkins, B.M. (1989) *Physical properties of biomass*, Gordon and breach, New York.

Jiang, X. and Ellis, N. (2010) Upgrading bio-oil through emulsification with biodesel: mixture production. *Energy & Fuels*, vol **24**, pp. 1358–1364.

Meier, D. and Faix, O. (1999) State of the art of applied fast pyrolysis of lignocellulosic materials – a review. *Bioresource Technology*, vol **68**, (1), pp. 71–77.

Natarajan, E. and Ganapathy, S.E. (2009) Pyrolysis of rice husk in a fixed bed reactor. *World Academy of Science, Engineering and Technology*, vol **32**, pp. 467–471.

Parikh, J., Channiwala, S.A. and Ghosal, G.K. (2005) A correlation for calculating hhv from proximate analysis of solid fuels. *Fuel*, vol **84**, pp. 487–494.

Ragauskas, A.J., Williams, C.K., Davison, B.H. et al. (2006) The path forward for biofuels and biomaterials. *Science*, vol **311** (5760), pp. 484–489.

Salleh, F., Samsuddin, R. and Husin, M. (2011) Bio-fuel source from combination feed of sewage sludge and rice waste. *International Conference on Environment Science and Engineering*, vol **8**, pp. 68–72.

Sharma, A. and Rao, T.R. (1999) Kinetics of pyrolysis of rice husk. *Bioresource Technology*, vol **67**, pp. 53–59.

Sukiran M.A., Chin C.M. and Bakar N.K.A. (2009) Bio-oils from pyrolysis of oil palm empty fruit bunches. *American Journal of Applied Sciences*, vol **6**, pp. 869–875.

Udomsap, P., Yapp, H.Y., Tiong, H.H., et al. (eds) (2011) Towards stabilization of bio-oil by addition of antioxidants and solvents, and emulsification with conventional hydrocarbon fuels. The International Conference & Utility Exhibition, 13–15 September 2011, Thailand.

Vlaev, L.T., Markovska, I.G. and Lyubchev, L.A. (2003) Non-isothermal kinetics of pyrolysis of rice husk. *Thermochimica Acta*, vol **406**, pp. 1–7.

Wanasundara, P.K.J.P.D. and Shahidi, F. (eds) (2005) Antioxidants: science, technology, and applications. *Bailey's Industrial Oil and Fat Products*, 6th ed., John Wiley & Sons, Inc., Hoboken, NJ.

Williams, P.T. and Nugranad, N. (2000) Comparison of products from the pyrolysis and catalytic pyrolysis of rice husks. *Energy*, vol **25**, pp. 493–513.

Xiu, S., Rojanala, H.K., Shahbazi, A., et al. (2012) Pyrolysis and combustion characteristics of bio-oil from swine manure. *Journal of Thermal Analysis and Calorimetry*, vol **107**, pp. 823–829.

Yaman, S. (2004) Pyrolysis of biomass to produce fuels and chemical feedstocks. *Energy Conversion and Management*, vol **45**, pp. 651–671.

Zheng, J.L. (2007) Bio-oil from fast pyrolysis of rice husk: yields and related properties and improvement of the pyrolysis system. *Journal of Analytical and Applied Pyrolysis*, vol **80**, pp. 30–35.

14

Overview of Safety and Health Assessment for Biofuel Production Technologies

Mimi H. Hassim[1], *Weng Hui Liew*[1], *and Denny K. S. Ng*[2]

[1] *Department of Chemical Engineering, Universiti Teknologi Malaysia, Malaysia*
[2] *Department of Chemical and Environmental Engineering/Centre of Sustainable Palm Oil Research (CESPOR), The University of Nottingham, Malaysia Campus, Selangor, Malaysia*

14.1 Introduction

Sustainability is defined as "meeting the needs of the present without compromising the ability of future generations to meet their own need" (Anon, 1987). It is now a necessity to adopt sustainability policies and practices in all types of process industry. Generally, for chemical process industries, several requirements related to sustainability have been imposed, either voluntarily or legally. For instance, Responsible Care which was introduced in 1988 commits the members of Chemical Manufacturers Association to improve safety, health, and environmental (SHE) performances of their processes (Hook, 1996). EU directives have also affected process development and design so that SHE aspects should be taken into consideration in the earlier phase of any new project lifecycle. For example, the Integrated Pollution Prevention and Control (IPPC) Directive was enacted to achieve a high level of protection of the environment for sustainable development. This is to be done on the basis of what is achievable with the best techniques available in the individual industrial sectors falling within the scope of the directive (O'Malley, 1999). The European Agency for Safety and Health at Work (EU-OSHA)

Process Design Strategies for Biomass Conversion Systems, First Edition. Edited by
Denny K. S. Ng, Raymond R. Tan, Dominic C. Y. Foo, and Mahmoud M. El-Halwagi.
© 2016 John Wiley & Sons, Ltd. Published 2016 by John Wiley & Sons, Ltd.

was set up in 1996 to make Europe's workplaces safer, healthier, and more productive. The European Risk Observatory was then set up in 2005 as an integral part of the EU-OSHA. It describes factors and anticipates changes in the working environment and their likely consequences to health and safety (EU-OSHA, 2010). It aims to identify new and emerging risks and to promote early preventive action. SHE considerations in process development and design have therefore become important because of legal requirements, company image, and economic reasons. Another very important regulation is the Registration, Evaluation, Authorisation and Restriction of Chemicals (REACH) legislation, which was enacted by the EU in 2007. Its ultimate aim is to protect lives and workers' health besides the environment from the risks posed by chemicals produced or imported into the EU.

A generally accepted division of sustainability is to divide it into economic, environmental, and social sustainability. From a company's point of view, corporate responsibility is the term used to cover these aspects. Safety and health are important parts of corporate responsibility which is integral for social sustainability. Therefore, workers' safety and health are among the sustainability indicators used (Al-Sharrah et al., 2010).

Traditionally, the factors that are given the highest priority when developing and designing a new process or system were economics and the technologies employed in terms of its feasibility, practicality, and effectiveness. However, following several disastrous cases of accidents in process industries, nowadays, other nonprofitable criteria such as safety and health have become more important considerations. Among the accidents related to process safety, the chemical explosion in a fertilizer factory in Toulouse, France, in September 2001 which caused 30 deaths including 21 employees is worth mentioning. A similar explosion also involved a fertilizer plant in Texas in 2013 which killed 15 people. As for health impacts, an area of concern is the long-term effect due to prolonged exposures to dioxin as a result of the explosion that took place in the Coalite TCP (2,4,5-trichlorophenol) plant in Bolsover, United Kingdom, in 1968. The chronic exposures to dioxin have led to 79 cases of chloracne among workers within 7 months of the occurrence of the accident. Another example is also involving dioxin at the Rhone-Poulenc TCP plant at Pont-de-Claix, France, which caused 100 cases of dioxin poisoning in the period of 1953–1970.

Following these accidents, the public has gradually become more aware of the importance of satisfying safety and health performance in process industries as one of the primary contributors to the protection of lives and the environment. In order to achieve the highest benefit of safety and health policy implementation, these aspects need to be considered at the very early point of the process lifecycle, that is, during the development and design phase. As previously mentioned, the aim of process design has now evolved into creating a safer, healthier, and environmentally friendlier process aside from being profitable and technically feasible. This is vital because of general legal requirements, company image, and economic reasons as well, since an unsafe plant is costly due to losses in production and capital and workers' compensation.

Various methods have been introduced to enhance process safety and health performances in the process industry. However, most of them were focusing on evaluating safety and health performance aspects in a proposed process design rather than incorporating those aspects when designing the process. Undoubtedly, hazard and risk assessment is an important element of safety and health management system—but the benefits will be greatest if the safety and health features can be embedded into the synthesis of the process itself. This is what inherent safety concept is all about. The next section discusses the idea behind inherently safer plant design in more details.

14.2 Inherent Safety in Process Design

Workers do not create hazards if the working guidelines are properly in place and strictly followed. In many cases, the hazards are built into the workplace (Kletz, 1991). It has been identified that most accidents involved design elements (Kidam, 2012). It is therefore important to make work safer by designing an inherently safer workplace rather than trying to get workers to adapt to unsafe conditions.

According to Kletz (1984), it is more effective to improve SHE performance by introducing inherent SHE principles earlier when developing and designing a plant. The traditional attitude in plant design is to rely much on the add-on safety systems. The plants are designed on a tight time schedule by using standards and so-called sound engineering practice (Heikkilä, 1999). On the contrary, inherent SHE is an effective and cost-optimal way for eliminating or reducing hazards using intrinsic means (e.g., focusing on chemical properties and process conditions), rather than controlling or managing them using external add-on systems (Kletz, 1984). In principle, an inherently safe, healthy, and environmentally friendly plant or activity cannot, under any circumstances, cause harm to people or the environment (Mansfield, 1996). Protective equipment may fail and humans may commit errors. Therefore, designing a fundamentally safer, healthier, and environmentally friendlier plant is more appealing and hence should be made as the first choice of designers and engineers compared to total reliance on extrinsic control system. The largest payoffs are achieved by verifying that inherent safety has been considered early and often in the process and engineering design phase (Lutz, 1997). This could be achieved through implementing inherently safer design (ISD) approach. ISD is a different way of thinking in designing chemical products and processes. Applying the concept at the very beginning of a project allows a safer product to be chosen instead of a hazardous one. A route that avoids the use of hazardous raw materials or intermediates can then be selected (Kletz, 1998). This sounds rather simple and revolves only around the chemistry level, yet it is actually critical since the basic decision on, for example, reaction chemistry affects the hazard potential of a plant more than the initial choice of technology (Anon, 1988). Manipulation of the chemistry and physics of the materials is more effective at preventing accidents than the dependence on additional elements to stop incipient incidents (CCPS, 1993). Once the chemistry has been decided, intensified equipment that does not require large inventories can be chosen during the flow sheet development (Kletz, 1998). Basically, the strategies to the inherently safer design of processes and plants discussed above can be grouped into four main inherent safety principles of (Kletz, 1984):

1. *Minimization* or *Intensification*
   Use smaller quantities of hazardous substances (either material or energy content).
2. Substitution
   Replace a hazardous material or process with a less hazardous one.

3. *Moderation* or *Attenuation*
   Use materials under less hazardous form, which can be accomplished either by physical (i.e., dilution) or by chemical (i.e., less severe process conditions) strategies (also called limitation of effects).
4. Simplification
   Design processes or facilities which eliminate unnecessary complexity, thereby reducing the opportunities for error, and which are forgiving of errors that are made (also called error tolerance).

More discussion about the inherent safety concept and its application is available from various references including Kletz (1984, 1985, 1991, 1998), Englund (1990), Hendershot (1991, 1995, 1997), Bollinger et al. (1996), Lawrence (1996), Heikkilä (1999), Khan and Amyotte (2003), Mannan and Lees (2005) and CCPS (2009).

14.3 Inherent Occupational Health in Process Design

The rationale of the inherent safety concept makes it attractive for adoption to the environmental and health aspects (Hassim, 2010). Kletz (1984) visualized the potential of also applying the concept to the prevention of pollution (environmental aspect) and the avoidance of small continuous leaks into the atmosphere of the workplace (occupational health/industrial hygiene aspect), but he did not evolve it further.

The adoption of the inherent concept to health started later than safety and environment due to its more complicated underlying principle. Its need to consider both toxicological and technical design disciplines makes the occupational health element receive much less interest in the design of chemical plants; rather, active works have been done dominantly from the medical point of view (Hassim, 2010). Health hazards are actually as threatening as, if not more threatening than, process safety hazards. Many do not realize the fact that each year, more people die from occupational-related diseases than by industrial accidents (Wenham, 2002).

Process plant industries are hazardous by nature as they involve various harmful chemicals as either raw materials or products and diverse risky activities that may expose workers to those chemicals, even though in most cases, process materials are well contained. Such potential risks to health must be clearly acknowledged and considered in the design of a facility. Although more is understood now about some occupational hazards than in the past, every year, new chemicals and new technologies are being introduced which present new and often unknown hazards to both workers and the community. Therefore, health property needs to be embedded into process design so that the process can be made fundamentally healthier, which brings us to the concept of inherent occupational health. Hassim and Hurme (2010a) were the first to define the concept of inherent occupational health although several works related to this subject have been conducted years before that (to name a few, those by INSIDE Project (2001), Johnson (2001), and Hassim and Edwards (2006)).

Inherent occupational health (IOH) is the prevention of occupational health hazards (i.e., chemical or physical condition) that have the potential to cause health damage to workers by trying to eliminate the use of hazardous chemicals, process conditions, and operating procedures that may cause occupational hazards to the employees. In this context, inherent occupational health hazards can be defined as a condition, inherent to the operation or use of material in a particular occupation or environment, that can cause death, injury, acute or chronic illness, disability, or reduced job performance of personnel by an acute or chronic exposure (Hassim and Hurme, 2010a).

Generally, there are twofold aims of inherent occupational health (Hassim and Hurme, 2010a). The first and most ideal strategy is to reduce the hazards due to the inherent properties of chemicals (such as toxicity and high vapor pressure) by using friendlier chemicals or the chemicals in safer physical condition (such as lower temperature) to eliminate the exposure. The second one is to reduce such process steps or procedures which involve inherent danger of exposure to the chemicals. Examples of such operations are some manual operations where the worker is in close contact with the material, such as manual handling and dosing of chemical, emptying, and cleaning of the equipment.

As a summary to the two previous sections, as widely known, inherent safety concept was proposed by Trevor Kletz in 1971. However, the idea started to rapidly propagate only after the major disastrous explosion that took place four years later involving a chemical plant in Flixborough, United Kingdom. Even though the idea emerged and was preached mostly from the view of chemical and petrochemical processes, the idea also does apply well to other industries including biofuel processing plants. In the next section of this chapter, an overview of the works that have been done related to safety and health assessment for biofuel production is presented. Future potential approaches to adopt inherent safety and inherent occupational health features in biofuel processes are also discussed. Before that, the benefits of considering safety and health in early design and the challenges in doing so are described in the next section.

14.4 Design Paradox

Before discussing available methods for safety and health assessment in biofuel production systems, it would be so much valuable to comprehend the idea behind design paradox. This section aims to bridge the understanding on the concept of inherent safety introduced earlier with the current and potential assessment approaches to be covered in the next sections.

Basically, a typical process plant goes through lifecycle stages of research and development, design, construction, operation, retrofitting, and finally decommissioning. The design stages can further be divided into process preliminary design, basic engineering, and detailed engineering (Hurme and Rahman, 2005). As the project embarks, taking chemical process as an example, the chemical synthesis route is selected during the *research and development* phase. In *preliminary design*, process structure is created, material and heat balances are calculated, and flow sheet diagrams are generated. Then the process piping, instrumentation diagrams, and so on are created in the *basic engineering* phase. Meanwhile in *detailed engineering* phase, detailed documents and drawings for procurement and construction are made. This is summarized as Table 14.1.

Even though inherent safety features can be incorporated at any stage of the process lifecycle, the best results will only be achieved if it is implemented during the earliest stages of process development. Therefore, the search for inherently safer process alternatives should begin early since many of the decisions on the process are actually conceptual and fundamental (Hassim, 2010). At the design stage, process designers and engineers

Table 14.1 *Information availability at different design stages*

R&D design	Process predesign	Basic engineering	Detailed engineering
First process concept	All in R&D stage	All in R&D and predesign stages	All in R&D, predesign, and basic engineering stages
Process block	Flow sheet (simple or detailed)		
Diagram		PI diagram	Detailed equipment, piping, and instrumentation
Reaction steps	Mass/energy balances	Process data on equipment, piping, and instrumentation	
Types of chemicals			
Physical/chemical/ toxicity properties	Operating conditions		Equipment sizing
	Major unit operations	Plant layout	Mechanical design/ engineering
Reaction conditions		Preliminary working procedures	
Stoichiometric equations			Structural, civil, and electrical engineering
Product yield			Design of ancillary services

Reproduced from Hassim and Hurme (2010a), with permission from Elsevier.

have maximum degrees of freedom in the process and plant specification (see Figure 14.1; CCPS, 1993). However, the lack of information especially in the early design generally complicates hazard assessments and decision-making for synthesizing a process. This is called the design paradox (Hurme and Rahman, 2005). Hazard and risk assessments should become step-by-step more quantitative and precise as the design becomes more detailed. This is because the knowledge on the process extends concurrently as the process design progresses. An assessment method claiming to be applicable during the whole rather than at a specific point of the design process therefore has neither a fixed region nor a fixed viewpoint (Koller, 2000). Therefore, it is ideal to have a specific method or tool for different stages of process design since the design phase itself comprises several stages as previously mentioned. This will be discussed in more detail in the next sections.

Early safety and health assessment will not only benefit from the safety and health performance but also contribute to lowering the overall plant costs (Edwards and Lawrence, 1993; Kletz, 1998; Shah et al., 2003). Besides, the cost of fixing a problem (e.g., making changes or modifications on the process) is lower when done at the earlier phases of the process lifecycle. Analyses made by Kletz (1988) revealed that the cost increases tenfold as one progresses through each phase. This is because hazard and risk assessments will eventually lead to decisions that require necessary preventive actions to either eliminate or reduce the hazards and risks as low as reasonably practicable (ALARP). The decisions cover diverse elements including alternative process routes, plant layout, and plant

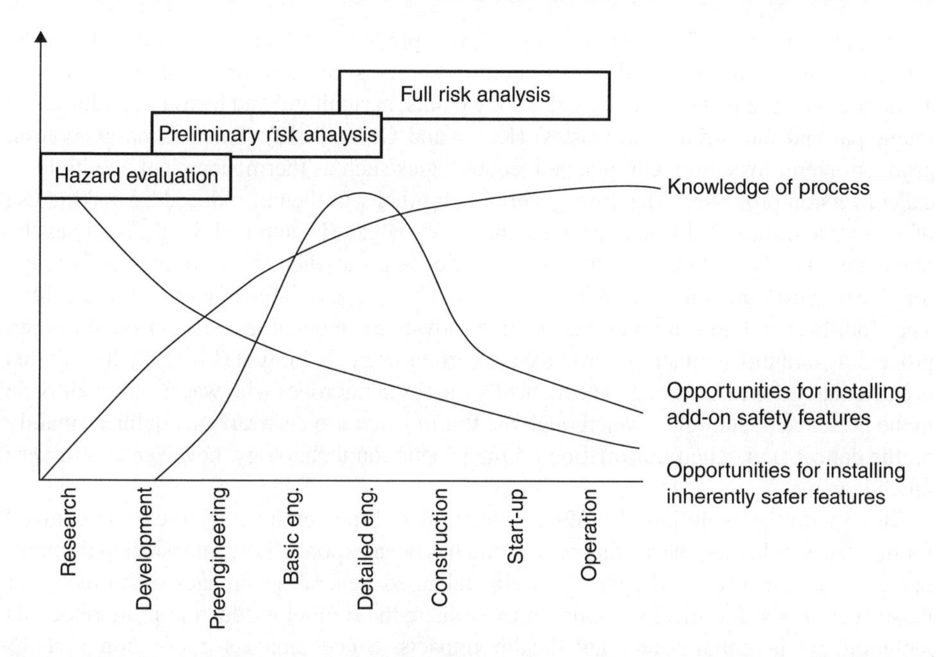

Figure 14.1 *The design paradox and inherently safer design. Reproduced from Hurme and Rahman (2005), with permission from Elsevier*

localization (Heikkilä, 1999). Performing modifications on those aforementioned aspects of a process plant are definitely involving huge financial resources unless they are made in early design phase. Therefore, the ideology of conducting early hazard assessment and designing safer and healthier plant should be taken as mandatory now rather than voluntary by all process industries including biofuel production.

14.5 Introduction to Biofuel Technologies

Unlike petrochemical and oil and gas industries which have been around for so long (approximately more than a century), biomass-based industry is comparably relatively new. Among diverse derivatives that can be produced from biomasses and their wastes, biofuel is regarded as one of the major breakthroughs that can solve a lot of problems caused by fossil fuels. Biofuel is recognized as an important and among the most promising source of renewable and sustainable energy to substitute fossil fuel. This is due to resources depletion and climate change problem, which in principle is a result of the rising emission of greenhouse gases (GHGs).

Generally, biofuel refers to solid, liquid, and gaseous fuels derived from renewable sources including biomass, animal fats, and waste oils (Demirbas, 2009). Biofuel production has evolved from the first to the fourth generation. Each generation primarily differs in terms of the feedstock and production technologies. For example, the first-generation biofuel is produced from food crops (e.g., sugar beet, corn, and oil seeds) via the transesterification process for biodiesel production and via the fermentation process for bioethanol production. These are the two primary products of the first-generation biofuels. The second-generation biofuel turns to nonfood-based sources of lignocellulosic biomass as the feedstock (e.g., herbaceous and woody plants, agricultural and forestry residues, and municipal and industrial solid wastes) (Kocar and Civas, 2013). This second-generation production employs more complicated technologies such as thermochemical and biological conversion processes. The third-generation biofuel was then introduced by making use of algae (Demirbas, 2011) and more recently microalgae (Nigam and Singh, 2011) as the raw materials. Microalgae have similar properties as algae; therefore, the same technology for oil extraction and transesterification applies to both types of algae for producing biofuel. The fourth-generation biofuel adopts the advanced technology of petroleum-based processing and biochemistry to produce the carbon-negative biofuel (Lü et al., 2011). This is particularly done by mixing the genetically modified microbes with waste carbon dioxide in the presence of sunlight. Even though the fourth generation is nearly acquainted (mainly on the concept), detailed information on the production technology, however, is still very much lacking.

Along with the evolution of biofuel generation and the continuous increase in demand for this renewable fuel, the biofuel production has been exponentially intensifying throughout the world. However, despite the well-acclaimed general advantages of biofuel over fossil fuel, it is still critically important to evaluate the biofuel production in more details including its potential safety and health impacts. Since biomass conversion systems (particularly biofuel) are relatively new compared to petrochemical industries, previous research efforts were focusing primarily on the chemistry, production technologies, and economic aspects of the system. Safety and health issues, which are also important elements of sustainability, have received much less attention especially from the process design point of view. Safety and health is just as significant as, if not more than, the other factors of biofuel production since the whole project can be simply scrapped off if the process does not meet safety and health requirements and standards.

Although efforts related to safety and health assessments in the biofuel production system are way behind those done on petrochemical industries, there are still some works that have to be done in this subject matter. Those existing works are discussed in the next two sections. Furthermore, new concepts and approaches for safety and health assessments of biofuel process development and design are also proposed.

14.6 Safety Assessment of Biofuel Production Technologies

It is mandatory that the safety of a process plant fulfills a certain required level and standards. This is because of general legal requirements, besides the indirect benefits to the image and economic performance of a company. Therefore, safety should influence process decisions from the first moments of the design project. One of the existing studies adopted the well-established hazard and operability (HAZOP) method for safety assessment of biofuel process (Jeerawongsuntorn et al., 2011).

Using the simulated biodiesel production process as case study, basically Jeerawongsuntorn and team (2011) integrated the human–machine interface with HAZOP analysis for safety protection. Safety instrumented system was developed to create interlocks such as automatic shutdown to prevent the system from being exposed to extreme or dangerous conditions. The HAZOP tool analyzes the deviation of critical variables in equipment or process streams. Subsequently, interlock actions were added to strengthen the existing protection means if they are insufficient (Cocchiara et al., 2001). The function of HAZOP in this particular work is more related to process control system and hence it does not contribute toward enhancing the safety protection capability of the process.

Gómex et al. (2013) took a step further by applying inherently safer design to the different types of reactor for biodiesel production using the integrated inherent safety index (I2SI), one of the available methods for inherent safety evaluation (Khan and Amyotte, 2005). They pointed out that even though safety risk is still very much uncertain in the design stage, it should already be assessed in detail.

The two works discussed above focused solely on the safety aspect of biodiesel production. However, before that, there is already an effort done to incorporate safety into the biodiesel synthesis processes (Narayanan et al., 2007). But the main interest of the study was more on process optimization, in which multisustainability criteria including safety and health are simultaneously optimized using the analytical hierarchical process (AHP) technique. The same goes with several works thereafter. Pokoo-Aikins et al. (2010) conducted economic and inherent safety analysis on biodiesel production process using sewage sludge as feedstock. They developed their own safety matrix tool, which was simplified from the existing inherent safety index-based methods (e.g., the prototype index of inherent safety (PIIS), inherent safety index (ISI), and *i*-Safe) for ease of application to their process under study. The matrix covers hazards due to both chemical properties (i.e., toxicity and vapor density) and process conditions (i.e., process temperature and pressure).

Jayswal et al. (2011) also conducted a similar kind of work on biodiesel production process at the early stages of design. Pareto analysis and fishbone diagram were used to analyze the complicated cause-and-effect relationship among several sustainability metrics which are inclusive of inherent safety, economics, environmental impact, and process efficiency. In their work, safety analysis is regarded as the main constituent of societal impact assessment. They proposed a new method called the enhanced inherent safety index (EISI), which was upgraded from the ISI (Heikkilä et al., 1996) to include additional factors of process complexity and chemicals' quantity into the assessment.

A more recent one is a study carried out by Ng et al. (2013) which also included inherent safety assessment in synthesizing and optimizing a so-called sustainable biorefinery system. Likewise Jayswal et al. (2011) used the ISI (Heikkilä et al., 1996) since the method is simple yet comprehensive enough to evaluate the inherent safety property of the biorefinery based on the information on process materials and unit operations available from the process flow sheet diagrams. Similarly, Liew et al. (2014) also considered inherent safety in assessing the sustainability of alternative process synthesis routes for biodiesel production. They calculated the inherent safety level of each route using the PIIS method developed by Edwards and Lawrence (1993). The PIIS fits their work well as the assessment was done during the R&D stage, when the only information available about the process is basic reaction chemistries data. The resulting inherent safety value is then used together with the values obtained from the environmental, health, and economic assessment to screen for the "best" route with optimized sustainability criteria.

Overall, from the review of these earlier works, attempts to incorporate safety in designing or operating a safer biofuel production system have already started. However, the works are still at a very beginning stage; hence, more aggressive efforts are needed to study this aspect more comprehensively and systematically.

14.7 Health Assessment of Biofuel Production Technologies

For the health aspect, the assessment is a bit trickier than safety since health assessment itself can be categorized mainly into public health, occupational health, and environmental health (Hassim, 2010). The three categories also do overlap each other. For example, public health and occupational health are mostly concerned with health impacts on the community and the workers respectively, as a result of process operations (emissions from the process itself which resulted in adverse health impacts), whereas environmental health refers to health impacts to the community as a result of environmental exposure. The health impact assessment in process industries has received much less interest from researchers compared to process safety as it has more complicated underlying principles; health typically deals with long-term exposures, whereas the impact of safety-related events can be immediately seen upon the occurrence of the cases. Besides, the impact of health-based incidents is less dramatic compared to safety, and this insidious nature of health effect is the reason why it rarely reaches the news and is not well publicized unlike the industrial accident cases (Hassim, 2010).

Realistically, health aspect of a process is worth as much as, if not more than, other concerns since health hazards directly affect human lives only unlike process safety hazards which also affect the plant, property, and cost (Hassim, 2010).

Even though health effects from chemical exposure can be both acute and chronic due to short-term and long-term exposure respectively, the main interest of health assessment is commonly on the chronic effects. Acute health effects are more relevant to process safety because of the large, accidental toxic releases such as in the Bhopal gas tragedy in 1984. However, there are diverse chronic health effects that may occur due to operations within the process industries. Such effects include respiratory and pulmonary problems, cancer, mutagenic effects, lung function defects, and dermal, eye, and nose irritation. The health impacts of process operation are not only an issue of petrochemical but also renewable fuel-based industries.

Many would assume that renewable fuel is confirmedly a healthy product. But this is not absolutely true particularly for biodiesel. For example, a study (Pearl, 2004) revealed that biodiesel may pose some health risks to humans for several reasons. Firstly, the biodiesel production process purifies methyl esters to eliminate insoluble products and other impurities. Secondly, the health risk to humans associated with potential airborne exposure to the combustion products when using biomass as a combustion fuel in diesel engines is also not negligible. For instance, by using biodiesel in diesel engine, its nature of higher oxygen content promotes significant reduction in the emission of particulate matter (PM), carbon monoxide, sulfur, poly-aromatic hydrocarbon (PAH), smoke, and noise (Zullaikah et al., 2005). However, in terms of NOx emission, biofuels, particularly biodiesel, generate up to 70% increase when diesel is replaced (Ravindranath et al., 2011). Sanz Requena et al. (2011) conducted a lifecycle assessment (LCA) on the overall life chain (from cultivation to consumption) of the first-generation biofuel produced from sunflower oil, rapeseed oil, and soybean oil. Their analysis reveals that carcinogenic impacts and inorganic respiratory problems are detected during the production of the seed. This is mainly due to the emission of heavy metal cadmium to soil which contributes to the carcinogenic effect, whereas the release of nitrogen oxide to air results in respiratory problems in human. Hence, it is important to study the whole chain when assessing the health impact of biomass-based products.

Another available study on the health assessment of biomass is associated with understanding the relationship between biomass smoke and child anemia. The study was conducted due to the high prevalence of child anemia in line with the increased usage of biomass for heating and cooking at home in 29 developing countries, which is inclusive of Azerbaijan, Armenia, Bolivia, Burkina Faso, Cambodia, Cameroon, Congo (Brazzaville), Congo (Democratic Republic), Egypt, Ethiopia, Ghana, Guinea, Haiti, Honduras, India, Jordan, Lesotho, Madagascar, Malawi, Mali, Moldova, Nepal, Niger, Rwanda, Senegal, Swaziland, Tanzania, Uganda, and Zimbabwe (Kyu et al., 2010). The result discloses high interactions between biomass smoke exposure and child age on mild and severe anemia.

It can be summarized that the earlier studies focused more on health impact to humans in conjunction with the usage of biofuel and fossil fuel. Not until recently, the idea of inherent occupational health is being adopted into biomass processes. The works are actually done in a company with inherent safety assessment by Ng et al. (2013) and Liew et al. (2014) as described before in Section 14.6. Both used the Inherent Occupational Health Index (IOHI), an index-based method for inherent occupational health assessment of chemical synthesis processes. The method will be discussed in further detail in the next section.

14.8 Proposed Ideas for Future Safety and Health Assessment in Biofuel Production Technologies

At present, as discussed in the two earlier sections, the available studies related to safety and health analysis on biofuel production are very limited. To date, neither holistic assessment method nor framework for such analysis has ever been developed. Therefore, in this section, potential approaches for safety and health analysis of biofuel production technologies are presented. These approaches are mainly adopted from those originally developed for petrochemical processes but deemed to be also suitable for biofuel application.

A safer and healthier design of a process always starts with hazard and risk assessment. Only when the root cause of the problem is understood can the problem be solved. The same goes with safety and health hazards. Therefore, safety and health assessment is an important element in synthesizing a safer and healthier process plant.

In an assessment, hazards in the process are identified and subsequently the magnitude of the hazards (severity/consequence) and likelihood of their occurrence are estimated to calculate the associated risk. Risk assessment is not always possible in all cases as it requires more detailed information on the process and the receptors. As described in detail in Section 14.4, assessment should be performed as early as possible to achieve the greatest benefit of incorporating safety and health features into the process design with a higher degree of freedom but at lower cost. But the challenge is on the limited process information available during the early lifecycle phases. In order to enhance the assessment so that it only makes use of the information available at that particular stage of assessment without compromising the comprehensiveness of the assessment, it is important to have a specific method of assessment for different stages. A method that claims to be applicable throughout the whole rather than at specific stages has neither a fixed region nor a fixed viewpoint (Koller, 2000).

Based on this perspective, Hassim and Hurme (2010a) introduced the idea of dividing the process design phase into several stages, namely the research and development (R&D), preliminary design, basic engineering, and detailed engineering. Each stage should have a dedicated method for safety and health analysis since each has different amounts and types of process information as summarized in Table 14.1.

For the R&D stage, not much data are accessible except for the process block diagram, reaction chemistries, and material properties which are obtained merely from handbooks or databanks. Therefore, safety and health assessment at this stage should focus more on the hazards posed by the chemical properties and reaction conditions. As for safety, process safety hazards (e.g., flammability, explosiveness, acute toxicity, and operating temperature and pressure) are among those that should be accounted for. As for health, toxicity properties (especially those that may cause long-term effects), chemical volatility, and process mode of operation (e.g., batch vs. semibatch vs. continuous) are of concern. Operating temperature and pressure should also be considered in terms of their susceptibility to cause chemical exposure to receptors and not for causing disastrous events (e.g., fire and explosion) which are more relevant to safety.

Such assessment, despite the fact that it sounds very fundamental and chemistry level-oriented, is actually significant as it is at this particular stage that ISD principles can be adopted (e.g., substituting highly flammable or toxic chemicals with safer ones or choosing a synthesis route that avoids intermediate with carcinogenic property). If chemical substitution is not possible, alternatives should be searched for to carry out the process operation under more moderate conditions—this is called "moderation/attenuation" strategy. Such strategies are highly needed in biofuel process synthesis since the current technologies do employ various highly hazardous, toxic, and volatile chemicals (e.g., alcohols in the pretreatment step). There is innovation to avoid the need for pretreatment processes. But this consequently caused reductions in yield and thus, the technology turns to supercritical operating conditions (can reach up to 350°C and 19 MPa (Lee et al., 2011), which does not at all solve the problem concerning safety and health). Another technology (e.g., thermochemical process) which is also used to convert biomass to energy, gas, and liquid products is even more extreme; the methods employed (liquefaction, pyrolysis, gasification, and direct combustion) in the process operate at high temperature ranges from 300 to 900°C (Zhang et al., 2010). Among the methods available for safety and health analysis during the R&D stage are the PIIS (Edwards and Lawrence, 1993) and IOHI (Hassim and Hurme, 2010a) for process safety and occupational health, respectively.

In preliminary design, more process data is available from the process flow diagrams (PFDs). At this stage, the conceptual process is created, offering more details about the process including mass and energy balances and unit operations. Such information allows more thorough safety and health analysis to be conducted. For example, as for safety, now the assessment can be extended to other units in the process besides reactor columns. Inventory and operating conditions of all unit operations in the whole process are important safety aspects to be looked into at this stage. The complexity of the process structure itself should also be considered.

On the other hand, the main concern from the health viewpoint at this stage is on the chemical exposure especially of workers within the process area. In process industries that deal mostly with airborne materials (including gas, vapor, dust, fumes, and aerosols), fugitive emissions are the main origin of the background exposures experienced by workers. Fugitive emissions are low level but continuous leaks that occur wherever there are discontinuities in the solid barriers that maintain containment. Health risk can now be calculated at this stage by estimating the amount of fugitive emissions in a process that can be potentially exposed to the workers and the volumetric flow rate within the process area. A method called the hazard quotient index (HQI) was developed by Hassim and Hurme (2010b) which describes in detail how the health risk due to exposure to fugitive emissions in a petrochemical process can be estimated during the preliminary design stage. The method also applies well to biofuel production process. As for process safety, the ISI (Heikkilä et al., 1996) and *i-Safe* (Palaniappan et al., 2002) are among the methods that can be used for assessment at this stage.

Various strategies can be taken to increase the safety and health performance of a process based on the data available here including intensification (i.e., using series of smaller capacity columns (reactor, distillation, storage tank) rather than a single very large one). Higee distillation is another example to enhance the safety feature of a process as it employs atmospheric pressure rather than storing the materials at compressed conditions. As for health, reactive distillation is regarded as one of the best innovations so as to reduce fugitive emissions. Reactive distillation is a state-of-the-art technology that combines distillation and chemical reaction in one single operating step, thus reducing the number of leaking points from piping fittings and components (simplification strategy).

In basic engineering, piping and instrumentation diagrams (P&IDs) and process plot plan are generated. From ISD perspective, basic engineering is the last step where process modifications can still be made at moderate cost and there are still large opportunities to adopt inherently SHE principles (Hurme and Rahman, 2005). More comprehensive assessment can be carried out based on the piping details from the P&IDs and the plant layout. Safety assessment can now cover more details about the process including fire and explosion protections, process control system, drainage and spill control, and spacing between units in the process. Dow Fire & Explosion Index (AIChE, 1994) is among the available methods capable of assessing safety thoroughly at this point of the process lifecycle.

As for health, the interest is always the same—on workers' potential exposure to fugitive emissions. Now exposure to releases due to manual operations (e.g., manual sampling, cleaning, or emptying vessels) can also be assessed based on the information from the P&IDs. Data on manual works also allow dermal risk exposure to be estimated besides inhalation-based risk. A method called the Occupational Health Index (OHI) was introduced to assess health risk of petrochemical processes at the basic engineering stage. Similar to the other methods mentioned earlier, the OHI can also be suitably used for biofuel processes since the nature of the process operations is similar even though they are both involving different types of feedstocks and unit operations.

ISD can be implemented at this stage by substituting conventional piping fittings and components with less or nonleaking ones. For example, when handling flammable liquids, drum pumps with well-equipped safety features (e.g., pump tubes made of conductive stainless steel) should be used. When handling highly toxic chemicals, monoblock pump or canned pump can be an alternative as they are designed for zero leakage. Flanges release very small amounts of fugitive emissions through gasket but the problem is there can be up to several thousands of flanges in one plant. Welded pipes represent an improvement over flanged pipes from the fugitive emissions' point of view. However, this is not always viable for various reasons. Since welded connections do eliminate leaks, they should be considered wherever possible.

Those described above are only a few and simple examples of ISD strategies to incorporate safety and health properties into the design of biofuel processes. Various other approaches can be applied depending on the suitability of the process operations and the availability of resources.

14.9 Conclusions

Biofuel is regarded as a potential substitute to fossil fuel as it is generated from sustainable renewable sources and it can potentially resolve the environmental issues associated with fossil fuel use. Considering the increasing demand on biofuel which concurrently will lead to vast production of biofuel, it is important to perform rigorous assessment on biofuel production technologies to assess their potential impacts in terms of the sustainability performance. Overall, this chapter focuses on safety and health criteria as part of sustainability metrics.

Extensive review on the existing related studies presented in this chapter shows that works associated with safety and health assessment of biofuel production technologies are still very much lacking. Since biomass-based industry is relatively new compared to petrochemical industry, to date analyses on particularly biofuel production processes revolved highly around improving the technologies and economics aspects. Only very few works are available which exclusively assess the safety or health aspect of biofuel processes. Other works are more on optimizing the process from multisustainability objective perspective, in which safety and/or health criteria are also included.

Early safety and health assessment during process development and design stages is very important in order to start incorporating safety and health features early on when developing and synthesizing new processes.

Different ISD strategies (i.e. substitution, minimization, attenuation, and simplification) can be greatly applied to eliminate or reduce safety and health hazards in the process plants. As conclusion to this chapter, concerted efforts on safety and health assessment of biofuel production technologies are very much in need to ensure sustainability of the process and industry as a whole in the long run.

References

AIChE. *Dow's Fire and Explosion Index Classification Guide*. Technical Manual, LC 80-29237. AIChE, New York, 1994.

Al-Sharrah, G., Elkamel, A., Almanssoor, A. Sustainability indicators for decisionmaking and optimization in the process industry: The case of the petrochemical industry, *Chemical Engineering Science* **65** (2010) 1452–1461.

Anon. *Our Common Future*. Oxford University Press, Oxford, 1987.

Anon. The design of inherently safer plants, *Chemical Engineering Progress* **84** (1988) 21.

Bollinger, R. E, Clark, D. G., Dowell III, A. M., Ewbank, R. M., Hendershot, D.C., Lutz, W. K., Meszaros, S. I., Park, D. E., Wixom, E. D. *Inherently Safer Chemical Processes – A Life Cycle Approach*. American, New York, 1996.

CCPS. Inherently Safer Chemical Processes: A Life Cycle Approach, 2nd Ed. AIChE, New York, 2009.

Cocchiara, M., Bartolozzi, V., Picciotto, A., Galluzzo, M. Integration of interlock system analysis with automated HAZOP analysis. *Reliability Engineering and System Safety* **74** (2001) 99–105.

Demirbas, A. Biofuels securing the planet's future energy needs. *Energy Conversion and Management* **50** (2009) 2239–2249.

Demirbas, M. F.Biofuels from algae for sustainable development. *Applied Energy* **88** (2011) 3473–3480.

Edwards, D.W., Lawrence, D., Assessing the inherent safety of chemical process routes: Is there a relation between plant costs and inherent safety? *Process Safety and Environment Protection* **71** (1993) 252–258.

Englund, S. M. Opportunities in the Design of Inherently Safer Chemical Plants. Academic Press, San Diego; *Advances in Chemical Engineering*, **15** (1990) 73–135.

EU-OSHA, 2010, European Agency for Safety and Health at Work. Available at: http://osha.europa.eu (accessed on August 1, 2015).

Gómex, G. E., Ramos, M. A., Cadena, J. E., Gómex, J. M., Munoz, F. Inherently safer design applied to the biodiesel production. *Chemical Engineering Transactions* **31** (2013) 619–624.

Hassim, M. H., 2010, Inherent Occupational Health Assessment in Chemical Process Development and Design [Doctoral Dissertation], Aalto University School of Science and Technology, Espoo.

Hassim, M. H., Edwards, D. W. Development of a methodology for assessing inherent occupational health hazards. *Process Safety and Environment Protectection* **84**(B5) (2006) 378–390.

Hassim, M. H., Hurme, M. Inherent occupational health assessment during process research and development stage. *Journal of Loss Prevention in the Process Industries* **23** (1) (2010a) 127–138.

Hassim, M. H., Hurme, M. Inherent occupational health assessment during preliminary design stage. *Journal of Loss Prevention in the Process Industries* **23** (3) (2010b) 476–482.

Heikkilä, A-M. 1999. Inherent Safety in Process Plant Design. An Index-Based Approach. VTT Publications, Espoo, 384.

Heikkilä, A-M., Hurme, M., Järveläinen, M. Safety considerations in process synthesis, *Computers& Chemical Engineering* **20** (A) (1996) S115–S120.

Hendershot, D. C. 1991. Design of Inherently Safer Process Facilities. Texas Chemical Council Safety Seminar, Session D, Inherently Safe Plant Design. pp. 2–22.

Hendershot, D. C. Conflicts and decisions in the search for inherently safer process options. *Process Safety Progress*. **14** (1) 1995 52–56.

Hendershot, D. C.Measuring inherent safety, health and environmental characteristics early in process development. *Process Safety Progress*. **16** (2) (1997) 78–79.

Hook, G. Responsible care and credibility. *Environmental Health Perspectives* **104** (11) (1996) 1.

Hurme, M., Rahman, M. Implementing inherent safety throughout process lifecycle, *Journal of Loss Preventionin the Process Industries* **18** (2005) 238–244.

INSIDE Project, The INSET toolkit. http://www.aeat-safety-andrisk com/html/inside.html (accessed on 3 July 2015), 2001.

Institute of Chemical Engineers, Centre for Chemical Process Hazards (CCPS). Guidelines for Engineering Design for Process Safety. Center for Chemical Process Safety (AIChE), New York, 1993.

Jayswal, A., Li, X., Zanwar, A., Lou, H. H., Huang, Y. A sustainability root cause analysis methodology and its application. *Computers and Chemical Engineering* **35** (2011) 2786–2798.

Jeerawongsuntorn, C., Sainyamsatit, N., Srinophakun, T. Integration of safety instrumented system with automated HAZOP analysis: An application for continuous biodiesel production. *Journal of Loss Prevention in the Process Industries* **24** (2011) 412–419.

Johnson, V. S. 2001 Occupational Health Hazard Index for Proposed Chemical Plant [MSc Thesis]. Loughborough University, Loughborough.

Khan, F. I., Amyotte, P. R. How to make inherent safety practice a reality, *Canadian Journal of Chemical Engineering* **81** 2003 2–16.

Khan, F. I., Amyotte, P. R. I2SI: A comprehensive quantitative tool for inherent safety and cost evaluation. *Journal of Loss Prevention in the Process Industries.* **18** 2005 310–326.

Kidam, K. 2012. Process Safety Enhancement in Chemical Plant Design by Exploiting Accident Knowledge. [Doctoral Dissertation]. Aalto University School of Science and Technology, Espoo.

Kletz, T.A. *Cheaper, Safer Plants, or Wealth and Safety at Work.* Institution of Chemical Engineers, Rugby, 1984.

Kletz, T. A., Inherently safer plants. *Plant Operations Progress* **4** (1985) 164–167.

Kletz, T.A. Seminar presentation. Union Carbide Corporation, 1988.

Kletz, T. A. *Plant Design for Safety: A User Friendly Approach.* Hemisphere Publishing Corporation, New York, 1991.

Kletz, T. A., *Process Plants: A Handbook for Inherently Safer Design.* Taylor & Francis, Philadelphia, PA, 1998.

Koçar, G., Civaş, N. An overview of biofuels from energy crops: Current status and future prospects. *Renewable and Sustainable Energy Reviews* **28** (2013) 900–916.

Koller, G. 2000. Identification and Assessment of Relevant Environmental, Health and Safety Aspects during Early Phases of Process Development. Doctoral Thesis (Diss. ETH Nr 13607). Swiss Federal Institute of Technology Zürich, Zurich.

Kyu, H.H., Georgiades, K., Boyle, M.H., 2010. Biofuel smoke and child anemia in 29 developing countries: a multilevel analysis. *Ann. Epidermiol.* **20**, 811–817.

Lawrence, D. 1996. Quantifying Inherent Safety of Chemical Process Routes [PhD Thesis]. Loughborough University, Loughborough.

Lee, S., Posarac, D., Ellis, N. Process simulation and economic analysis of biodiesel production processes using fresh and waste vegetable oil and supercritical methanol. *Chemical Engineering Research and Design.* **89** (2011) 2626–2642.

Liew, W. H., Hassim, M. H., Ng, D. K. S. *Screening of sustainable biodiesel production pathways during process research and development (R&D) stage using fuzzy optimization, clean technology and environmental policy (CTEP)*, 2014**16**(7), 1431–1444.

Lü, J., Sheahan, C., Fu, P. Metabolic engineering of algae for fourth generation biofuels production. *Energy and Environmental Science* **4** (2011) 2451–2466.

Lutz, W. K. Advancing inherent safety into methodology. *Process Safety Progress* **16** (2) (1997) 86–88.

Mannan, S., Lees, F. P. *Lee's Loss Prevention in the Process Industries*, vol. **2**. Elsevier Butterworth-Heinemann, Burlington, MA, 2005.

Mansfield, D. P. 1996. The development of an integrated toolkit for inherent SHE. International Conference and Workshop on Process Safety Management and Inherently Safer Processes (AIChE), Orlando, FL, October 8–11, 1996, American Institute of Chemical Engineers, New York, pp. 103–117.

Narayanan, D., Zhang, Y. Mannan, M. S. Engineering for sustainable development (ESD) in biodiesel production. *Process Safety and Environmental Protection* **85** (2007) 349–359.

Ng, R. T. L., Hassim, M. H., Ng, D. K. S. Process synthesis and optimization of a sustainable integrated biorefinery via fuzzy optimization. *AIChE Journal* **59** (2013) 4212–4227.

Nigam, P. S., Singh, A. Production of liquid biofuels from renewable sources. *Progress in Energy and Combustion Science* **37** (2011) 52–68.

O'Malley, V. The Integrated Pollution Prevention and Control (IPPC) Directive and its implications for the environment and industrial activities in Europe. *Sensors and Actuators B* **59** (1999) 78–82.

Palaniappan, C., Srinivasan, R., Tan, R. Expert system for the design of inherently safer processes: 1. Route selection stage, *Industrial & Engineering Chemistry Research* **41** (2002) 6698–6710.

Pearl, G. G. *Biodiesel and BSE, Director's Digest.* Fats and Proteins Research Foundation, Inc., Chicago, IL, pp. 1–3, 2004.

Pokoo-Aikins, G., Heath, A., Mentzer, R. A., Sam Mannan, M., Rogers, W. J., El-Halwagi, M. M. A multi-criteria approach to screening alternatives for converting sewage sludge to biodiesel. *Journal of Loss Prevention in the Process Industries* **23** (2010) 412–420.

Ravindranath, N. H., Sita Lakshmi, C., Manuvie, R., Balachandra, P. Biofuel production and implications for land use, food production and environment in India. *Energy Policy* **39** (2011) 5737–5745.

Sanz Requena, J. F., Guimaraes, A. C., Quiros Alpera, S., Relea Gangas, E., Hernandez-Navarro, S., Navas Gracia, L. M., Martin-Gil, J., Fresneda Cuesta, H. Life cycle assessment (LCA) of the biofuel production process from sunflower oil, rapeseed oil and soybean oil, *Fuel Processing Technology* **92** (2011) 190–199.

Shah, S., Fischer, U., Hungerbühler, K. A hierarchical approach for the evaluation of chemical process aspects from the perspective of inherent safety. *Process Safety and Environment Protectection* **81** (2003) 430–443.

Wenham, D. 2002 Occupational Health and Safety Management Course Module. Centre for Hazard and Risk Management (CHaRM), Loughborough.

Zhang, L., Xu, C., Champagne, P. Overview of recent advances in thermo-chemical conversion of biomass. *Energy Conversion and Management* **51** (2010) 969–982.

Zullaikah, S., Lai, C. C., Vali, S. R., Ju, Y. H. A two-step acid-catalyzed process for the production of biodiesel from rice bran oil. *Bioresource Technology* **96** (2005) 1889–1896.

Index

Process Design Strategies for Biomass Conversion Systems, First Edition. Edited by Denny K. S. Ng, Raymond R. Tan, Dominic C. Y. Foo, and Mahmoud M. El-Halwagi.
© 2016 John Wiley & Sons, Ltd. Published 2016 by John Wiley & Sons, Ltd.